T0292814

COARSE GEOMETRY OF TOPOLOGICAL GROUPS

This book provides a general framework for doing geometric group theory for many non-locally compact topological transformation groups that arise in mathematical practice, including homeomorphism and diffeomorphism groups of manifolds, isometry groups of separable metric spaces and automorphism groups of countable structures. Using Roe's framework of coarse structures and spaces, the author defines a natural coarse geometric structure on all topological groups. This structure is accessible to investigation, especially in the case of Polish groups, and often has an explicit description, generalising well-known structures in familiar cases including finitely generated discrete groups, compactly generated locally compact groups and Banach spaces. In most cases, the coarse geometric structure is metrisable and may even be refined to a canonical quasi-metric structure on the group. The book contains many worked examples and sufficient introductory material to be accessible to beginning graduate students. An appendix outlines several open problems in this young and rich theory.

Christian Rosendal is Professor of Mathematics at the University of Maryland. He received a Simons Fellowship in Mathematics in 2012 and is Fellow of the American Mathematical Society.

CAMBRIDGE TRACTS IN MATHEMATICS

GENERAL EDITORS

J. BERTOIN, B. BOLLOBÁS, W. FULTON, B. KRA, I. MOERDIJK, C. PRAEGER,
P. SARNAK, B. SIMON, B. TOTARO

A complete list of books in the series can be found at www.cambridge.org/mathematics.
Recent titles include the following:

Coarse Geometry of Topological Groups

CHRISTIAN ROSENDAL
University of Maryland

CAMBRIDGE
UNIVERSITY PRESS

CAMBRIDGE
UNIVERSITY PRESS

University Printing House, Cambridge CB2 8BS, United Kingdom

One Liberty Plaza, 20th Floor, New York, NY 10006, USA

477 Williamstown Road, Port Melbourne, VIC 3207, Australia

314–321, 3rd Floor, Plot 3, Splendor Forum, Jasola District Centre,
New Delhi – 110025, India

103 Penang Road, #05–06/07, Visioncrest Commercial, Singapore 238467

Cambridge University Press is part of the University of Cambridge.

It furthers the University's mission by disseminating knowledge in the pursuit of
education, learning, and research at the highest international levels of excellence.

www.cambridge.org
Information on this title: www.cambridge.org/9781108842471
DOI: 10.1017/9781108903547

First published 2021

A catalogue record for this publication is available from the British Library.

ISBN 978-1-108-84247-1 Hardback

Contents

Preface

The present book is the product of research initiated during an extended sabbatical financed in part by a fellowship by the Simons Foundation. Although the basic discovery of an appropriate coarse structure on Polish groups really materialised around 2013, the initial seeds were sown in several prior works and, in my own case, in the two studies [77] and [78], where various concepts of boundedness in topological groups were developed.

The main objects of this study are *Polish groups*, that is, separable and completely metrisable topological groups. Although these can of course be treated abstractly, I am principally interested in them as they appear in applications, namely as transformation groups of various mathematical structures and even as the additive groups of separable Banach spaces. This class of groups has received a substantial amount of attention over the past two decades and the results here are my attempt at grappling with possible geometric structures on them. In particular, this is a response to the question of how to apply the language and techniques of geometric group theory, abstract harmonic and functional analysis to their study.

Because this project has been a long time coming, my ideas on the subject have evolved over time and have been influenced by a number of people. Evidently, the work of Mikhail Gromov pervades all of geometric group theory and hence also the ideas presented here. But another specific reference is John Roe's lectures on coarse geometry [75] and whose framework of coarse spaces allowed me to extend the definition of a geometric structure to all topological groups and not just the admittedly more interesting subclass of locally bounded Polish groups.

Although this book contains the first formal presentation of the theory, parts of it have already found their way into other publications. In particular, in [80] I made a systematic study of equivariant geometry of amenable Polish

vii

groups, including separable Banach spaces, and made use of some of the theory presented here. Similarly, in a collaboration with Kathryn Mann [**57**], I investigated the large-scale geometry of homeomorphism groups of compact manifolds within the present framework, while other authors [**19, 38, 99**] have done so for other groups. Finally, [**81**] contains a characterisation of the small-scale geometry of Polish groups and its connection to the large scale.

Acknowledgements

Over the years, I have greatly benefitted from all my conversations on this topic with a number of people. Although I am bound to exclude many, these include Uri Bader, Bruno Braga, Michael Cohen, Yves de Cornulier, Marc Culler, Alexander Dranishnikov, Alexander Furman, William Herndon, Kathryn Mann, Julien Melleray, Emmanuel Militon, Nicolas Monod, Justin Moore, Vladimir Pestov, Konstantin Slutsky, Sławomir Solecki, Andreas Thom, Simon Thomas, Todor Tsankov, Phillip Wesolek, Kevin Whyte, Takamitsu Yamauchi and Joseph Zielinski.

I also take this occasion to thank my colleagues and the department at the University of Illinois at Chicago for giving me great flexibility in my work, and Chris Laskowski and the mathematics department at the University of Maryland for hosting me during the preparation of this book. My research benefitted immensely from the continuous support of the National Science Foundation and a fellowship from the Simons Foundation.

Finally, I wish to thank Valentin Ferenczi, Gilles Godefroy, Alexander Kechris, Alain Louveau and Stevo Todorčević for their support over the years and, most of all, my entire family for indulging and occasionally even encouraging my scientific interests.

My research benefitted immensely from the continuous support of the National Science Foundation (DMS 1201295, DMS 1464974 and DMS 1764247) and a fellowship from the Simons Foundation.

1

Introduction

1.1 Motivation

Geometric group theory, or the large-scale geometry of finitely generated discrete groups and of compactly generated locally compact groups, is by now a well-established theory (see [**22, 68**] for recent accounts). In the finitely generated case, the starting point is the elementary observation that the word metrics ρ_Σ on a discrete group Γ given by finite symmetric generating sets $\Sigma \subseteq \Gamma$ are all mutually quasi-isometric, and thus any such metric may be said to define the large-scale geometry of Γ. This has led to a very rich theory weaving together combinatorial group theory, geometry, topology and functional analysis stimulated by the impetus of M. Gromov (see, e.g., [**36**]).

To fix the language, let us recall that a map $(X, d_X) \xrightarrow{\phi} (Y, d_Y)$ between two metric spaces is a *quasi-isometry* provided that there is a constant K so that, for all $x, x' \in X$,

$$\frac{1}{K} d_X(x, x') - K \leqslant d_Y(\phi(x), \phi(x')) \leqslant K \cdot d_X(x, x') + K$$

and also

$$\sup_{y \in Y} \inf_{x \in X} d_Y(y, \phi(x)) \leqslant K.$$

The existence of a quasi-isometry between metric spaces defines an equivalence relation on the class of metric spaces, and hence the large-scale geometry of a finitely generated group Γ is well defined up to this notion of equivalence.

In the locally compact setting, matters have not progressed equally swiftly, even though the basic tools have been available for quite some time. Indeed, by a result of R. Struble [**84**] dating back to 1951, every locally compact second countable group admits a compatible left-invariant *proper* metric, i.e., so that the closed balls are compact. Struble's theorem was based on an earlier well-known result due independently to G. Birkhoff [11] and S. Kakutani [43] characterising the metrisable topological groups as the first countable topological groups and, moreover, stating that every such group admits a compatible left-invariant metric. However, as is evident from the construction underlying the Birkhoff–Kakutani theorem, if one begins with a compact symmetric generating set Σ for a locally compact second countable group G, then one may obtain a compatible left-invariant metric d that is quasi-isometric to the word metric ρ_Σ induced by Σ. By applying the Baire category theorem and arguing as in the discrete case, one sees that any two such word metrics ρ_{Σ_1} and ρ_{Σ_2} are quasi-isometric, which shows that the compatible left-invariant metric d is uniquely defined up to quasi-isometry by this procedure.

Thus far, there has been no satisfactory general method of studying large-scale geometry of topological groups beyond the locally compact groups, although, of course, certain subclasses such as Banach spaces arrive with a naturally defined geometry. This state of affairs may be largely the result of the presumed absence of canonical generating sets in general topological groups as opposed to the finitely or compactly generated ones. In certain cases, substitute questions have been considered, such as the boundedness or unboundedness of specific metrics [27] or of all metrics [77]; growth type and distortion of individual elements or subgroups [34, 72]; equivariant geometry [70] and specific coarse structures [67].

In the present book, we offer a solution to this problem that, in many cases, allows one to isolate and compute a canonical word metric on a topological group G and thus to identify a unique quasi-isometry type of G. Moreover, this quasi-isometry type agrees with that obtained in the finitely or compactly generated settings and also verifies the main characteristics encountered there, namely that it is a topological isomorphism invariant of G capturing all possible large-scale behaviour of G. Furthermore, under mild additional assumptions on G, this quasi-isometry type may also be implemented by a compatible left-invariant metric on the group.

Although applicable to all topological groups, our main interest is in the class of *Polish groups*, i.e., separable completely metrisable topological groups. These include most interesting topological transformation groups, e.g.,

$$\text{Homeo}(\mathcal{M}), \qquad \text{Diff}^k(\mathcal{M}),$$

for $\mathcal{M}$ a compact (smooth) manifold, and

$$\text{Aut}(\mathbf{A}),$$

for a countable discrete structure $\mathbf{A}$, along with all separable Banach spaces and locally compact second countable groups. Another class that has recently received much attention by geometric topologists is the mapping class groups of *infinite-type* surfaces; that is, so that the mapping class group is not finitely generated. In this case, the mapping class group can be viewed as the automorphism group of an associated countable graph and thus falls into the framework of automorphism groups of countable discrete structures. However, it should be stressed that the majority of our results are directly applicable in the greater generality of *European* groups, i.e., Baire topological groups, countably generated over every identity neighbourhood. This includes, for example, all σ-compact locally compact Hausdorff groups and all (potentially non-separable) Banach spaces.

One central technical tool is the notion of coarse structure due to J. Roe [74, 75], which may be viewed as the large-scale counterpart to uniform spaces. Indeed, given an écart (also known as pre- or pseudo-metric) d on a group G, let $\mathcal{E}_d$ be the coarse structure on G generated by the entourages

$$E_\alpha = \{(x, y) \in G \times G \mid d(x, y) < \alpha\},$$

for $\alpha < \infty$. That is, $\mathcal{E}_d$ is the ideal of subsets of $G \times G$ generated by the E_α. In analogy with A. Weil's result [94] that the left-uniform structure $\mathcal{U}_L$ on a topological group G can be written as the union

$$\mathcal{U}_L = \bigcup_d \mathcal{U}_d$$

of the uniform structures $\mathcal{U}_d$ induced by the family of continuous left-invariant écarts d on G, we define the *left-coarse structure* $\mathcal{E}_L$ on G by

$$\mathcal{E}_L = \bigcap_d \mathcal{E}_d.$$

This definition equips every topological group with a left-invariant coarse structure, which, like a uniformity, may or may not be metrisable, i.e., be the coarse structure associated to a metric on the group. To explain when that happens, we say that a subset $A \subseteq G$ is *coarsely bounded in G* if A has finite diameter with respect to every continuous left-invariant écart on G. This may be viewed as an appropriate notion of 'geometric compactness' in topological groups and, in the case of a Polish group G, has the following combinatorial reformulation. Namely, $A \subseteq G$ is coarsely bounded in G if, for every identity neighbourhood V, there is a finite set $F \subseteq G$ and a k so that $A \subseteq (FV)^k$.

Theorem 1.1 *The following conditions are equivalent for a Polish group G:*

(1) *the left-coarse structure $\mathcal{E}_L$ is metrisable;*
(2) *G is* locally bounded, *i.e., has a coarsely bounded identity neighbourhood;*
(3) *$\mathcal{E}_L$ is generated by a compatible left-invariant metric d, i.e., $\mathcal{E}_L = \mathcal{E}_d$;*
(4) *a sequence (g_n) eventually leaves every coarsely bounded set in G if and only if there is some compatible left-invariant metric d on G for which $d(g_n, 1) \xrightarrow[n]{} \infty$.*

In analogy with proper metrics on locally compact groups, the metrics appearing in condition (3) above are said to be *coarsely proper*. Indeed, these are exactly the compatible left-invariant metrics all of whose bounded sets are coarsely bounded. Moreover, by Struble's result, on a locally compact second countable group these are the proper metrics.

The category of coarse spaces may best be understood by its morphisms, namely, the bornologous maps. In the case where $(X, d_X) \xrightarrow{\phi} (Y, d_Y)$ is a map between pseudo-metric spaces, then ϕ is *bornologous* if there is an increasing modulus $\theta \colon \mathbb{R}_+ \to \mathbb{R}_+$ so that, for all $x, x' \in X$,

$$d_Y\big(\phi(x), \phi(x')\big) \leqslant \theta\big(d_X(x, x')\big).$$

By using this, we may quasi-order the continuous left-invariant écarts on G by setting $\partial \lll d$ if the identity map $(G, d) \to (G, \partial)$ is bornologous. One then shows that a metric is coarsely proper when it is the maximum element of this ordering. Although seemingly most familiar groups are locally bounded, counterexamples exist, such as the infinite direct product of countably infinite groups, e.g., $\mathbb{Z}^{\mathbb{N}}$.

However, just as the word metric on a finitely generated group is well defined up to quasi-isometry, we may obtain a similar canonicity provided that the group G is actually *generated by a coarsely bounded* set A, that is, every element of G can be written as a product of elements of $A \cup A^{-1} \cup \{1\}$. In order to do this, we refine the quasi-ordering $\lll$ on continuous left-invariant écarts on G above by letting $\partial \ll d$ if there is a constant K so that $\partial \leqslant K \cdot d + K$. Again, if d is maximum in this ordering, we say that d is *maximal*. Obviously, two maximal écarts are quasi-isometric, whence these induce a canonical *quasi-isometry type* on G. Moreover, as it turns out, the maximal écarts are exactly those that are quasi-isometric to the word metric

$$\rho_\Sigma(x,y) = \min(k \mid \exists z_1, \ldots, z_k \in \Sigma \colon x = yz_1 \cdots z_k)$$

given by a coarsely bounded generating set $\Sigma \subseteq G$.

Theorem 1.2 *The following are equivalent for a Polish group G:*

(1) *G admits a compatible left-invariant maximal metric;*
(2) *G is generated by a coarsely bounded set;*
(3) *G is locally bounded and not the union of a countable chain of proper open subgroups.*

A reassuring fact about our definition of coarse structure and quasi-isometry type is that it is a conservative extension of the existing theory. Namely, as the coarsely bounded sets in a σ-compact locally compact group coincide with the relatively compact sets, one sees that our definition of the quasi-isometry type of a compactly generated locally compact group coincides with the classical definition given in terms of word metrics for compact generating sets. The same argument applies to the category of finitely generated groups when these are viewed as discrete topological groups. Moreover, as will be shown, if $(X, \|\cdot\|)$ is a Banach space, then the norm metric will be maximal on the underlying additive group $(X,+)$, whereby $(X,+)$ will have a well-defined quasi-isometry type, namely, that of $(X, \|\cdot\|)$. But even in the case of homeomorphism groups of compact manifolds M, as shown in [**57, 62**], the maximal metric on the group $\mathrm{Homeo}_0(M)$ of isotopically trivial homeomorphisms of M is quasi-isometric to the fragmentation metric originating in the work of R. D. Edwards and R. C. Kirby [**26**].

1.2 A Word on the Terminology

Some of the basic results presented here have previously been included in the pre-print [**79**], which now is fully superseded by this book. Under the impetus of T. Tsankov, we have changed the terminology from [**79**] to become less specific and more in line with the general language of geometric group theory. Thus, the coarsely bounded sets were originally called *relatively (OB) sets* to keep in line with the terminology from [**77**]. Similarly, locally bounded groups were denoted *locally (OB)* and groups generated by coarsely bounded sets were called *(OB) generated*. For this reason, other papers based on [**79**],

such as [19], [57], [80] and [98], also use the language of relatively (OB)
sets. The translation between the two is straightforward and involves no
change in theory.

1.3 Summary of Findings

To aid the reader in the navigation of the new concepts appearing here,
we include Figure 1.1, depicting the main classes of Polish groups and a
few simple representative examples from some of these. Observe that in the
diagram the classes increase going up and from left to right.

Note that the shaded areas reflect the fact that every Polish group of bounded
geometry is automatically locally bounded, and that coarsely bounded groups
trivially have bounded geometry.

1.3.1 Coarse Structure and Metrisability

Chapter 2 introduces the basic machinery of coarse structures with its associated morphisms of bornologous maps and analyses these in the setting of
topological groups. We introduce the canonical left-invariant coarse structure

		$\mathrm{Isom}(\mathbb{U})$	$\mathrm{Isom}(\mathbb{U})$ $\times \mathbb{F}_\infty$	$\prod_n \mathbb{Z}$
Bounded geometry $\cup$	$\mathrm{Homeo}(\mathbb{S}^n)$	$\mathrm{Homeo}_\mathbb{Z}(\mathbb{R})$	$\mathrm{Homeo}_\mathbb{Z}(\mathbb{R})$ $\times \mathbb{F}_\infty$	
Locally compact $\cup$	Compact groups	Compactly generated	$\mathbb{R} \times \mathbb{F}_\infty$	
Discrete	Finite groups	Finitely generated	$\mathbb{F}_\infty$	

$$\underset{\substack{\text{Coarsely}\\\text{bounded}}}{} \subset \underset{\substack{\text{Generated by}\\\text{bounded set}}}{} \subset \underset{\substack{\text{Locally}\\\text{bounded}}}{}$$

Figure 1.1 Main classes of Polish groups organised according to their coarse
geometry.

$\mathcal{E}_L$ with its ideal of coarsely bounded sets and compare this with other coarse structures such as the group-compact coarse structure $\mathcal{E}_K$.

The main results of the chapter concern the identification of coarsely proper and maximal metrics along with Theorems 1.1 and 1.2 characterising the existence of these. This also leads to a version of the Milnor–Schwarz Lemma [**64, 83**] adapted to our setting; this is the central tool in the computation of actual quasi-isometry types of groups.

1.3.2 Basic Structure Theory

In Chapter 3 we provide some of the basic tools for the geometric study of Polish groups and present a number of computations of the geometry of specific groups. The simplest class to consider is that of the 'metrically compact' groups, i.e., those quasi-isometric to a one-point space. These are exactly those coarsely bounded in themselves. This class of groups was extensively studied in [**77**] and includes a large number of topological transformation groups of highly homogeneous mathematical structures such as homeomorphism groups of spheres and the unitary group of separable infinite-dimensional Hilbert space.

The locally Roelcke pre-compact groups comprise another particularly interesting class. This class includes examples such as the automorphism group of the countably regular tree $\text{Aut}(\mathsf{T}_\infty)$ and the isometry group of the Urysohn metric space $\text{Isom}(\mathbb{U})$ that turn out to be quasi-isometric to the tree T_∞ and the Urysohn space $\mathbb{U}$, respectively. Because, by a recent result of J. Zielinski [**99**], the locally Roelcke pre-compact groups have locally compact Roelcke completions, they also provide us with an important tool for the analysis of Polish groups of bounded geometry in Chapter 5.

Indeed, a closed subgroup H of a Polish group G is said to be *coarsely embedded* if the inclusion map is a coarse embedding, or equivalently, a subset $A \subseteq H$ is coarsely bounded in H if and only if it is coarsely bounded in G. Because, in a locally compact group, the coarsely bounded sets are simply the relatively compact sets, every closed subgroup is coarsely embedded, although not necessarily quasi-isometrically embedded in the compactly generated case. However, this fails dramatically for Polish groups. Indeed, every Polish group is isomorphic to a closed subgroup of the coarsely bounded group $\text{Homeo}([0, 1]^{\mathbb{N}})$. So this subgroup is coarsely embedded only if coarsely bounded itself. This difference, along with the potential non-metrisability of the coarse structure, accounts for a great deal of the additional difficulties arising when investigating general Polish groups.

Theorem 1.3 *Every locally bounded Polish group G is isomorphic to a coarsely embedded closed subgroup of the locally Roelcke pre-compact group* $\mathrm{Isom}(\mathbb{U})$.

Via this embedding, every locally bounded Polish group can be seen to act continuously on a locally compact space preserving its geometric structure.

The main structural theory of Chapter 3 is a byproduct of the analysis of the coarse geometry of product groups. Indeed, we show that a subset A of a product $\prod_i G_i$ is coarsely bounded if and only if each projection $\mathrm{proj}_i(A)$ is coarsely bounded in G_i. From this, we obtain a universal representation of all Polish groups.

Theorem 1.4 *Every Polish group G is isomorphic to a coarsely embedded closed subgroup of the countable product* $\prod_n \mathrm{Isom}(\mathbb{U})$.

This can be viewed as providing a product resolution of the coarse structure on non-locally bounded Polish groups.

1.3.3 Coarse Geometry of Group Extensions

In Chapter 4 we address the fundamental and familiar problem of determining the coarse geometry of a group G from those of a closed normal subgroup K and the quotient group G/K. That is, we will reconstruct the coarse geometry of the middle term G from those of K and G/K in the short exact sequence

$$1 \longrightarrow K \longrightarrow G \longrightarrow G/K \longrightarrow 1.$$

Although certain things can be said about the general situation, we mainly focus on a more restrictive setting, which includes that of central extensions. Namely, we suppose that K is a closed normal subgroup of a Polish group G, where the latter is generated by K and the centraliser $C_G(K) = \{g \in G \mid \forall k \in K \ gk = kg\}$ of K in G, i.e., such that $G = K \cdot C_G(K)$. Note that, in this case, we also have that

$$G/K = C_G(K)/Z(K),$$

where $Z(K) = \{h \in K \mid \forall k \in K \ hk = kh\}$ is the centre of K. Assume furthermore that K is coarsely embedded in G and that $G/K \xrightarrow{\phi} C_G(K)$ is a section for the quotient map that is bornologous as a map $G/K \xrightarrow{\phi} G$. Then the map

$$K \times G/K \to G, \qquad (k,h) \mapsto k\phi(h)$$

defines a coarse equivalence between $K \times G/K$ and G.

A common instance of this setup is seen when G is generated by a discrete normal subgroup $K = \Gamma$ and a connected closed subgroup F.

Theorem 1.5 *Suppose G is a Polish group generated by a discrete normal subgroup Γ and a connected closed subgroup F. Assume also that $\Gamma \cap F$ is coarsely embedded in F and that $G/\Gamma \xrightarrow{\phi} F$ is a bornologous section for the quotient map. Then G is coarsely equivalent to $G/\Gamma \times \Gamma$.*

In connection with these problems, several fundamental issues emerge:

- When is K coarsely embedded in G?
- When does the quotient map $G \xrightarrow{\pi} G/K$ admit a bornologous section $G/K \xrightarrow{\phi} G$?
- Is G locally bounded provided that K and G/K are?

Indeed, to determine whether K is coarsely embedded in G or whether a section $G/K \xrightarrow{\phi} G$ is bornologous both require some advance knowledge of the coarse structure on G itself. However, the latter is exactly what we are trying to determine. To circumvent this conumdrum, we study the cocycle associated with a section ϕ. Indeed, given a section $G/K \xrightarrow{\phi} C_G(K)$ for the quotient map, one obtains an associated cocycle $G/K \times G/K \xrightarrow{\omega_\phi} Z(K)$ by the formula

$$\omega_\phi(h_1, h_2) = \phi(h_1 h_2)^{-1} \phi(h_1) \phi(h_2).$$

Assuming that ϕ is Borel and G/K locally bounded, the coarse qualities of the map $G/K \xrightarrow{\phi} G$ and whether K is coarsely embedded in G now become intimately tied to the coarse qualities of ω_ϕ. Let us state this for the case of central extensions.

Theorem 1.6 *Suppose K is a closed central subgroup of a Polish group G so that G/K is locally bounded and that $G/K \xrightarrow{\phi} G$ is a Borel measurable section of the quotient map. Assume also that, for every coarsely bounded set $B \subseteq G/K$, the image*

$$\omega_\phi[G/K \times B]$$

is coarsely bounded in K. Then G is coarsely equivalent to $K \times G/K$.

The main feature here is, of course, that the assumptions make no reference to the coarse structure of G, only to those of K and G/K.

We then apply our analysis to covering maps of manifolds or more general locally compact spaces, which builds on a specific subcase from our joint work with K. Mann [57]. Our initial setup is a proper, free and cocompact action

$$\Gamma \curvearrowright X$$

of a finitely generated group Γ on a path-connected, locally path-connected and semi-locally simply connected, locally compact metrisable space X. Then the normaliser $N_{\mathrm{Homeo}(X)}(\Gamma)$ of Γ in the homeomorphism group $\mathrm{Homeo}(X)$ is the group of all lifts of homeomorphisms of $M = X/\Gamma$ to X, whereas the centraliser $C_{\mathrm{Homeo}(X)}(\Gamma)$ is an open subgroup of $N_{\mathrm{Homeo}(X)}(\Gamma)$. Let

$$N_{\mathrm{Homeo}(X)}(\Gamma) \xrightarrow{\ \pi\ } \mathrm{Homeo}(M)$$

be the corresponding quotient map and let

$$Q_0 = \pi\big[C_{\mathrm{Homeo}(X)}(\Gamma)\big]$$

be the subgroup of $\mathrm{Homeo}(M)$ consisting of homeomorphisms admitting lifts in $C_{\mathrm{Homeo}(X)}(\Gamma)$. We show that Q_0 is open in $\mathrm{Homeo}(M)$. Also, assume H is a subgroup of Q_0 that is Polish in a finer group topology, say H is the transformation group of some additional structure on M, e.g., a diffeomorphism or symplectic group. Then the group of lifts $G = \pi^{-1}(H) \leqslant N_{\mathrm{Homeo}(X)}(\Gamma)$ carries a canonical lifted Polish group topology and is related to H via the exact sequence

$$1 \to \Gamma \to G \xrightarrow{\ \pi\ } H \to 1.$$

By using only assumptions on the structure of Γ, we can relate the geometry of G to those of H and Γ.

Theorem 1.7 *Suppose* $\Gamma/Z(\Gamma) \xrightarrow{\ \psi\ } \Gamma$ *is a bornologous section for the quotient map, that* $H \leqslant Q_0$ *is Polish in some finer group topology and that* $G = \pi^{-1}(H)$. *Then* G *is coarsely equivalent to* $H \times \Gamma$.

Observe here that ψ is a section for the quotient map from the discrete group Γ to its quotient by the centre, which a priori has little to do with H and G. Nevertheless, a main feature of the proof is the existence of a bornologous section $H \xrightarrow{\ \phi\ } C_G(\Gamma)$ for the quotient map π, which is extracted from ψ.

Also, applying our result to the universal cover $X = \tilde{M}$ of a compact manifold M, we arrive at the following result.

Theorem 1.8 *Suppose that M is a compact manifold, H is a subgroup of* $\mathrm{Homeo}_0(M)$, *which is Polish in some finer group topology, and let G be the group of all lifts of elements in H to homeomorphisms of the universal cover $\tilde{M}$. Assume that the quotient map*

$$\pi_1(M) \longrightarrow \pi_1(M)/Z\big(\pi_1(M)\big)$$

admits a bornologous section. Then G is coarsely equivalent to $\pi_1(M) \times H$.

1.3.4 Polish Groups of Bounded Geometry

Chapter 5 concerns perhaps the geometrically most well-behaved class of Polish groups beyond the locally compact ones, namely those of bounded geometry. To get a foretaste of this concept, we note that a metric space (X,d) is said to have *bounded geometry* if there is some number α with the property that, for every β, there is an integer $k = k(\beta)$ so that every subset of diameter β in X can be covered by k sets of diameter α. J. Roe [**75**] extended this definition to all coarse spaces and, by using his definition, we may therefore investigate the class of Polish groups of bounded geometry. As it turns out, these groups are all locally bounded and hence, when equipped with a corresponding coarsely proper metric, are just metric spaces of bounded geometry. Furthermore, relying on work of J. Zielinski [**99**] on locally Roelcke pre-compact groups, we obtain a dynamical characterisation of these.

Theorem 1.9 *The following conditions are equivalent for a Polish group G:*

(1) *G has bounded geometry;*
(2) *G is coarsely equivalent to a metric space of bounded geometry;*
(3) *G is coarsely equivalent to a proper metric space;*
(4) *G admits a continuous, coarsely proper, modest and cocompact action $G \curvearrowright X$ on a locally compact metrisable space X;*
(5) *G is locally bounded and every continuous, coarsely proper and modest action $G \curvearrowright X$ on a locally compact Hausdorff space X is cocompact.*

Here, a continuous action $G \curvearrowright X$ on a locally compact metrisable space X is said to be *coarsely proper* if

$$\{g \in G \mid gK \cap K \neq \emptyset\}$$

for every compact set $K \subseteq X$. Furthermore, we say that the action is *modest* if $\overline{B \cdot K}$ is compact for all coarsely bounded subsets $B \subseteq G$ and compact $K \subseteq X$.

Of course, every locally compact Polish group has bounded geometry, but beyond that the primordial example is that of $\mathrm{Homeo}_{\mathbb{Z}}(\mathbb{R})$, which is the group of all homeomorphisms of $\mathbb{R}$ commuting with integral translations. One reason for the group's importance is its appearance in dynamics and topology, namely as the group of all lifts of orientation preserving homeomorphisms of the circle $\mathbb{S}^1$ to the universal cover $\mathbb{R}$, and thus its inclusion in the exact sequence

$$\mathbb{Z} \to \mathrm{Homeo}_{\mathbb{Z}}(\mathbb{R}) \to \mathrm{Homeo}_{+}(\mathbb{S}^1).$$

Other examples can be obtained in a similar manner or be built from these. For example, if we define a cocycle $\omega \colon \mathbb{Z}^2 \times \mathbb{Z}^2 \to \mathrm{Homeo}_{\mathbb{Z}}(\mathbb{R})$ by $\omega\big((x_1, x_2), (y_1, y_2)\big) = \tau_{x_1 y_2}$, where τ_n is the translation by n, then we obtain an extension

$$\mathrm{Homeo}_{\mathbb{Z}}(\mathbb{R}) \times_\omega \mathbb{Z}^2$$

of $\mathbb{Z}^2$ by $\mathrm{Homeo}_{\mathbb{Z}}(\mathbb{R})$ that is quasi-isometric to the Heisenberg group $H_3(\mathbb{Z})$.

Having established a dynamical criterion for bounded geometry of Polish groups, we turn our attention to a seminal result of geometric group theory due to M. Gromov. Namely, in Theorem $0.2.C_2'$ in [**36**], Gromov provides the following dynamical reformulation of quasi-isometry of finitely generated groups.

Theorem 1.10 (M. Gromov) *Two finitely generated groups Γ and Λ are quasi-isometric if and only if they admit commuting, continuous, proper and cocompact actions*

$$\Gamma \curvearrowright X \curvearrowleft \Lambda$$

on a locally compact Hausdorff space X.

This result also provides a model for other well-known notions of equivalence of groups such as measure equivalence involving measure-preserving actions on infinite measure spaces. Although Gromov's result easily generalises to arbitrary countable discrete groups, only recently, a collaboration with U. Bader [**3**] established the theorem for locally compact groups. Combining this analysis with the mechanics entering into the proof of Theorem 1.9, we succeed in finding the widest possible generalisation.

Theorem 1.11 *Let G and H be Polish groups of bounded geometry. Then the following hold:*

(1) *G coarsely embeds into H if and only if G and H admit commuting, continuous, coarsely proper, modest actions*

$$G \curvearrowright X \curvearrowleft H$$

on a locally compact Hausdorff space X where the H action is cocompact.

(2) *G and H are coarsely equivalent if and only if G and H admit commuting, cocompact, coarsely proper, modest and continuous actions*

$$G \curvearrowright X \curvearrowleft H$$

on a locally compact Hausdorff space X.

The commuting cocompact actions $G \curvearrowright X \curvearrowleft H$ above are said to be a *topological coupling* of G and H. An almost tautological example is given by $\text{Homeo}_\mathbb{Z}(\mathbb{R})$ in its coupling with $\mathbb{Z}$,

$$\mathbb{Z} \curvearrowright \mathbb{R} \curvearrowleft \text{Homeo}_\mathbb{Z}(\mathbb{R}).$$

The preceding results indicate the proximity of Polish groups of bounded geometry to the class of locally compact groups. However, these classes also display significant differences regarding their harmonic analytical properties. For example, by virtue of [**59**], $\text{Homeo}_\mathbb{Z}(\mathbb{R})$ admit no non-trivial continuous linear isometric representations on reflexive Banach spaces, whereas, N. Brown and E. Guentner [**14**] have shown that every locally compact Polish group has a *proper continuous affine isometric action* on a separable reflexive space. We obtain their result for bounded geometry groups under the added assumption of amenability.

Theorem 1.12 *Let G be an amenable Polish group of bounded geometry. Then G admits a coarsely* proper continuous affine isometric action *on a reflexive Banach space.*

Turning to topological dynamics, a result of W. Veech [**92**] states that every locally compact group acts freely on its *universal minimal flow*, which is a compact G-flow of which every other compact G-flow is a factor. For general Polish groups, however, this fails dramatically. In fact, there are examples of so-called *extremely amenable* Polish groups, that is, groups whose universal minimal flow reduces to a single point. Nevertheless, we show that Veech's theorem has purely geometric content by providing an appropriate generalisation of Veech's theorem to Polish groups of bounded geometry. In fact, by drawing on recent work of I. Ben Yaacov, J. Melleray and T. Tsankov [**7**], we are also able to generalise a result due to A. S. Kechris, V. G. Pestov and S. Todorčević [**48**] from the setting of locally compact groups to Polish groups of bounded geometry. Indeed, these authors show that if a locally compact Polish group has a metrisable universal minimal flow, then it must be compact. In the coarse setting instead, we find the following result.

Proposition 1.13 *Let G be a Polish group of bounded geometry whose universal minimal flow is metrisable. Then G is coarsely bounded.*

The Polish groups of bounded geometry also furnish the first example of a reasonably complete structural classification. Namely, a well-known result of H. Hopf [40] states that every finitely generated group quasi-isometric to $\mathbb{R}$ contains a finite index cyclic subgroup. We extend this to groups of bounded geometry in the following result.

Theorem 1.14 *Let G be a Polish group coarsely equivalent to $\mathbb{R}$. Then there is an open subgroup H of index at most 2 in G and a coarsely bounded set $A \subseteq H$ so that every $h \in H \setminus A$ generates a cobounded undistorted infinite cyclic subgroup.*

1.3.5 The Geometry of Automorphism Groups

In Chapter 6 we turn our attention to the class of non-Archimedean Polish groups or, equivalently, the automorphism groups Aut(**M**) of countable first-order structures **M**. These are of particular interest in mathematical logic and provide interesting links to model theory.

Given a structure **M**, the automorphism group Aut(**M**) acts naturally on finite tuples $\bar{a} = (a_1, \ldots, a_n)$ in **M** via

$$g \cdot (a_1, \ldots, a_n) = (ga_1, \ldots, ga_n)$$

and, with this notation, the pointwise stabiliser subgroups

$$V_{\bar{a}} = \{g \in \text{Aut}(\mathbf{M}) \mid g \cdot \bar{a} = \bar{a}\},$$

where $\bar{a}$ ranges over all finite tuples in **M**, form a neighbourhood basis at the identity in Aut(**M**). So, if $A \subseteq \mathbf{M}$ is the finite set enumerated by $\bar{a}$ and $\mathbf{A} \subseteq \mathbf{M}$ is the substructure generated by A, we have $V_A = V_A = V_{\bar{a}}$. An *orbital type* $\mathcal{O}$ in **M** is simply the orbit $\mathcal{O}(\bar{a}) = \text{Aut}(\mathbf{M}) \cdot \bar{a}$ of some tuple $\bar{a}$. Therefore, in the case where **M** is ω-homogeneous, $\mathcal{O}(\bar{a})$ is the set of realisations of the type $\text{tp}^{\mathbf{M}}(\bar{a})$ in **M**. Also, a collection $\mathcal{S}$ of orbital types $\mathcal{O} \subseteq \mathcal{O}(\bar{a}) \times \mathcal{O}(\bar{a})$ is *symmetric* if $\mathcal{O}(\bar{c}, \bar{b}) \in \mathcal{S}$ whenever $\mathcal{O}(\bar{b}, \bar{c}) \in \mathcal{S}$.

A geometric tool that will be used throughout this chapter is the graph $\mathbb{X}_{\bar{a}, \mathcal{S}}$ associated to a tuple $\bar{a}$ and a finite symmetric set $\mathcal{S}$ of orbital types $\mathcal{O} \subseteq \mathcal{O}(\bar{a}) \times \mathcal{O}(\bar{a})$. Here the vertex set of $\mathbb{X}_{\bar{a}, \mathcal{S}}$ is just $\mathcal{O}(\bar{a})$, while

$$(\bar{b}, \bar{c}) \in \text{Edge } \mathbb{X}_{\bar{a}, \mathcal{S}} \iff \bar{b} \neq \bar{c} \ \& \ \mathcal{O}(\bar{b}, \bar{c}) \in \mathcal{S}.$$

In particular, if $\mathbf{M}$ is atomic, then the edge relation is type $\emptyset$-definable, that is, it is type definable without parameters. We equip $\mathbb{X}_{\bar{a},\mathcal{S}}$ with the shortest-path metric $\rho_{\bar{a},\mathcal{S}}$ (which may take the value ∞ if $\mathbb{X}_{\bar{a},\mathcal{S}}$ is not connected).

Our first result provides a necessary and sufficient criterion for when an automorphism group has a well-defined quasi-isometric type and associated tools allowing for concrete computations of this same type.

Theorem 1.15 *Let $\mathbf{M}$ be a countable structure in a countable language. Then* $\mathrm{Aut}(\mathbf{M})$ *is monogenic if and only if there is a tuple $\bar{a}$ in $\mathbf{M}$ satisfying the following two requirements.*

(1) *For every tuple $\bar{b}$, there is a finite symmetric family $\mathcal{S}$ of orbital types $\mathcal{O} \subseteq \mathcal{O}(\bar{b}) \times \mathcal{O}(\bar{b})$ for which the orbit*

$$V_{\bar{a}} \cdot \bar{b} = \{\bar{c} \mid \mathcal{O}(\bar{c},\bar{a}) = \mathcal{O}(\bar{b},\bar{a})\}$$

has finite $\rho_{\bar{b},\mathcal{S}}$-diameter,

(2) *there is a finite symmetric family $\mathcal{R}$ of orbital types $\mathcal{O} \subseteq \mathcal{O}(\bar{a}) \times \mathcal{O}(\bar{a})$ so that $\mathbb{X}_{\bar{a},\mathcal{R}}$ is connected.*

Moreover, if $\bar{a}$ and $\mathcal{R}$ are as in (2), then the mapping

$$g \in \mathrm{Aut}(\mathbf{M}) \mapsto g \cdot \bar{a} \in \mathbb{X}_{\bar{a},\mathcal{R}}$$

is a quasi-isometry between $\mathrm{Aut}(\mathbf{M})$ and $(\mathbb{X}_{\bar{a},\mathcal{R}}, \rho_{\bar{a},\mathcal{R}})$.

Furthermore, condition (1) alone gives a necessary and sufficient criterion for $\mathrm{Aut}(\mathbf{M})$ being locally bounded and thus having a coarsely proper metric.

With this at hand, we can subsequently relate the properties of the theory $T = \mathrm{Th}(\mathbf{M})$ of the model $\mathbf{M}$ with properties of its automorphism group. First, a metric d on a set X is said to be *stable* in the sense of B. Maurey and J.-L. Krivine [53] provided that, for all bounded sequences (x_n), (y_n) and ultrafilters $\mathcal{U}, \mathcal{V}$,

$$\lim_{n \to \mathcal{U}} \lim_{m \to \mathcal{V}} d(x_n, y_m) = \lim_{m \to \mathcal{V}} \lim_{n \to \mathcal{U}} d(x_n, y_m).$$

We may then combine stability with a result from [80] to produce affine isometric actions on Banach spaces.

Theorem 1.16 *Suppose $\mathbf{M}$ is a countable atomic model of a stable theory T so that $\mathrm{Aut}(\mathbf{M})$ is locally bounded. Then $\mathrm{Aut}(\mathbf{M})$ admits a coarsely proper continuous affine isometric action on a reflexive Banach space.*

Although Theorem 1.15 furnishes an equivalent reformulation of admitting a metrically proper or maximal compatible left-invariant metric, it is often

useful to have more concrete instances of this. A particular case of this is when
M admits an orbital A-independence relation, $\perp_A$, that is, an independence
relation over a finite subset $A \subseteq \mathbf{M}$ satisfying the usual properties of symmetry,
monotonicity, existence and stationarity (see Definition 6.21 for a precise
rendering). In particular, this applies to the Boolean algebra of clopen subsets
of Cantor space with the dyadic probability measure and to the countably
regular tree.

Theorem 1.17 *Suppose A is a finite subset of a countable structure* **M** *and* $\perp_A$
is an orbital A-independence relation. Then the pointwise stabiliser subgroup
V_A *is coarsely bounded. Thus, if $A = \emptyset$, the automorphism group* Aut(**M**) *is
coarsely bounded and, if $A \neq \emptyset$,* Aut(**M**) *is locally bounded.*

Now, model theoretical independence relations arise, in particular, in
models of ω-stable theories. Although stationarity of the independence relation
may fail, we nevertheless arrive at the following result.

Theorem 1.18 *Suppose that* **M** *is a saturated countable model of an ω-stable
theory. Then* Aut(**M**) *is coarsely bounded.*

In this connection, we should mention an earlier observation by P. Cameron
[15], namely, that automorphism groups of countable $\aleph_0$-categorical structures
are Roelcke pre-compact and thus coarsely bounded.

A particular setting giving rise to orbital independence relations, which has
been studied by K. Tent and M. Ziegler [86], is Fraïssé classes admitting a
canonical amalgamation construction. For our purposes, we need a stronger
notion than that considered in [86], and say that a Fraïssé class $\mathcal{K}$ admits a
functorial amalgamation over some $\mathbf{A} \in \mathcal{K}$ if there is a map Θ that to every
pair of embeddings $\eta_1 \colon \mathbf{A} \hookrightarrow \mathbf{B}_1$ and $\eta_2 \colon \mathbf{A} \hookrightarrow \mathbf{B}_2$ with $\mathbf{B}_i \in \mathcal{K}$ produces
an amalgamation of $\mathbf{B}_1$ and $\mathbf{B}_2$ over these embeddings so that Θ is symmetric
in its arguments and commutes with embeddings (see Definition 6.29 for full
details).

Theorem 1.19 *Suppose $\mathcal{K}$ is a Fraïssé class with limit* **K** *admitting a functorial
amalgamation over some* $\mathbf{A} \in \mathcal{K}$. *Then* Aut(**K**) *is locally bounded. Moreover,
if* **A** *is generated by the empty set, then* Aut(**K**) *is coarsely bounded.*

1.3.6 Zappa–Szép Products

Zappa–Szép products of Polish groups appear frequently throughout our study.
These are groups G containing closed subgroups H and K so that $G = H \cdot K$

and $H \cap K = \{1\}$. For example, if H is a locally compact Polish group, the homeomorphism group $\mathrm{Homeo}(H)$ admits a Zappa–Szép decomposition

$$\mathrm{Homeo}(H) = H \cdot K,$$

where H is identified with the group of left-translations of H itself and

$$K = \{g \in \mathrm{Homeo}(H) \mid g(1_H) = 1_H\}$$

is the isotropy subgroup at the identity of H.

A closely related example is given by the automorphism group $\mathrm{Aut}(\mathsf{T}_n)$ of the n-regular tree, $n = 2, 3, \ldots, \aleph_0$. Because T_n is the Cayley graph of the non-abelian free group $\mathbb{F}_n$ on n generators with respect to its free generating set, one has a similar Zappa–Szép decomposition

$$\mathrm{Aut}(\mathsf{T}_n) = \mathbb{F}_n \cdot \mathrm{Aut}(\mathsf{T}_n, e),$$

where $\mathrm{Aut}(\mathsf{T}_n, e)$ is the isotropy subgroup at the identity element $e \in \mathbb{F}_n$.

As familiar from the special cases of direct and semi-direct products, in Chapter 7 we show that if a Polish group G is the Zappa–Szép product of two closed subgroups H and K, then G is actually homeomorphic with $H \times K$.

Theorem 1.20 *Suppose a Polish group G is the Zappa–Szép product of two closed subgroups H and K. Then the group multiplication*

$$H \times K \xrightarrow{\phi} G, \quad (h,k) \mapsto hk$$

defines a homeomorphism between the cartesian product $H \times K$ and G.

We also embark on a detailed study of their coarse structure and, in particular, investigate when the multiplication map is a coarse equivalence between the cartesian product $H \times K$ and G. For this, again let

$$H \times K \xrightarrow{\phi} G$$

be the group multiplication and let $G \xrightarrow{\pi_H} H$ and $G \xrightarrow{\pi_K} K$ denote the corresponding projections defined by

$$g = \pi_H(g) \cdot \pi_K(g).$$

Observe that neither ϕ, π_H nor π_K are homomorphisms in general. Also, if $X \subseteq H$, we let

$$X^K = \{khk^{-1} \mid h \in X \ \& \ k \in K\}.$$

Theorem 1.21 *Suppose a Polish group G is the Zappa–Szép product of closed subgroups H and K, where H is locally bounded. Then ϕ is a coarse equivalence if and only if, for every coarsely bounded subset $X \subseteq H$, the images $\pi_H[X^K]$ and $\pi_K[X^K]$ are coarsely bounded in H and K respectively.*

Note that, despite the apparent symmetry between writing $G = HK$ and $G = KH$, the groups H and K take on different roles in Theorem 1.21. This is because of the fact that requiring the multiplication map $H \times K \xrightarrow{\phi} G$ to be bornologous is not equivalent to demanding the multiplication map

$$K \times H \xrightarrow{\psi} G, \quad (k,h) \mapsto kh$$

to be bornologous.

2

Coarse Structure and Metrisability

2.1 Coarse and Uniform Spaces

We begin our introduction to large-scale geometry of groups by reviewing a few basic facts about uniform and coarse spaces. The first topic is, of course, classical by this point, whereas coarse spaces are not yet widely known. Let us start with some notation, terminology and simple facts.

Suppose X is a set and E, F are subsets of $X \times X$. We let

$$E^{\mathsf{T}} = \{(y, x) \mid (x, y) \in E\}$$

denote the *transpose* of E and let

$$E \circ F = \{(x, z) \mid \exists y \ (x, y) \in E \text{ and } (y, z) \in F\}$$

denote the *composition* of the two. Also, $\Delta(X)$ or often just Δ will denote the *diagonal* of X,

$$\Delta(X) = \{(x, x) \mid x \in X\}.$$

Occasionally, if $\mathcal{F}$ is a collection of subsets of $X \times X$, we will write $\mathcal{F}^{\mathsf{T}}$ for the collection of transpositions $\{E^{\mathsf{T}} \mid E \in \mathcal{F}\}$. One immediately sees that composition is associative, so we can unambiguously write $E_1 \circ E_2 \circ \cdots \circ E_n$ for iterated compositions.

Remark 2.1 For any sets $E, F, E_i, F_i \subseteq X \times X$, we have

$$(E \cup F)^{\mathsf{T}} = E^{\mathsf{T}} \cup F^{\mathsf{T}}, \qquad (E \circ F)^{\mathsf{T}} = F^{\mathsf{T}} \circ E^{\mathsf{T}}$$

along with the following distributive law

$$(E_1 \cup E_2) \circ (F_1 \cup F_2) = (E_1 \circ F_1) \cup (E_1 \circ F_2) \cup (E_2 \circ F_1) \cup (E_2 \circ F_2).$$

19

Intersections, on the other hand, are slightly more complicated:

$$(E_1 \cap E_2) \circ (F_1 \cap F_2) \subseteq (E_1 \circ F_1) \cap (E_1 \circ F_2) \cap (E_2 \circ F_1) \cap (E_2 \circ F_2).$$

Note that there is, in general, only an inclusion from left to right.

Uniform spaces first appear in the work of A. Weil [94]. A *uniform structure* or *uniformity* on a set X is a family $\mathcal{U}$ of subsets $E \subseteq X \times X$ called *entourages* verifying the following conditions.

(1) $\mathcal{U}$ is a *filter of sets*, that is, $\mathcal{U}$ is non-empty and is closed under taking supersets

$$E \subseteq F \text{ and } E \in \mathcal{U} \Rightarrow F \in \mathcal{U}$$

and finite intersections

$$E, F \in \mathcal{U} \Rightarrow E \cap F \in \mathcal{U},$$

(2) every $E \in \mathcal{U}$ contains the diagonal Δ of X,
(3) $\mathcal{U}$ is closed under transposition

$$E \in \mathcal{U} \Rightarrow E^{\mathsf{T}} \in \mathcal{U},$$

(4) for any $E \in \mathcal{U}$, there is some $F \in \mathcal{U}$ so that

$$F \circ F \subseteq E.$$

A *uniform space* is simply a set X equipped with a uniform structure $\mathcal{U}$ on X.

Whereas the concept of uniform spaces captures the idea of being 'uniformly close' in a topological space, J. Roe [75] (see also [74] for an earlier definition) provided a corresponding axiomatisation of being at 'uniformly bounded distance', namely the notion of a coarse space.

Definition 2.2 (Coarse structures and spaces) A *coarse structure* on a set X is a collection $\mathcal{E}$ of subsets $E \subseteq X \times X$ called *entourages* satisfying the following conditions.

(1) $\mathcal{E}$ is an *ideal of sets*, that is, $\mathcal{E}$ is non-empty and closed under taking subsets

$$E \supseteq F \text{ and } E \in \mathcal{U} \Rightarrow F \in \mathcal{U}$$

and finite unions

$$E, F \in \mathcal{U} \Rightarrow E \cup F \in \mathcal{U},$$

(2) the diagonal Δ belongs to $\mathcal{E}$,
(3) $\mathcal{E}$ is closed under transposition,

(4) $\mathcal{E}$ is closed under composition

$$E, F \in \mathcal{E} \Rightarrow E \circ F \in \mathcal{E}.$$

A *coarse space* is just a set X equipped with a coarse structure $\mathcal{E}$.

Example 2.3 (Pseudo-metric spaces) The canonical example of both a coarse space and a uniform space is when (X, d) is a metric or, more generally, a pseudo-metric space. Recall here that a *pseudo-metric* space is a set X equipped with an *écart*, that is, a map $d \colon X \times X \to \mathbb{R}_+$ so that, for all $x, y, z \in X$, we have

(1) $d(x, y) = d(y, x)$,
(2) $d(x, x) = 0$,
(3) $d(x, y) \leqslant d(x, z) + d(z, y)$.

In this case, we may, for every $\alpha > 0$, construct an entourage by

$$E_\alpha = \{(x, y) \mid d(x, y) < \alpha\}$$

and define a uniformity $\mathcal{U}_d$ on X by

$$\mathcal{U}_d = \{E \subseteq X \times X \mid \exists \alpha > 0 \ E_\alpha \subseteq E\}.$$

Similarly, a coarse structure $\mathcal{E}_d$ is obtained by

$$\mathcal{E}_d = \{E \subseteq X \times X \mid \exists \alpha < \infty \ E \subseteq E_\alpha\}.$$

Alternatively, a set $E \subseteq X \times X$ belongs to $\mathcal{E}_d$ if and only if

$$\sup_{(x, y) \in E} d(x, y) < \infty.$$

As is evident from the definition, the intersection $\bigcap_{i \in I} \mathcal{E}_i$ of an arbitrary family $\{\mathcal{E}_i\}_{i \in I}$ of coarse structures on a set X is again a coarse structure. This observation gives rise to the following example.

Example 2.4 (The coarse structure generated by a family of écarts) Suppose $\{d_i\}_{i \in I}$ is a family of écarts on a set X. Then the intersection

$$\mathcal{E} = \bigcap_{i \in I} \mathcal{E}_{d_i}$$

of the associated coarse structures is again a coarse structure on X. Note also that a set $E \subseteq X \times X$ belongs to $\mathcal{E}$ if and only if

$$\sup_{(x, y) \in E} d_i(x, y) < \infty$$

for all $i \in I$. We say that $\mathcal{E}$ is *generated* by the family of écarts $\{d_i\}_{i \in I}$.

Example 2.5 (Generation, subbases and bases) Another noteworthy fact is that, if $\mathcal{F}$ is any collection of subsets of $X \times X$, then there is a smallest coarse structure $\mathcal{E}$ on X containing $\mathcal{F}$. Indeed, note first that there is at least one coarse structure containing $\mathcal{F}$, namely the powerset $\mathcal{P}(X \times X)$. So the smallest coarse structure containing $\mathcal{F}$ is simply the intersection of all coarse structures that contain $\mathcal{F}$. We say that $\mathcal{E}$ is *generated* by $\mathcal{F}$ and that $\mathcal{F}$ is a *subbasis* for $\mathcal{E}$.

Also, a collection $\mathcal{F}$ of subsets of $X \times X$ is a *basis* for a coarse structure $\mathcal{E}$ on X provided that

$$\mathcal{E} = \{E \subseteq X \times X \mid E \subseteq F \text{ for some } F \in \mathcal{F}\}.$$

We can also explicitly describe the coarse structure $\mathcal{E}$ generated by any collection $\mathcal{F}$ as follows. Namely, given $\mathcal{F}$, let $\tilde{\mathcal{F}}$ denote the family consisting of all finite unions of compositions

$$F_1 \circ \cdots \circ F_n,$$

where $F_1, \ldots, F_n \in \{\Delta\} \cup \mathcal{F} \cup \mathcal{F}^\mathsf{T}$. Then because, by Remark 2.1, T commutes with $\cup$ and anti-commutes with $\circ$, we see that $\tilde{\mathcal{F}}$ is closed under T. As $\circ$ distributes over $\cup$, we have that $\tilde{\mathcal{F}}$ is closed under $\circ$ and, of course, is also closed under finite unions. Because each of the operators T, $\circ$ and $\cup$ are monotone with respect to inclusion $\subseteq$, it follows that

$$\mathcal{E} = \{E \subseteq X \times X \mid E \subseteq F \text{ for some } F \in \tilde{\mathcal{F}}\}$$

is an ideal stable under T and $\circ$; that is, $\mathcal{E}$ is the smallest coarse structure on X containing $\mathcal{F}$. In particular, $\tilde{\mathcal{F}}$ is a basis for $\mathcal{E}$.

We observe that the family $\mathfrak{C}$ of coarse structures on a set X forms a complete lattice. Indeed, if $\{\mathcal{E}_i\}_{i \in I} \subseteq \mathfrak{C}$ is any subcollection, then the meet is given by

$$\bigwedge_{i \in I} \mathcal{E}_i = \bigcap_{i \in I} \mathcal{E}_i$$

whereas the join $\bigvee_{i \in I} \mathcal{E}_i$ is the smallest coarse structure containing all of the $\mathcal{E}_i$.

For uniform structures the situation is slightly more complicated. Although, in general, neither an arbitrary union nor an arbitrary intersection of uniform structures is a uniform structure, it is easy to see that the union of a directed system of uniform structures is again a uniform structure. That is, if $\{\mathcal{U}_i\}_{i \in I}$ is a family of uniform structures on X so that, for all $i, j \in I$, there is some $k \in I$ with $\mathcal{U}_i \cup \mathcal{U}_j \subseteq \mathcal{U}_k$, then $\bigcup_{i \in I} \mathcal{U}_i$ is again a uniform structure.

Example 2.6 (The uniform structure generated by a directed family of écarts) Suppose $\{d_i\}_{i \in I}$ is a directed family of écarts on a set X, that is, for all $i, j \in I$ there is some $k \in I$ so that $d_i \leqslant d_k$ and $d_j \leqslant d_k$. Then the associated family of uniform structures $\{\mathcal{U}_{d_i}\}_{i \in I}$ is also directed and so the union $\mathcal{U} = \bigcup_{i \in I} \mathcal{U}_{d_i}$ is also a uniform structure on X. We note that a set $E \subseteq X \times X$ belongs to $\mathcal{U}$ if and only if

$$\inf_{(x,y) \notin E} d_i(x,y) > 0$$

for some $i \in I$.

Still, if $\{\mathcal{U}_i\}_{i \in I}$ is an arbitrary family of uniformities on a set X, then there is a smallest uniformity containing all of them. Concretely, we set

$$\bigvee_{i \in I} \mathcal{U}_i = \left\{ \bigcap_{E \in \mathcal{S}} E \;\middle|\; \mathcal{S} \subseteq \bigcup_{i \in I} \mathcal{U}_i \text{ and } |\mathcal{S}| < \infty \right\}.$$

It is easy to check that this is a filter all of whose elements contain Δ and which is closed under transposes $E \mapsto E^{\mathsf{T}}$. Also, suppose $E_1, \ldots, E_n \in \bigcup_{i \in I} \mathcal{U}_i$. Then there are $F_1, \ldots, F_n \in \bigcup_{i \in I} \mathcal{U}_i$ with $F_k \circ F_k \subseteq E_k$, whereby

$$(F_1 \cap \cdots \cap F_n) \circ (F_1 \cap \cdots \cap F_n) \subseteq E_1 \cap \cdots \cap E_n$$

and $F_1 \cap \cdots \cap F_n \in \bigvee_{i \in I} \mathcal{U}_i$. So $\bigvee_{i \in I} \mathcal{U}_i$ is indeed the smallest uniformity containing all the $\mathcal{U}_i$, and hence is the join of the $\mathcal{U}_i$ in the family $\mathfrak{U}$ of uniformities on X. The meet $\bigwedge_{i \in I} \mathcal{U}_i$ in $\mathfrak{U}$ is then the join of all uniformities that are contained in the intersection $\bigcap_{i \in I} \mathcal{U}_i$. So also $\mathfrak{U}$ is a complete lattice.

Suppose $(X, \mathcal{U})$ is a uniform space and d is an écart on X. We say that d is $\mathcal{U}$-*uniformly continuous* or just *uniformly continuous* provided that $\mathcal{U}_d \subseteq \mathcal{U}$, that is, provided that

$$\{(x,y) \mid d(x,y) < \epsilon\} \in \mathcal{U}$$

for all $\epsilon > 0$. The écart d is also said to be *compatible* with $\mathcal{U}$ if it induces the uniformity $\mathcal{U}$, that is, if actually $\mathcal{U} = \mathcal{U}_d$.

The family of uniformly continuous écarts is directed and, as shown by Weil [94], every uniform structure $\mathcal{U}$ on a set is generated by its family of uniformly continuous écarts.

2.2 Coarse and Uniform Structures on Groups

Throughout this text, we shall use operations on sets in groups. So let us fix the notation once and for all.

Notation 2.7 Suppose A and B are subsets of a group G and $n \geqslant 1$. Then we set

$$A^{-1} = \{a^{-1} \mid a \in A\},$$

$$AB = A \cdot B = \{ab \mid a \in A \text{ and } b \in B\}$$

and

$$A^n = \{a_1 a_2 \cdots a_n \mid a_i \in A\}.$$

To avoid confusion with cartesian products, for example in the notation G^2, we will always write the cartesian product using $\times$, i.e., $G \times G$.

Also, in order to facilitate the discussion of uniformities and coarse structures on group G, note that, if $A \subseteq G$ is an arbitrary subset, we may define a subset of $G \times G$ by

$$E_A = \{(x,y) \in G \times G \mid x^{-1}y \in A\}.$$

Note that E_A is *left-invariant* in the sense that

$$(x,y) \in E_A \iff (zx, zy) \in E_A$$

for all $x, y, z \in G$. We observe that, while $y^{-1}x = (x^{-1}y)^{-1}$, the product xy^{-1} cannot, in general, be computed from $x^{-1}y$. Thus, in the definition of E_A, the exact formula $x^{-1}y$ is important. Using xy^{-1} instead would lead to a right-invariant set.

Remark 2.8 It is worth pointing a few basic facts that will be used repeatedly. Namely, suppose $A, B \subseteq G$. Then

$$E_A^{\mathsf{T}} = E_{A^{-1}}, \quad E_{A \cup B} = E_A \cup E_B, \quad E_A \circ E_B = E_{AB}$$

and

$$E_A[B] := \{x \in G \mid \exists b \in B \ (x,b) \in E_A\} = BA^{-1}.$$

Moreover, evidently, $A \subseteq B$ if and only if $E_A \subseteq E_B$.

As is well known, a topological group G has a number of naturally defined uniformities (cf. [76] for a deeper study). The following is of particular importance to us.

Example 2.9 (Left-uniform structure on a topological group) Suppose G is a topological group. The *left-uniformity* $\mathcal{U}_L$ on G is that generated by the family of entourages

$$E_V = \{(x,y) \in G \times G \mid x^{-1}y \in V\},$$

where V varies over identity neighbourhoods in G.

As with any other uniformity, $\mathcal{U}_L$ is generated by the family of $\mathcal{U}_L$-uniformly continuous écarts. More interestingly, the left-uniformity $\mathcal{U}_L$ is actually generated by the directed family of continuous *left-invariant* écarts on G, i.e., continuous écarts d so that $d(xy, xz) = d(y, z)$ for all $x, y, z \in G$. In other words,

$$\mathcal{U}_L = \bigcup \{\mathcal{U}_d \mid d \text{ is a continuous left-invariant écart on } G\}.$$

Note also that left-invariant continuous écarts on G are automatically $\mathcal{U}_L$-uniformly continuous.

The left-uniformity is the main uniformity on topological groups we shall consider and, unless explicitly stated otherwise, all notions of uniform continuity will always refer to this structure. So uniformly or left-uniformly continuous functions will always mean uniformly continuous with respect to $\mathcal{U}_L$.

If $\mathcal{U}$ and $\mathcal{V}$ are uniformities on a set X, we say that $\mathcal{U}$ is *finer than* or *refines* $\mathcal{V}$ if

$$\mathcal{U} \supseteq \mathcal{V}.$$

On the other hand, for coarse structures $\mathcal{E}$ and $\mathcal{F}$ on X, we say that $\mathcal{E}$ is *finer than* $\mathcal{F}$ if

$$\mathcal{E} \subseteq \mathcal{F}.$$

By the previous example, $\mathcal{U}_L$ is the smallest common refinement of all the $\mathcal{U}_d$, where d varies over continuous left-invariant écarts on G. However, the common refinement of a class of coarse structures is given by their intersection rather than by their union. Therefore, in analogy with the description of the left-uniform structure given above, we define the left-coarse structure on a topological group as follows.

Definition 2.10 (Left-coarse structure on a topological group) For a topological group G, we define its *left-coarse structure* $\mathcal{E}_L$ by

$$\mathcal{E}_L = \bigcap \{\mathcal{E}_d \mid d \text{ is a continuous left-invariant écart on } G\}.$$

From the preceding discussion, we have the following basic but important reformulation.

Lemma 2.11 *Let G be a topological group. Then a subset $E \subseteq G \times G$ is a coarse entourage, that is, belongs to $\mathcal{E}_L$, if and only if*

$$\sup_{(x, y) \in E} d(x, y) < \infty$$

for every continuous left-invariant écart d on G.

The definition of the left-coarse structure, although completely analogous to the description of the left-uniformity, is highly impredicative because it involves quantification over the class of all continuous left-invariant écarts on G. It is therefore instructive to seek alternative descriptions and approaches to it, which we will do in the following.

2.3 Coarsely Bounded Sets

With every coarse structure comes a notion of coarsely bounded sets, which, in the case of a metric space, would be the sets of finite diameter.

Definition 2.12 A subset $A \subseteq X$ of a coarse space $(X, \mathcal{E})$ is said to be *coarsely bounded* or *$\mathcal{E}$-bounded* if $A \times A \in \mathcal{E}$.

Our next task is to provide an informative reformulation of the class of coarsely bounded subsets of a topological group equipped with its left-coarse structure. This will be based on the classical metrisation theorem of G. Birkhoff [11] and S. Kakutani [43] or, more precisely, on the following lemma underlying Birkhoff's construction in [11] (see also [41]).

Lemma 2.13 *Let G be a topological group and $(V_n)_{n \in \mathbb{Z}}$ a increasing chain of symmetric open identity neighbourhoods satisfying $G = \bigcup_{n \in \mathbb{Z}} V_n$ and $V_n^3 \subseteq V_{n+1}$ for all $n \in \mathbb{Z}$. Define, for $g, f \in G$,*

$$\delta(g, f) = \inf \left(2^n \mid g^{-1} f \in V_n \right)$$

and put

$$d(g, f) = \inf \left(\sum_{i=0}^{k-1} \delta(h_i, h_{i+1}) \mid h_0 = g, h_k = f \right).$$

Then

$$\frac{1}{2} \delta(g, f) \leqslant d(g, f) \leqslant \delta(g, f)$$

and d is a continuous left-invariant écart on G. Moreover, if $(V_n)_{n \in \mathbb{Z}}$ is actually a neighbourhood basis at the identity, then d induces the topology on G.

Note that, if d is a left-invariant *compatible* metric on G, that is, one that induces the topology of G, then d is also a *compatible* metric for the left-uniformity, i.e., $\mathcal{U}_L = \mathcal{U}_d$. With this observation, the following metrisation theorem of Birkhoff and Kakutani is an immediate consequence. We shall use both Lemma 2.13 and the formulation below several times in our study.

Theorem 2.14 (Birkhoff–Kakutani metrisation theorem) *Let G be a Hausdorff topological group. Then the following conditions are equivalent.*

(1) *The topology of G is first countable,*
(2) *the topology on G is metrisable,*
(3) *there is a compatible left-invariant metric for the left-uniformity on G.*

Our aim is eventually to provide a similar characterisation for the (left-invariant) metrisability of the left-coarse structure on Polish groups in terms of the unidimensional criterion as in condition (1) of Theorem 2.14. However, first we must study the coarsely bounded sets by using Lemma 2.13.

Proposition 2.15 *Let G be a topological group equipped with its left-coarse structure. Then the following conditions are equivalent for a subset $A \subseteq G$.*

(1) *A is coarsely bounded,*
(2) *for every continuous left-invariant écart d on G,*

$$\mathrm{diam}_d(A) < \infty,$$

(3) *for every continuous isometric action on a metric space $G \curvearrowright (X, d)$ and every $x \in X$, we have*

$$\mathrm{diam}_d(A \cdot x) < \infty,$$

(4) *for every increasing exhaustive sequence $V_1 \subseteq V_2 \subseteq \cdots \subseteq G$ of open subsets with $V_n^2 \subseteq V_{n+1}$, we have $A \subseteq V_n$ for some n.*

Moreover, suppose G is countably generated over every identity neighbourhood, *that is, for every identity neighbourhood V there is a countable set $C \subseteq G$ so that G is algebraically generated by $V \cup C$. Then (1)–(4) are equivalent to*

(5) *for every identity neighbourhood V, there is a finite set $F \subseteq G$ and a $k \geqslant 1$ so that $A \subseteq (FV)^k$.*

Proof (1)⇔(2) By Lemma 2.11, observe that $A \times A \in \mathcal{E}_L$ if and only if

$$\mathrm{diam}_d(A) = \sup_{(x,y) \in A \times A} d(x, y) < \infty$$

for every continuous left-invariant écart d. So (1) and (2) are equivalent.

(3)⇒(2) If d is a continuous left-invariant écart on G, we let X be the corresponding metric quotient of G, i.e., X is the quotient of G by the equivalence relation of having d-distance 0 and equipped with the metric induced by d. Then the left-shift action of G on itself factors through to a continuous transitive isometric action on the metric space X. Therefore, if every A-orbit in X is bounded, A is also d-bounded in G.

(2)$\Rightarrow$(3) Conversely, suppose $G \curvearrowright (X, d)$ is a continuous isometric action on some metric space. Then, for any fixed $x \in X$, the formula

$$\partial(g, f) = d(g \cdot x, f \cdot x)$$

defines a continuous left-invariant écart ∂ on G. Moreover, if A is ∂-bounded, then the A-orbit $A \cdot x$ is d-bounded and the same is true for any other A-orbit on X.

(4)$\Rightarrow$(2) Note that, if d is a continuous left-invariant écart on G, then

$$d(1, xy) \leqslant d(1, x) + d(x, xy) = d(1, x) + d(1, y)$$

for all $x, y \in G$. It follows that

$$V_n = \{x \in G \mid d(1, x) < 2^n\}$$

defines an increasing exhaustive chain of open subsets of G so that $V_n^2 \subseteq V_{n+1}$. Moreover, the d-bounded sets in G are exactly those that are contained in some V_n. Thus, if A satisfies (4), it must have finite diameter with respect to every continuous left-invariant écart on G.

(2)$\Rightarrow$(4) Conversely, suppose there is some increasing exhaustive chain of open subsets $W_1 \subseteq W_2 \subseteq \cdots \subseteq G$ so that $W_n^2 \subseteq W_{n+1}$, while $A \nsubseteq W_n$ for all n. Then $U_n = W_n \cap W_n^{-1}$ defines an increasing exhaustive sequence of *symmetric* open subsets of G still satisfying $U_n^2 \subseteq U_n$, while $A \nsubseteq U_n$ for all n. By passing to a tail subsequence, we may already suppose that $1 \in U_1$. Also, pick symmetric open identity neighbourhoods $V_k \subseteq U_1$ for all $k \leqslant 0$ so that $V_{k-1}^3 \subseteq V_k$ and set $V_k = U_{2k+2}$ for $k > 0$. Then $(V_k)_{k \in \mathbb{Z}}$ is an increasing and exhaustive bi-infinite sequence of symmetric open identity neighbourhoods in G satisfying

$$V_k^3 \subseteq V_{k+1}$$

for all $k \in \mathbb{Z}$. Therefore, by Lemma 2.13, there is a continuous left-invariant écart d on G so that, for all $k \geqslant 1$ and $g \in G$,

$$d(g, 1) < 2^k \Rightarrow g \in V_k.$$

As A is not contained in any of the V_k, it follows that $\mathrm{diam}_d(A) = \infty$, showing $\neg(4) \Rightarrow \neg(2)$.

$(5) \Rightarrow (4)$ That (5) implies (4) is obvious. Suppose (5) holds and

$$V_1 \subseteq V_2 \subseteq \cdots \subseteq G$$

are as in (4). Then we may find some finite F and $k \geqslant 1$ for which $A \subseteq (FV_1)^k$, whereby

$$A \subseteq (FV_1)^k \subseteq (V_m V_1)^k \subseteq (V_{m+1})^k \subseteq (V_{m+2})^{k-1} \subseteq \cdots \subseteq V_{m+k}$$

provided that m is large enough so that $F \subseteq V_m$.

$(4) \Rightarrow (5)$ Conversely, assume now that, in addition, G is countably generated over every identity neighbourhood. Suppose that A is coarsely bounded and V is an identity neighbourhood. Pick a countable set $C = \{x_n\}_{n=1}^{\infty}$ generating G over V and let $V_n = \left(V \cup \{x_1, \ldots, x_n\}\right)^{2^n}$. Then the V_n are as in (4), whence $A \subseteq V_n = \left(V \cup \{x_1, \ldots, x_n\}\right)^{2^n}$ for some n, showing $(4) \Rightarrow (5)$. $\square$

Definition 2.16 Let G be a topological group. Then $\mathcal{CB}$ denotes the family of coarsely bounded sets in G when G is equipped with its left-coarse structure.

As a consequence of Proposition 2.15, we have the following.

Corollary 2.17 *Let G be a topological group equipped with its left-coarse structure. Then $\mathcal{CB}$ is an ideal of sets stable under the operations*

$$A \mapsto \mathsf{cl}\, A, \quad A \mapsto A^{-1}, \quad (A, B) \mapsto AB.$$

Proof That $\mathcal{CB}$ is closed under taking subsets, finite unions and topological closures follows directly from condition (2) of Proposition 2.15. Stability of $\mathcal{CB}$ under products AB, on the other hand, follows from condition (4) of the same proposition. Also, if d is a continuous left-invariant écart on G, then $d(a^{-1}, 1) = d(1, a)$ for all a and, hence, if A has finite d-diameter, so does A^{-1}. $\square$

Note that, if G is countably generated over every identity neighbourhood, then, by condition (5) of Proposition 2.15, in order to detect the coarse boundedness of a subset A, we need only quantify over basic identity neighbourhoods V rather than over all continuous left-invariant écarts d on G. In practice, the first condition is often the easiest to consider and, in particular, has the following consequence.

Corollary 2.18 *Let G be a Polish group and $\mathcal{F}(G)$ the Effros–Borel space of closed subsets of G. Then*

$$\mathcal{CB} \cap \mathcal{F}(G)$$

is a Borel set in $\mathcal{F}(G)$.

We recall that the *Effros–Borel space* of a Polish space X is the measurable space consisting of the collection $\mathcal{F}(X)$ of closed subsets of X equipped with the σ-algebra generated by sets

$$\{C \in \mathcal{F}(G) \mid C \cap U \neq \emptyset\},$$

where U varies over open subsets of X. Because this σ-algebra is actually the Borel algebra associated with a Polish topology on $\mathcal{F}(G)$, its elements are called the Borel sets of $\mathcal{F}(G)$.

Proof Let $\{U_n\}_{n \in \mathbb{N}}$ be a neighbourhood basis at the identity in G and let $\{x_n\}_{n \in \mathbb{N}}$ be a countable dense subset of G with $x_1 = 1$. We let

$$V_{n,k} = G \setminus \overline{\big(\{x_1, \ldots, x_k\}U_n U_n\big)}^k$$

and claim that a closed subset $C \subseteq G$ is coarsely bounded if and only if

$$\forall n \; \exists k \quad C \cap V_{n,k} = \emptyset,$$

which is clearly a Borel condition in $\mathcal{F}(G)$.

To see the equivalence, assume first that C is coarsely bounded and fix some n. Then there is a finite set $F \subseteq G$ and some m so that $C \subseteq (FU_n)^m$. Because $\{x_i\}_{i \in \mathbb{N}}$ is dense in G, we have $G = \bigcup_{i \in \mathbb{N}} x_i U_n$ and so $F \subseteq \{x_1, \ldots, x_k\}U_n$ for some $k \geqslant m$ that is large enough. Therefore,

$$C \subseteq (FU_n)^m \subseteq \big(\{x_1, \ldots, x_k\}U_n U_n\big)^m \subseteq \overline{\big(\{x_1, \ldots, x_k\}U_n U_n\big)}^k$$

and hence $C \cap V_{n,k} = \emptyset$ as required.

Conversely, assume that $\forall n \; \exists k \; C \cap V_{n,k} = \emptyset$. To see that C is coarsely bounded, let an identity neighbourhood W be given and find n so that $U_n \subseteq W$. Choose k so that $C \cap V_{n,k} = \emptyset$. Then

$$C \subseteq \overline{\big(\{x_1, \ldots, x_k\}U_n U_n\big)}^k \subseteq \big(\{x_1, \ldots, x_k\}U_n U_n\big)^k \cdot W \subseteq \big(\{x_1, \ldots, x_k\}W^2\big)^{k+1},$$

thus verifying that C is coarsely bounded. $\qquad\square$

Also, in σ-compact locally compact groups we get a simple characterisation of the coarsely bounded sets.

Corollary 2.19 *A subset A of a σ-compact locally compact group G is coarsely bounded in the left-coarse structure if and only if it is relatively compact.*[1]

---

[1] We say that A is *relatively compact* if it has compact topological closure in G.

Proof Assume first that G is locally compact, but not necessarily countably generated over every identity neighbourhood. Fix a compact identity neighbourhood V in G. Then $(FV)^k$ is also compact for all finite sets $F \subseteq G$ and $k \geqslant 1$. Also, V covers any compact set by finitely many left-translates. Therefore, we see that $A \subseteq G$ is relatively compact if and only if $A \subseteq (FV)^k$ for some finite F and $k \geqslant 1$.

Note now that a locally compact group G is σ-compact exactly when G is countably generated over every identity neighbourhood. Indeed, if G is σ-compact, we write $G = \bigcup_{n \in \mathbb{N}} K_n$ for a sequence of compact sets K_n. Then, if V is an identity neighbourhood, we can for each n find some finite set F_n so that $K_n \subseteq F_n V$, whereby G is generated by $\bigcup_{n \in \mathbb{N}} F_n$ over V. Conversely, if G is generated by some countable set $\{x_n\}_{n \in \mathbb{N}}$ over a compact identity neighbourhood V, then G will also be the union of the countable many compact sets

$$K_n = \overline{\left(\{1, x_1^{\pm}, \dots, x_n^{\pm}\} V^{\pm}\right)^n}$$

and thus will be σ-compact.

The equivalence of the corollary now follows from Proposition 2.15. □

Corollary 2.20 *A subset A of a Banach space $(X, \|\cdot\|)$ is coarsely bounded in the left-coarse structure of the underlying additive group $(X, +)$ if and only if A is norm bounded.*

Proof Since the norm defines a compatible invariant metric on X, the coarsely bounded sets are automatically norm bounded. Thus, it suffices to show that a norm-bounded set A is also coarsely bounded. But, every identity neighbourhood in $(X, +)$ contains a norm ball $\epsilon B_X = \{x \in X \mid \|x\| < \epsilon\}$ for some $\epsilon > 0$ and

$$A \subseteq (n\epsilon) \cdot B_X = \underbrace{\epsilon B_X + \cdots + \epsilon B_X}_{n \text{ times}}$$

for some sufficiently large n, showing that A is coarsely bounded. □

Let us note that whereas relative compactness of a subset A of a topological group G depends only on the closure cl A in G as a topological space, but not on the remaining part of the ambient topological group G, the situation is very different for coarse boundedness. Indeed, although A is coarsely bounded in G if and only if cl A is, it is possible that A is coarsely bounded in G, while failing to be coarsely bounded in some intermediate closed subgroup $A \subseteq H \leqslant G$. So, one must always stress in which ambient group G a subset A is coarsely bounded. We will see many examples of this going forward.

2.4 Comparison with Other Left-Invariant Coarse Structures

In the same way as a left-invariant uniformity on a group can be described in terms of a certain filter, one may reformulate left-invariant coarse structure on groups as ideals of subsets. This connection is explored in greater detail by A. Nicas and D. Rosenthal in [**67**] and we shall content ourselves with some elementary observations here.

Recall that, if G is a group, we call a subset $F \subseteq G \times G$ left-invariant if, for all $x, y, z \in G$, we have

$$(x, y) \in F \Leftrightarrow (zx, zy) \in F.$$

Note that, if F is left-invariant, then it can be recovered from the one-dimensional set

$$A_F = \{x^{-1}y \mid (x, y) \in F\}$$

by noting that $F = \{(x, y) \in G \times G \mid x^{-1}y \in A_F\}$. Indeed, if $x^{-1}y \in A_F$, then $x^{-1}y = z^{-1}u$ for some pair $(z, u) \in F$, whereby also

$$(x, y) = (xz^{-1} \cdot z, xz^{-1} \cdot u) \in F.$$

Conversely, if B is any subset of G, then B may be recovered from the left-invariant set

$$E_B = \{(x, y) \in G \times G \mid x^{-1}y \in B\}$$

by observing that $B = \{x^{-1}y \mid (x, y) \in E_B\}$. Thus, $B \mapsto E_B$ defines a bijection between subsets of G and left-invariant subsets of $G \times G$ with inverse $F \mapsto A_F$.

Observe also that, if E, F are left-invariant, then so are E^{T}, $E \cup F$ and $E \circ F$. It follows that, if $\mathcal{F}$ is a collection of left-invariant subsets $E \subseteq G \times G$, then the closure $\tilde{\mathcal{F}}$ of $\mathcal{F} \cup \{\Delta\}$ under T, $\circ$ and $\cup$ (see Example 2.5) is a basis consisting of left-invariant sets for a coarse structure $\mathcal{E}$.

If $E \subseteq G \times G$ is any set, then the *left-saturation*

$$\hat{E} = \{(zx, zy) \in G \times G \mid (x, y) \in E \text{ and } z \in G\}$$

is the smallest left-invariant set containing E. We say that a coarse structure $\mathcal{E}$ on a group G is *left-invariant* if it is stable under left-saturation, that is, if

$$E \in \mathcal{E} \implies \hat{E} \in \mathcal{E}.$$

Alternatively, $\mathcal{E}$ is left-invariant if the left-invariant entourages form a basis for $\mathcal{E}$. Such coarse structures are also called *compatible* in [**67**]. For example, being the intersection of a family of left-invariant coarse structures, the left-coarse structure $\mathcal{E}_L$ on a topological group G is itself left-invariant.

Lemma 2.21 *Suppose G is a group equipped with a left-invariant coarse structure $\mathcal{E}$. Then*

$$\mathcal{I}_{\mathcal{E}} = \{A_F \mid F \in \mathcal{E}\} = \{B \subseteq G \mid E_B \in \mathcal{E}\}$$

is the ideal of coarsely bounded sets. Furthermore, this ideal contains $\{1\}$ and is closed under inversion $B \mapsto B^{-1}$ and products of sets $(A, B) \mapsto AB$.

Proof To see that

$$\{A_F \mid F \in \mathcal{E}\} = \{B \subseteq G \mid E_B \in \mathcal{E}\},$$

suppose first that $C \in \{A_F \mid F \in \mathcal{E}\}$. Then $C = A_F$ for some $F \in \mathcal{E}$, whereby $E_C = E_{(A_F)} = F \in \mathcal{E}$ and so $C \in \{B \subseteq G \mid E_B \in \mathcal{E}\}$. Conversely, if $C \in \{B \subseteq G \mid E_B \in \mathcal{E}\}$, then $E_C \in \mathcal{E}$ and so $C = A_{(E_C)} \in \{A_F \mid F \in \mathcal{E}\}$.

Also $\{1\} = A_\Delta \in \mathcal{I}_{\mathcal{E}}$. Furthermore, by Remark 2.8, we see that, if $B \in \mathcal{I}_{\mathcal{E}}$, then $E_B \in \mathcal{E}$ and $E_{B^{-1}} = (E_B)^{\mathsf{T}} \in \mathcal{E}$, whence $B^{-1} \in \mathcal{I}_{\mathcal{E}}$ too. Finally, if $B, C \in \mathcal{I}_{\mathcal{E}}$, then $E_{B \cup C} = E_B \cup E_C \in \mathcal{E}$, whereby $B \cup C \in \mathcal{I}_{\mathcal{E}}$. That $\mathcal{I}_{\mathcal{E}}$ is closed under taking subsets follows from the monotonicity of the operator $B \mapsto E_B$ and the fact that $\mathcal{E}$ is an ideal. Thus, $\mathcal{I}_{\mathcal{E}}$ is an ideal containing $\{1\}$ and closed under inversion and products of sets.

Now, to see that $\mathcal{I}_{\mathcal{E}}$ is the ideal of coarsely bounded sets, assume first that B is coarsely bounded. Because $\{1\} \times \{1\} \subseteq \Delta \in \mathcal{E}$, also $\{1\}$ and $B \cup \{1\}$ are coarsely bounded and so $\{1\} \times B \subseteq (B \cup \{1\}) \times (B \cup \{1\}) \in \mathcal{E}$. It follows that

$$B = \{x^{-1}y \mid (x, y) \in \{1\} \times B\} = A_{\{1\} \times B} \in \mathcal{I}_{\mathcal{E}}.$$

Conversely, if $B \in \mathcal{I}_{\mathcal{E}}$, then $B^{-1}B \in \mathcal{I}_{\mathcal{E}}$ and so

$$B \times B \subseteq \{(x, y) \in G \times G \mid x^{-1}y \in B^{-1}B\} = E_{B^{-1}B} \in \mathcal{E},$$

i.e., B is coarsely bounded. $\square$

Suppose conversely that $\mathcal{I}$ is an ideal on a group G containing $\{1\}$ and closed under inversion and products of sets. Then, by Remark 2.8,

$$\mathcal{E}_{\mathcal{I}} = \{F \mid F \subseteq E_B \text{ for some } B \in \mathcal{I}\}$$

is a left-invariant coarse structure on G. The following proposition is now immediate.

Proposition 2.22 *Let G be a group. Then the map*

$$\mathcal{E} \mapsto \mathcal{I}_{\mathcal{E}}$$

taking each left-invariant coarse structure to its ideal of coarsely bounded sets is a $1-1$-correspondence between between the left-invariant coarse structures on G and ideals $\mathcal{I}$ that contain $\{1\}$ and are closed under inversion and products of sets. Furthermore, the inverse correspondence is given by

$$\mathcal{I} \mapsto \mathcal{E}_{\mathcal{I}}.$$

The gist of this result is, of course, that the left-invariant coarse structures on a group G are completely determined by the associated ideal of coarsely bounded sets and, furthermore, that each ideal as above determines an invariant coarse structure of which it is the ideal of coarsely bounded sets.

As the left-coarse structure $\mathcal{E}_L$ is left-invariant, we thus have the following corollary.

Corollary 2.23 *For every topological group G, we have $\mathcal{E}_L = \mathcal{E}_{CB}$.*

As, by Corollary 2.20, the coarsely bounded sets in a Banach space are the norm-bounded sets, the following is immediate.

Corollary 2.24 *The left-coarse structure on the underlying additive topological group $(X,+)$ of a Banach space $(X, \|\cdot\|)$ is that induced by the norm metric.*

The study of Banach spaces under their coarse structure is a main pillar in geometric non-linear functional analysis, cf. the treatise [**9**] or the more recent survey [**45**].

In addition to the left-coarse structure, we could consider the left-invariant coarse structures on a topological group G defined by the following ideals:

$$\mathcal{K} = \{A \subseteq G \mid A \text{ is relatively compact in } G\},$$
$$\mathcal{V} = \{A \subseteq G \mid \forall V \ni 1 \text{ open } \exists k \ A \subseteq V^k\},$$
$$\mathcal{F} = \{A \subseteq G \mid \forall V \ni 1 \text{ open } \exists k \ \exists F \subseteq G \text{ finite } A \subseteq (FV)^k\}.$$

As can be easily checked, all of these ideals contain $\{1\}$ and are closed under inversion and products. They therefore define left-invariant coarse structures on G. Note also that, because for every set $A \subseteq G$ and identity neighbourhood V we have $\mathrm{cl}A \subseteq AV$, each of these ideals is stable under the topological closure operation $A \mapsto \mathrm{cl}A$. Furthermore, the following inclusions are obvious.

$$\mathcal{K} \ \subseteq \ \underset{\underset{\mathcal{V}}{\cup\mathsf{I}}}{\mathcal{F}} \ \subseteq \ \mathcal{CB}$$

Observe also that $\mathcal{K} = \mathcal{F}$ in every locally compact group. And, furthermore, if this latter is σ-compact, then $\mathcal{K} = \mathcal{F} = \mathcal{CB}$.

As can easily be seen, $\bigcup \mathcal{V}$ equals the intersection of all of open subgroups of G. From this, we readily verify that the following conditions are equivalent on a topological group G.

(1) G has no proper open subgroups,
(2) $G = \bigcup \mathcal{V}$,
(3) $\mathcal{V} = \mathcal{F}$,
(4) $\mathcal{V} = \mathcal{F} = \mathcal{CB}$,
(5) the associated coarse structure $\mathcal{E}_\mathcal{V}$ is *connected*, that is, $\{(x, y)\} \in \mathcal{E}_\mathcal{V}$ for all $x, y \in G$.

Because the notion of coarse structure $\mathcal{E}$ on a space X is supposed to capture what it means for points to a bounded distance apart, it is natural to require that it is connected in the above sense. Also, the coarse structure $\mathcal{E}_\mathcal{V}$ essentially lives only on the intersection of all open subgroups of G and therefore solely constitutes a non-trivial structure on that part of the group. For these reasons, $\mathcal{K}$, $\mathcal{F}$ and $\mathcal{CB}$ are better adapted to illuminate the geometric features of a topological group than $\mathcal{V}$.

In as much as our aim is to study geometry of topological groups, as opposed to topology, metrisability or écartability of a coarse structure is highly desirable. From that perspective, Corollary 2.23, which shows that the coarse structure $\mathcal{E}_{\mathcal{CB}} = \mathcal{E}_L$ is the common refinement of the continuously écartable coarse structures, gives prominence to the left-coarse structure as opposed to $\mathcal{E}_\mathcal{K}$, $\mathcal{E}_\mathcal{V}$ and $\mathcal{E}_\mathcal{F}$. Moreover, in some of the classical cases, namely, finitely generated, countable discrete or σ-compact locally compact groups, Corollary 2.19 also shows that the left-coarse structure coincides with the one classically studied, i.e., $\mathcal{E}_\mathcal{K}$.

Another indication that the left-coarse structure is more suitable from a geometric perspective is that, in contradistinction to $\mathcal{E}_\mathcal{K}$, it induces the correct coarse structure on Banach spaces, namely, that given by the norm.

One may, for example, consult the monograph [**22**] by Y. de Cornulier and P. de la Harpe for the coarse geometry of locally compact σ-compact groups and the paper [**67**] for more information about the *group-compact coarse structure* $\mathcal{E}_\mathcal{K}$ on general topological groups. The left-coarse structure appears to be new, but has its roots in earlier work in [**77**].

Remark 2.25 Suppose $\mathcal{I}$ is an ideal on a topological group G closed under inversion and products and containing $\{1\}$. Then G is *locally in $\mathcal{I}$*, that is, G has an identity neighbourhood V belonging to $\mathcal{I}$, if and only if there is a set $E \subseteq G \times G$, which is simultaneously an entourage for the left-uniformity $\mathcal{U}_L$ on G and the coarse structure $\mathcal{E}_\mathcal{I}$, i.e., if $\mathcal{U}_L \cap \mathcal{E}_\mathcal{I} \neq \emptyset$.

Indeed, given such an $E \in \mathcal{U}_L \cap \mathcal{E}_{\mathcal{I}}$, there is an identity neighbourhood W so that $E_W \subseteq E$, whence $E_W \in \mathcal{E}_{\mathcal{I}}$ and thus $W \in \mathcal{I}$. Conversely, if G is locally in $\mathcal{I}$, as witnessed by some identity neighbourhood $V \in \mathcal{I}$, then $E_V \in \mathcal{U}_L \cap \mathcal{E}_{\mathcal{I}}$.

Our next examples indicate that there is no simple characterisation of when $\mathcal{K} = \mathcal{F}$ or $\mathcal{F} = \mathcal{CB}$.

Example 2.26 Let $\mathrm{Sym}(\mathbb{N})$ denote the group of all (not necessarily finitely supported) permutations of $\mathbb{N}$ equipped with the discrete topology. Then, as $V = \{1\}$ is an identity neighbourhood, we see that $\mathcal{K} = \mathcal{F}$ is the ideal of finite subsets of $\mathrm{Sym}(\mathbb{N})$, while $\mathcal{V}$ has a single element, namely $\{1\}$.

On the other hand, it follows from the main result of G. M. Bergman's paper [**10**] that, for any exhaustive chain $V_1 \subseteq V_2 \subseteq \cdots \subseteq \mathrm{Sym}(\mathbb{N})$ of subsets with $V_n^2 \subseteq V_{n+1}$, we have $\mathrm{Sym}(\mathbb{N}) = V_k$ for some k. In other words, $\mathcal{CB}$ contains all subsets of $\mathrm{Sym}(\mathbb{N})$.

Example 2.27 A subset A of a topological group G is said to be *bounded in the left-uniformity* $\mathcal{U}_L$ if, for every identity neighbourhood V, there is some n and a finite set $F \subseteq G$ so that $A \subseteq FV^n$. As is well known in the theory of uniform spaces, this is equivalent to demanding that every left-uniformly continuous function $\phi \colon G \to \mathbb{R}$ is bounded on A. Clearly, the class of sets bounded in the left-uniformity contains $\mathcal{V}$ and is contained in $\mathcal{F}$. Thus, when G has no proper open subgroups, a set is bounded in the left-uniformity if and only if it belongs to $\mathcal{CB}$.

On the other hand, let S_∞ be the group of all permutations of $\mathbb{N}$ equipped with the Polish topology obtained by declaring pointwise stabilisers of finite sets to be open. Then the open subgroups form a neighbourhood basis at the identity and thus a subset $A \subseteq S_\infty$ is bounded in the left-uniformity exactly when A intersects only finitely many left cosets of every open subgroup $H \leqslant G$. Letting H vary over pointwise stabilisers of finite sets, we see that A is bounded in the left-uniformity if and only if

$$A(n) = \{f(n) \mid f \in A\}$$

is finite for every $n \in \mathbb{N}$.

For example, let $A \subseteq S_\infty$ be the following collection of cycles

$$A = \{(1\ 2\ 3\ 4 \ldots n) \mid n \in \mathbb{N}\}.$$

Then we see that $A(n) = \{1, n, n+1\}$ for all $n \in \mathbb{N}$ and hence A is bounded in the left-uniformity. On the other hand, it is straightforward to see that the sequence of cycles $\sigma_n = (1\ 2\ 3\ 4\dots n)$ has no accumulation point in S_∞ and thus A fails to be relatively compact.

Example 2.28 Consider the free non-abelian group $\mathbb{F}_{\aleph_1}$ on $\aleph_1$ generators $(a_\xi)_{\xi < \aleph_1}$ ($\aleph_1$ is the first uncountable cardinal number) equipped with the discrete topology. Then $\mathcal{K} = \mathcal{F}$ is the class of finite subsets, whereas $\mathcal{V}$ consists of $\{1\}$.

We claim that any element of $\mathcal{CB}$ is finite and thus $\mathcal{K} = \mathcal{F} = \mathcal{CB}$, despite the fact that $\mathbb{F}_{\aleph_1}$ is not generated over the identity neighbourhood $\{1\}$ by a countable set. Indeed, if $A \subseteq \mathbb{F}_{\aleph_1}$ is infinite, it either contains elements with unbounded word length in $(a_\xi)_{\xi < \aleph_1}$ or it must use an infinite number of generators. In the first case, we let V_n denote the set of $x \in \mathbb{F}_{\aleph_1}$ of word length at most 2^n and see that $A \nsubseteq V_n$ for all n. In the second case, suppose a_{ξ_n} is a sequence of generators all appearing in elements of A. We then find an increasing exhaustive chain $C_1 \subseteq C_2 \subseteq \cdots \subseteq \{a_\xi\}_{\xi < \aleph_1}$ so that $a_{\xi_n} \notin C_n$ and let $V_n = \langle C_n \rangle$. Again, $A \nsubseteq V_n$ for all n. So, in either case, $A \notin \mathcal{CB}$.

Example 2.29 For each n, let Γ_n be a copy of the discrete group $\mathbb{Z}$ and let $G = \prod_n \Gamma_n$ be equipped with the product topology. Because the coordinate projections proj_{Γ_n} are all continuous, we see that $A \subseteq G$ is relatively compact if and only if $\mathsf{proj}_{\Gamma_k}(A)$ is finite for all k. Thus, if A is not relatively compact, i.e., $A \notin \mathcal{K}$, fix k so that $\mathsf{proj}_{\Gamma_k}(A)$ is infinite. Then, using $V_n = \{x \in G \mid \mathsf{proj}_{\Gamma_k}(x) \in [-2^n, 2^n]\}$, we see that $A \notin \mathcal{CB}$. In other words, $\mathcal{K} = \mathcal{F} = \mathcal{CB}$ even though G is not locally compact.

If, instead, Γ_n is a discrete copy of $\mathbb{F}_{\aleph_1}$ for each n, we obtain an example $G = \prod_n \Gamma_n$ that is neither locally compact nor countably generated over every identity neighbourhood. Nevertheless, by repeating the argument of Example 2.28, we have $\mathcal{K} = \mathcal{F} = \mathcal{CB}$.

2.5 Metrisability and Monogenicity

In what follows, on a topological group, we shall be considering only the left-coarse structure $\mathcal{E}_L$ and the left-uniformity $\mathcal{U}_L$. All concepts will refer to these.

As is the case with uniform structure, the simplest case to understand is the metrisable coarse spaces.

Definition 2.30 A coarse space $(X, \mathcal{E})$ is said to be *metrisable* if it is of the form $\mathcal{E}_d$ for some generalised metric $d \colon X \times X \to [0, \infty]$ (thus possibly talking the value ∞).

Observe that, if d takes the value ∞, say $d(x,y) = \infty$, then $\{(x,y)\} \notin \mathcal{E}$ and thus the coarse space $(X, \mathcal{E})$ is disconnected. So, because $\mathcal{E}_L$ is connected, we will not need to be concerned about metrics taking the value ∞.

Let us also note that, because we do not require any continuity here, the difference between (generalised) écarts and metrics is not important. Indeed, if d is an écart on X, then $\partial(x,y) = d(x,y) + 1$ for all $x \neq y$ defines a metric on X inducing the same coarse structure as d.

In analogy with the characterisation of metrisable uniformities as those that are countably generated, Roe (Theorem 2.55 in [75]) characterises the metrisable coarse structures $\mathcal{E}$ on a set X as those that are *countably generated*, i.e., that have a countable subbasis. Observe that, if $\mathcal{F}$ is a countable subbasis for $\mathcal{E}$, then the closure $\tilde{\mathcal{F}}$ of $\mathcal{F} \cup \{\Delta\}$ under T, $\circ$ and $\cup$ (see Example 2.5 for details) will be a countable basis for $\mathcal{E}$.

Lemma 2.31 *The following are equivalent for a topological group G.*

(1) *The left-coarse structure $\mathcal{E}_L$ is metrisable,*
(2) *the ideal $\mathcal{CB}$ is countably generated,*
(3) *$\mathcal{E}_L$ is metrised by a left-invariant (possibly discontinuous) metric on G.*

Proof (1)$\Rightarrow$(2) Suppose $\mathcal{E}_L$ is given by some metric d on G. Then a subset A is coarsely bounded if and only if it has finite d-diameter, that is, exactly when it is contained in some finite radius ball

$$B_d(1,n) = \{x \in G \mid d(1,x) < n\}.$$

These countably many balls thus generate the ideal $\mathcal{CB}$.

(2)$\Rightarrow$(3) Assume that $\mathcal{CB}$ is countably generated and let $\{A_n\}_n \subseteq \mathcal{CB}$ be a countable set of generators. Set $V_0 = V_{-1} = \cdots = \{1\}$ and

$$V_{n+1} = \{0\} \cup A_{n+1} \cup A_{n+1}^{-1} \cup V_n^3$$

for $n \geqslant 0$. Then $(V_n)_{n \in \mathbb{Z}}$ is an increasing exhaustive chain in G consisting of coarsely bounded symmetric sets satisfying $V_n^3 \subseteq V_{n+1}$. Because the V_n are increasing and $A_n \subseteq V_n$, we see that a set is coarsely bounded if and only if it is contained in some V_n.

Applying Lemma 2.13 to the underlying discrete group of G, we obtain a left-invariant metric d on G whose bounded sets are exactly those contained in one of the V_n, i.e., the coarsely bounded sets. Thus, $\mathcal{E}_d$ and $\mathcal{E}_L$ are two left-invariant coarse structures with the same coarsely bounded sets and therefore must coincide by Proposition 2.22.

(3)$\Rightarrow$(1) This is immediate. $\qquad\qquad\qquad\qquad\qquad\qquad\qquad\qquad$ $\square$

It is worth pointing out that G may be the union of a countable sequence $A_1 \subseteq A_2 \subseteq \cdots \subseteq G$ of coarsely bounded sets even though the ideal $\mathcal{CB}$ is not countably generated. An instance of this will be given in Example 2.49.

Definition 2.32 A topological group G is *locally bounded* if and only it has a coarsely bounded identity neighbourhood.

Observe that, by Remark 2.25, this happens if and only if

$$\mathcal{U}_L \cap \mathcal{E}_L \neq \emptyset.$$

The importance of this condition will become even clearer from the results below.

Lemma 2.33 *Let G be a topological group and suppose that $\mathcal{E}_L$ is induced by a continuous left-invariant écart d on G. Then G is locally bounded.*

Proof This is trivial because the continuous left-invariant écart d must be bounded on some identity neighbourhood, which thus must be coarsely bounded in G. □

Before stating the next lemma, let us recall that a topological group G is *Baire* if it satisfies the Baire category theorem, i.e., if the intersection of a countable family of dense open sets is dense in G. Prime examples of Baire groups are, of course, the locally compact and the completely metrisable groups.

Lemma 2.34 *Let G be a Baire topological group. Suppose that G is the union of a sequence of coarsely bounded sets, e.g., if G has metrisable left-coarse structure $\mathcal{E}_L$. Then G is locally bounded.*

Proof If $G = \bigcup_n A_n$ for any sequence of coarsely bounded sets, then, as G is Baire, some $\overline{A_n}$ must be non-meagre and thus have non-empty interior W. It follows that the identity neighbourhood $V = WW^{-1}$ is coarsely bounded in G. □

Lemma 2.35 *Suppose G is a topological group countably generated over every identity neighbourhood. Then, for every symmetric open identity neighbourhood V, there is a continuous left-invariant écart d so that a subset $A \subseteq G$ is d-bounded if and only if there are a finite set F and an n so that $A \subseteq (FV)^n$.*

Proof Fix a symmetric open identity neighbourhood V and choose $x_1, x_2, \ldots \in G$ so that $G = \langle V \cup \{x_1, x_2, \ldots\} \rangle$. We then let

$$V_n = \left(V \cup \{x_1, x_1^{-1}, \ldots, x_n, x_n^{-1}\} \right)^{3^n}$$

and note that the V_n form an increasing exhaustive chain of open symmetric identity neighbourhoods satisfying $V_n^3 \subseteq V_{n+1}$ for all n. We now complement the chain by symmetric open identity neighbourhoods

$$V_0 \supseteq V_{-1} \supseteq V_{-2} \supseteq \cdots$$

so that $V_n^3 \subseteq V_{n+1}$ holds for all $n \in \mathbb{Z}$.

Applying Lemma 2.13, we obtain a continuous left-invariant écart d on G whose balls are each contained in some V_n and so that each V_n has finite d-diameter. It follows that a subset $A \subseteq G$ is d-bounded if and only if $A \subseteq V_n$ for some n. Also, if $F \subseteq G$ is finite, then $F \subseteq V_n$ for some $n \geqslant 1$, whereby $(FV)^k \subseteq V_{n+k}$ must have finite diameter for all $k \geqslant 1$. This shows that a subset $A \subseteq G$ is d-bounded if and only if there are $F \subseteq G$ finite and $k \geqslant 1$ so that $A \subseteq (FV)^k$. $\qquad\square$

Lemma 2.36 *Let G be a locally bounded topological group and assume that G is countably generated over every identity neighbourhood. Then $\mathcal{E}_L$ is induced by a continuous left-invariant écart d on G.*

Proof Fix a symmetric open identity neighbourhood V coarsely bounded in G and let d be a continuous left-invariant écart as in Lemma 2.35. Then a subset $A \subseteq G$ is d-bounded if and only if A is coarsely bounded in G, whence d induces the left-coarse structure $\mathcal{E}_L$ on G. $\qquad\square$

As the combination of being Baire and countably generated over identity neighbourhoods will reappear several times, it will be useful to have a name for this.

Definition 2.37 A topological group G is *European* if it is Baire and countably generated over every identity neighbourhood.

Observe that the class of European groups is a proper extension of the class of Polish groups; hence the name. For example, all connected completely metrisable groups, e.g., all Banach spaces, and all locally compact σ-compact Hausdorff groups are European, but may not be Polish. Recall also that, by Proposition 2.15, the ideals $\mathcal{CB}$ and $\mathcal{F}$ from Section 2.4 coincide in a European topological group.

Combining the preceding lemmas, we obtain the following characterisation of the metrisability of the coarse structure.

Theorem 2.38 *The following are equivalent for a European topological group G.*

(1) *The left-coarse structure $\mathcal{E}_L$ is metrisable,*
(2) *G is covered by a sequence of coarsely bounded sets,*
(3) *G is locally bounded,*
(4) *$\mathcal{E}_L$ is induced by a continuous left-invariant écart d on G.*

Theorem 2.38 can be seen as an analogue of well-known facts about locally compact groups. Indeed, S. Kakutani and K. Kodaira [**44**] showed that, if G is locally compact, σ-compact, then for any sequence U_n of identity neighbourhoods there is a compact normal subgroup $K \subseteq \bigcap_n U_n$ so that G/K is metrisable.

Moreover, R. Struble [**84**] showed that any metrisable (i.e., second countable) locally compact group admits a compatible left-invariant proper metric, i.e., so that all bounded sets are relatively compact. To construct such a metric, one can simply choose the V_n of Lemma 2.13 to be relatively compact.

Combining these two results, one sees that every locally compact σ-compact group G admits a continuous proper left-invariant écart, thus inducing the group-compact coarse structure $\mathcal{E}_L = \mathcal{E}_{\mathcal{K}}$ on G.

The next concept from [**75**] will help us to delineate, for example, the finitely generated groups in the class of all countable discrete groups.

Definition 2.39 A coarse structure $(X, \mathcal{E})$ is *monogenic* if $\mathcal{E}$ is generated by a single entourage E.

Note that, by replacing the generator $E \in \mathcal{E}$ by $E \cup E^{\mathsf{T}} \cup \Delta$, one sees that $\mathcal{E}$ is monogenic if and only if there is some entourage $E \in \mathcal{E}$ so that $\{E^n\}_n$ is a basis for $\mathcal{E}$, where $E^n = \underbrace{E \circ \cdots \circ E}_{n}$. Also, if $\mathcal{E}_L$ is the left-coarse structure on a group G, this E can be taken of the form E_A for some coarsely bounded symmetric set A containing 1. Now, recall from Remark 2.8 that $E_A^n = E_{A^n}$. By using this, we see that the coarse structure $\mathcal{E}_L$ is monogenic if and only if there is some coarsely bounded set A so that $\{A^n\}_n$ is a basis for the ideal $\mathcal{CB}$.

Theorem 2.40 *The following are equivalent for a European topological group G.*

(1) *The left-coarse structure $\mathcal{E}_L$ is monogenic,*
(2) *G is generated by a coarsely bounded set, i.e., there is some $A \in \mathcal{CB}$ algebraically generating G,*
(3) *G is locally bounded and not the union of a countable chain of proper open subgroups.*

Proof (2)⇒(1) Suppose that G is generated by a coarsely bounded symmetric subset A and so $G = \bigcup_n A^n$. By the Baire category theorem, some A^n must be somewhere dense and thus $B = \overline{A^n}$ is a coarsely bounded set with non-empty interior generating G. To see that $\{B^n\}_n$ is cofinal in $\mathcal{CB}$ and thus that $\mathcal{E}_{CB}$ is monogenic, observe that if $C \subseteq G$ is coarsely bounded, then, as $\mathsf{int}(B) \neq \emptyset$, there is a finite set $F \subseteq G$ and a k so that $C \subseteq (FB)^k$. As furthermore B generates G, one has $C \subseteq B^m$ for some sufficiently large m.

(1)⇒(3) If $\mathcal{E}_L$ is monogenic, it is countably generated and thus metrisable. Thus, by Theorem 2.38, G is locally bounded. Also, if $H_1 \leqslant H_2 \leqslant \cdots \leqslant G$ is a countable chain of open subgroups exhausting G, then, by definition of the ideal $\mathcal{CB}$, every coarsely bounded set is contained in some H_n. Thus, as G is generated by a coarsely bounded set, it must equal some H_n.

(3)⇒(2) If G is coarsely bounded as witnessed by some identity neighbourhood V, let $\{x_n\}_n$ be a countable set generating G over V. If, moreover, G is not the union of a countable chain of proper open subgroups, it must be generated by some $V \cup \{x_1, \ldots, x_n\}$ and hence be generated by a coarsely bounded set.   □

Observe that Theorem 2.40 applies, in particular, to locally compact σ-compact Hausdorff groups where the coarsely bounded and relative compact sets coincide. Thus, among these, the compactly generated groups are exactly those whose coarse structure is monogenic.

Example 2.41 (Banach spaces with the weak topology) Let X be a Banach space and consider the additive group $(X, +)$ equipped with the weak topology w. Recall that, by the Banach–Steinhaus uniform boundedness principle, a subset $A \subseteq X$ is norm-bounded if and only if it is weakly bounded, i.e.,

$$\sup_{x \in A} |\phi(x)| < \infty$$

for all $\phi \in X^*$. Note also that every functional ϕ defines a w-continuous invariant écart via

$$d_\phi(x, y) = |\phi(x - y)|.$$

Therefore, we see that the norm-bounded sets are exactly the coarsely bounded sets in $(X, +, \mathrm{w})$. In other words, the left-coarse structures of the topological groups $(X, +, \|\cdot\|)$ and $(X, +, \mathrm{w})$ have the same ideals of coarsely bounded sets and hence the left-coarse structures actually coincide by Proposition 2.22. To recapitulate, we have two distinct group topologies on the same group $(X, +)$ that generate the same left-coarse structure.

Observe also that $(X, +, w)$ is generated by a coarsely bounded set, namely, any open ball of finite diameter, and has metrisable coarse structure. However, if X is finite-dimensional, then no weakly open set has finite diameter. So we see that $(X, +, w)$ fails to be locally bounded except when X is finite-dimensional. The apparent contradiction with Theorems 2.38 and 2.40 is resolved when we observe that $(X, +, w)$ is not Baire, and thus not European, when X is infinite-dimensional.

Stretching the use of language, we introduce the following terminology.

Definition 2.42 A topological group G is *monogenic* if its left-coarse structure is monogenic.

Thus, by Theorem 2.40, a Polish or, more generally, European group is monogenic exactly when it is generated by a coarsely bounded set. Although it is true that every monogenic group is generated by a coarsely bounded set, the opposite fails in general. An example of this will be given in Example 2.49.

2.6 Bornologous Maps

Let us begin by fixing some notation. If $X \xrightarrow{\phi} Y$ is a map, we get a naturally defined map $\mathcal{P}(X \times X) \xrightarrow{\phi \times \phi} \mathcal{P}(Y \times Y)$ by

$$\left(\phi \times \phi\right)E = \{(\phi(x_1), \phi(x_2)) \mid (x_1, x_2) \in E\}$$

and, similarly, a map $\mathcal{P}(Y \times Y) \xrightarrow{(\phi \times \phi)^{-1}} \mathcal{P}(X \times X)$ by

$$\left(\phi \times \phi\right)^{-1}F = \{(x_1, x_2) \mid \left(\phi(x_1), \phi(x_2)\right) \in F\}.$$

Recall now that, if $X \xrightarrow{\phi} Y$ is a map between uniform spaces $(X, \mathcal{U})$ and $(Y, \mathcal{V})$, then ϕ is *uniformly continuous* if, for all $F \in \mathcal{V}$, there is $E \in \mathcal{U}$ so that

$$(x_1, x_2) \in E \Rightarrow \left(\phi(x_1), \phi(x_2)\right) \in F,$$

or, equivalently, if

$$E \subseteq (\phi \times \phi)^{-1}F,$$

which again implies that $(\phi \times \phi)^{-1}F \in \mathcal{U}$. Thus, ϕ is uniformly continuous exactly when $(\phi \times \phi)^{-1}[\mathcal{V}] \subseteq \mathcal{U}$. This motivates the definition of bornologous maps between coarse spaces.

Definition 2.43 Let $(X,\mathcal{E})$ and $(Y,\mathcal{F})$ be coarse spaces, Z a set and $X \xrightarrow{\phi} Y$, $Z \xrightarrow{\alpha,\beta} X$ mappings. We say that

(1) ϕ is *bornologous* if $(\phi \times \phi)[\mathcal{E}] \subseteq \mathcal{F}$,
(2) ϕ is *expanding* if $(\phi \times \phi)^{-1}[\mathcal{F}] \subseteq \mathcal{E}$,
(3) ϕ is *modest* if $\phi[A]$ is $\mathcal{F}$-bounded whenever $A \subseteq X$ is $\mathcal{E}$-bounded,
(4) ϕ is *coarsely proper* if $\phi[A]$ is $\mathcal{F}$-unbounded whenever $A \subseteq X$ is $\mathcal{E}$-unbounded,
(5) ϕ is a *coarse embedding* if it is both bornologous and expanding,
(6) $A \subseteq X$ is *cobounded* in X if there is an entourage $E \in \mathcal{E}$ so that

$$X = E[A] := \{x \in X \mid \exists y \in A \ (x,y) \in E\},$$

(7) ϕ is a *cobounded map* if $\phi[X]$ is cobounded in Y,
(8) α and β are *close* if there is some $E \in \mathcal{E}$ so that $(\alpha(z),\beta(z)) \in E$ for all $z \in Z$,
(9) ϕ is a *coarse equivalence* if it is bornologous and there is a bornologous map $Y \xrightarrow{\psi} X$ so that $\psi \circ \phi$ is close to id_X, whereas $\phi \circ \psi$ is close to id_Y.

There are a number of comments that are in order here.

(a) First, the map ϕ is coarsely proper if and only if $\phi^{-1}(B)$ is coarsely bounded for every coarsely bounded $B \subseteq Y$.

(b) A map ϕ between coarse spaces $(X,\mathcal{E})$ and $(Y,\mathcal{F})$ is bornologous if and only if, for every entourage $E \in \mathcal{E}$, there is an entourage $F \in \mathcal{F}$ so that

$$(x_1,x_2) \in E \Rightarrow (\phi x_1, \phi x_2) \in F.$$

(c) Similarly, ϕ is expanding if and only if, for all $F \in \mathcal{F}$, there is $E \in \mathcal{E}$ so that

$$(x_1,x_2) \notin E \Rightarrow (\phi x_1, \phi x_2) \notin F.$$

(d) In particular, if $(X,d_X) \xrightarrow{\phi} (Y,d_Y)$ is a map between pseudo-metric spaces, then ϕ is bornologous with respect to the induced coarse structures if and only if, for every $t < \infty$, there is a $\theta(t) < \infty$ so that

$$d_X(x_1,x_2) < t \Rightarrow d_Y(\phi x_1, \phi x_2) < \theta(t)$$

and is expanding if and only if, for every $t < \infty$, there is a $\kappa(t) < \infty$ so that

$$d_X(x_1,x_2) > \kappa(t) \Rightarrow d_Y(\phi x_1, \phi x_2) > t.$$

(e) An expanding map is always coarsely proper, whereas a bornologous map is always modest.

(f) If $X \xrightarrow{\phi} Y$ is a coarse equivalence as witnessed by $Y \xrightarrow{\psi} X$, then also ψ is a coarse equivalence. Furthermore, if $Y \xrightarrow{\eta} Z$ is a coarse equivalence into a coarse space Z, then $\eta \circ \phi$ is a coarse equivalence. So the existence of a coarse equivalence between coarse spaces defines an equivalence relation.

(g) Since $E_{B^{-1}}[A] = AB$, a subset A of a topological group G is cobounded if and only if there is a coarsely bounded set B so that $G = A \cdot B$.

Let us note the following elementary fact.

Lemma 2.44 *Let $(X,\mathcal{E})$ and $(Y,\mathcal{F})$ be coarse spaces and $X \xrightarrow{\phi} Y$, $Y \xrightarrow{\psi} X$ mappings so that $\psi \circ \phi$ is close to id_X. It follows that*

(1) *if $Y \xrightarrow{\psi} X$ is bornologous, then ϕ is expanding,*

(2) *if $Y \xrightarrow{\psi} X$ is expanding, then ϕ is bornologous and $\phi \circ \psi$ is close to id_Y.*

Proof Because $\psi \circ \phi$ is close to id_X, we may fix some $E_1 \in \mathcal{E}$ so that $(\psi\phi(x),x) \in E_1$ for all $x \in X$.

Suppose first that $Y \xrightarrow{\psi} X$ is bornologous and let $F \in \mathcal{F}$ be given. As ψ is bornologous, we have $E_2 = (\psi \times \psi)F \in \mathcal{E}$. Thus, if $(x_1,x_2) \in (\phi \times \phi)^{-1}F$, we have $(\psi\phi(x_1),\psi\phi(x_2)) \in E_2$ and so $(x_1,x_2) \in E_1^{-1} \circ E_2 \circ E_1$. That is, $(\phi \times \phi)^{-1}F \subseteq E_1^{-1} \circ E_2 \circ E_1 \in \mathcal{E}$, showing that ϕ is expanding.

Assume instead that $Y \xrightarrow{\psi} X$ is expanding. Given $E_2 \in \mathcal{E}$, we have $F = (\psi \times \psi)^{-1}\big(E_1 \circ E_2 \circ E_1^{-1}\big) \in \mathcal{F}$. Thus, if $(x_1,x_2) \in E_2$, we have $(\psi\phi(x_1),\psi\phi(x_2)) \in E_1 \circ E_2 \circ E_1^{-1}$ and so also $(\phi(x_1),\phi(x_2)) \in F$, showing that $(\phi \times \phi)E_2 \in \mathcal{F}$. In other words, ϕ is bornologous. Moreover, for every $y \in Y$, we have $(\psi\phi\psi(y),\psi y) \in E_1$ and thus $(\phi\psi(y),y) \in (\psi \times \psi)^{-1}E_1 \in \mathcal{F}$, whereby $\phi \circ \psi$ is close to id_Y. $\square$

This in turn leads to the following well-known fact.

Lemma 2.45 *A map $X \xrightarrow{\phi} Y$ between coarse spaces $(X,\mathcal{E})$ and $(Y,\mathcal{F})$ is a coarse equivalence if and only if ϕ is bornologous, expanding and cobounded.*

Proof Suppose first that ϕ is a coarse equivalence as witnessed by some bornologous $\psi \colon Y \to X$ so that $\psi \circ \phi$ is close to id_X and $\phi \circ \psi$ is close to id_Y. Then ϕ is expanding by Lemma 2.44 (1). Also, as $\phi \circ \psi$ is close to id_Y, we see that $\phi[X]$ must be cobounded in Y.

Conversely, if ϕ is bornologous, expanding and cobounded, pick an entourage $F \in \mathcal{F}$ so that $\phi[X]$ is F-cobounded and let $Y \xrightarrow{\psi} X$ be defined by

$$\psi(y) = x \text{ for some } x \in X \text{ so that } (y,\phi x) \in F.$$

Then, by construction, we have $(y, \phi\psi(y)) \in F$ for all $y \in Y$ and thus $\phi \circ \psi$ is close to id_Y. Applying Lemma 2.44 (2) (with the roles of ϕ and ψ reversed), we see that ψ is bornologous and $\psi \circ \phi$ is close to id_X, i.e., that ϕ is a coarse equivalence.   □

Regarding maps between groups, note that, if $G \xrightarrow{\phi} H$ is a group homomorphism and $A \subseteq G$ and $B \subseteq H$ are subsets, then

$$(\phi \times \phi)^{-1} E_B = E_{\phi^{-1}(B)}$$

and

$$(\phi \times \phi) E_A \subseteq E_{\phi[A]}.$$

Also, if $G \xrightarrow{\psi} H$ is an arbitrary map and $(\psi \times \psi) E_A \subseteq E_B$ for some $A \subseteq G$ and $B \subseteq H$, then

$$(\psi \times \psi) E_{A^n} = (\psi \times \psi) E_A^n \subseteq E_B^n = E_{B^n}$$

for all $n \geqslant 1$.

Lemma 2.46 *Let* $G \xrightarrow{\phi} H$ *be a continuous homomorphism between topological groups. Then* ϕ *is bornologous. Moreover, the following conditions on* ϕ *are equivalent.*

(1) ϕ *is a coarse embedding,*
(2) ϕ *is expanding,*
(3) ϕ *is coarsely proper.*

Proof To see that ϕ is bornologous, let E be a coarse entourage in G. Then there is a coarsely bounded set $A \subseteq G$ so that $E \subseteq E_A$, whence $(\phi \times \phi) E \subseteq (\phi \times \phi) E_A \subseteq E_{\phi[A]}$. It thus suffices to see that $\phi[A]$ is coarsely bounded in H, whereby $E_{\phi[A]}$ is an entourage in H. But this follows from the fact that, if d is a continuous left-invariant écart on H, then $d(\phi(\cdot), \phi(\cdot))$ defines a continuous left-invariant écart on G with respect to which A has finite diameter and hence $\phi[A]$ has finite d-diameter. As d was arbitrary, $\phi[A]$ is coarsely bounded in H.

To verify the set of equivalences, because every expanding map is coarsely proper, it suffices to see that ϕ is expanding provided it is coarsely proper. But, if F is an entourage in H, find a coarsely bounded set $B \subseteq H$ so that $F \subseteq E_B$. Then, because ϕ is coarsely proper, $\phi^{-1}(B)$ is coarsely bounded in G, whence $(\phi \times \phi)^{-1} E_B = E_{\phi^{-1}(B)}$ is an entourage in G. That is, ϕ is expanding.   □

Definition 2.47 Let H be a subgroup of a topological group G. We say that H is *coarsely embedded in* G if the inclusion map $H \xrightarrow{\iota} G$ is a coarse embedding, i.e., if the coarse structure on H coincides with the coarse structure on G restricted to H.

Note that, by Lemma 2.46, the subgroup H is coarsely embedded in G exactly when every subset A of H, which is coarsely bounded in G, is also coarsely bounded in H. In the case where H is a closed subgroup of a locally compact group, the coarsely bounded sets of H and G are the relatively compact sets, which thus does not depend on whether they are seen as subsets of H or of G. In other words, a closed subgroup of a locally compact group is automatically coarsely embedded. On the other hand, in a Polish group, this is very far from being true. In fact, as we shall see there is a universal Polish group G, i.e., in which every other Polish group can be embedded as a closed subgroup, so that G is coarsely bounded in itself. It thus follows that any closed subgroup H will be coarsely bounded in G, but need not be coarsely bounded in itself and thus may not be coarsely embedded. This fact is a source of much of the additional complexity compared with the coarse geometry of locally compact groups.

For the next result, observe that if H is a dense subgroup of a topological group G, then

$$(UV) \cap H = (U \cap H)(V \cap H)$$

for all open subsets $U, V \subseteq G$. Indeed, suppose $x \in G$. Then, since $U^{-1}x \cap V$ is open and H is dense in G, we have

$$x \in UV \iff U^{-1}x \cap V \neq \emptyset \iff U^{-1}x \cap V \cap H \neq \emptyset \iff x \in U(V \cap H),$$

i.e., $UV = U(V \cap H)$. Therefore, if $z \in (UV) \cap H$ is any element, we may write $z = uv$ for some $u \in U$ and $v \in V \cap H$, whereby $u = zv^{-1} \in H$ and thus $z \in (U \cap H)(V \cap H)$. In other words,

$$(UV) \cap H = (U \cap H)(V \cap H),$$

as claimed.

Proposition 2.48 *Let G be a topological group and let H be a dense subgroup equipped with the induced topology. Then H is coarsely embedded in G. Moreover, if the coarse structure of H is metrisable and G is first countable, then G is the union of a countable family of coarsely bounded sets.*

Proof Suppose some subset $B \subseteq H$ is coarsely bounded in G. We must show that B is also coarsely bounded in H. For this, suppose $V_1 \subseteq V_2 \subseteq \cdots \subseteq H$ is an exhaustive sequence of open subsets. As the topology of H is that induced from G, there are open sets $U_1 \subseteq U_2 \subseteq \cdots \subseteq G$ so that $V_n = H \cap U_n$. Again, as H is dense in G, we find that $\bigcup_n U_n$ is dense in G and thus $G = \bigcup_n U_n^2$. Because B is coarsely bounded in G, there are k, n so that $B \subseteq U_n^k$. However, as H is dense in G, we have, by the calculation above,

$$B \subseteq U_n^k \cap H = (U_n^{k-1} \cap H)(U_n \cap H) = \cdots = (U_n \cap H)^k = V_n^k,$$

which shows that B is coarsely bounded in H too.

For the moreover part, suppose H has metrisable coarse structure, say induced by some left-invariant metric d. Let $B_n \subseteq H$ denote the d-ball of radius n centred at the identity. We show that $G = \bigcup_n \overline{B_n}$. Indeed, let $g \in G$ and pick a sequence (h_n) in H with $h_n \to g$. This is possible as G is first countable. Then $\{h_n\}_n$ is relatively compact and thus coarsely bounded in G and hence also in H. It follows that $\{h_n\}_n \subseteq B_k$ for some k and thus that $g \in \overline{B_k}$ as claimed. Finally, because the B_k are coarsely bounded in H and thus in G, it follows that the closures $\overline{B_k}$ remain coarsely bounded. $\qquad \square$

Example 2.49 (Generation by coarsely bounded sets without local boundedness) Let G be a non-locally bounded Polish group and Γ a countable dense subgroup. Equip Γ with the induced topology from G. Because Γ is countable, it is certainly the union of a countable family $\{F_n\}_n$ of coarsely bounded subsets, namely, the finite subsets. However, Γ cannot have metrisable coarse structure. For, if it did, then by Proposition 2.48 also G would be the union of a countable sequence of coarsely bounded sets and, being Polish, Lemma 2.34 would imply that G is locally bounded.

Similarly, Γ cannot be cobounded in G. For, if $B \subseteq G$ is coarsely bounded with $G = \Gamma \cdot B$, then $G = \Gamma \cdot B = \bigcup_n F_n B$ and so again G would be locally bounded.

Taking an example where Γ is moreover finitely generated, we see that Γ could be generated by a coarsely bounded set, while failing to be monogenic.

Proposition 2.50 *Suppose $G \xrightarrow{\phi} H$ is a continuous homomorphism between European topological groups so that the image $\phi[G]$ is dense in H. Assume that, for some identity neighbourhood $V \subseteq H$, the pre-image $\phi^{-1}(V)$ is coarsely bounded in G. Then ϕ is a coarse equivalence between G and H.*

Proof Without loss of generality, we may assume that V is open. Let us first see that ϕ is a coarse embedding. So assume that $A \subseteq G$ and that $\phi[A]$ is coarsely bounded in H. Then, as $\phi[G]$ is dense in H, there are a finite set $F \subseteq G$ and a k so that $\phi[A] \subseteq (\phi[F]V)^k$. Now, observe that given

$a \in A$, by density of $\phi[G] \cap V$ in the open set V, we can find $g_1, \ldots, g_{k-1} \in \phi^{-1}(V)$ and $f_1, \ldots, f_k \in F$ so that $\phi(a) \in \phi(f_1)\phi(g_1)\cdots\phi(f_{k-1})\phi(g_{k-1})\phi(f_k)V$. But then, because $\phi[G]$ is a subgroup of H, it follows that actually $\phi(a) \in \phi(f_1)\phi(g_1)\cdots\phi(f_{k-1})\phi(g_{k-1})\phi(f_k) \cdot (\phi[G] \cap V)$ and therefore that $\phi(a) \in \big(\phi[F] \cdot (\phi[G] \cap V)\big)^k = \phi\big[(F\phi^{-1}(V))^k\big]$. In other words, $A \subseteq (F\phi^{-1}(V))^k$, showing that A is coarsely bounded in G. So ϕ is coarsely proper and thus a coarse embedding.

Because $\phi^{-1}(V)$ is a coarsely bounded identity neighbourhood, G is locally bounded. We can therefore find a sequence $A_1 \subseteq A_2 \subseteq \cdots \subseteq G$ of coarsely bounded sets forming a basis for the ideal $\mathcal{CB}$. We claim that $H = \bigcup_n \overline{\phi[A_n]}$. Indeed, if $h \in H$, choose $g_n \in G$ so that $\phi(g_n) \to h$. Then $\{\phi(g_n)\}_n$ is relatively compact and thus coarsely bounded in H. As ϕ is a coarse embedding, it follows that also $\{g_n\}_n$ is coarsely bounded in G and thus is contained in some set A_k. In other words, $h \in \overline{\phi[A_k]}$. It follows that the ideal $\mathcal{CB}$ on H is countably generated and thus that H is locally bounded. Choosing $U \subseteq H$ to be a coarsely bounded identity neighbourhood, we have $H = \phi[G]U$ and hence $\phi[G]$ is cobounded in H. So ϕ is a coarse equivalence. $\qquad\square$

2.7 Coarsely Proper Écarts

With the above concepts at hand, we may now return to issues regarding metrisability.

Definition 2.51 An continuous left-invariant écart d on a topological group G is said to be *coarsely proper* if d induces the left-coarse structure on G, i.e., if $\mathcal{E}_L = \mathcal{E}_d$.

Thus, by Theorem 2.38, a European group admits a coarsely proper écart if and only if it is locally bounded. On the other hand, the identification of coarsely proper écarts is also of interest.

Lemma 2.52 *The following are equivalent for a continuous left-invariant écart d on a topological group G:*

(1) *d is coarsely proper,*
(2) *a set $A \subseteq G$ is coarsely bounded if and only if it is d-bounded,*
(3) *for every continuous left-invariant écart ∂ on G, the mapping*

$$(G, d) \xrightarrow{\text{id}} (G, \partial)$$

is bornologous.

Proof Because

$$\mathcal{E}_L = \bigcap \{\mathcal{E}_\partial \mid \partial \text{ is a continuous left-invariant écart on } G\},$$

coarse properness of the écart d simply means that $\mathcal{E}_d \subseteq \mathcal{E}_\partial$ for every other continuous left-invariant écart ∂, that is, that the map

$$(G,d) \xrightarrow{\text{id}} (G,\partial)$$

is bornologous.

Note that, in general, if d and ∂ are left-invariant écarts on G so that any d-bounded set is ∂-bounded, then $(G,d) \xrightarrow{\text{id}} (G,\partial)$ is bornologous. For, if the d-ball of radius R is contained in the ∂-ball of radius S, then

$$d(x,y) = d(1,x^{-1}y) < R \Rightarrow \partial(x,y) = \partial(1,x^{-1}y) < S.$$

Thus, as the coarsely bounded sets are those bounded in every continuous left-invariant écart, this shows that d is coarsely proper if and only if a set $A \subseteq G$ is coarsely bounded exactly when it is d-bounded. $\square$

The following criterion is useful for identifying coarsely proper metrics.

Lemma 2.53 *Suppose that d is a compatible left-invariant metric on a topological group G without proper open subgroups. Then d is coarsely proper if and only if, for all constants Δ and $\delta > 0$, there is a k so that, for any $x \in G$ with $d(x,1) < \Delta$, there are $y_0 = 1, y_1, \ldots, y_{k-1}, y_k = x$ so that $d(y_i, y_{i+1}) < \delta$.*

In particular, every compatible geodesic metric is coarsely proper.

Proof To see this, suppose first that d is coarsely proper. Then, for any Δ, the open ball $B_d(\Delta) = \{x \in G \mid d(x,1) < \Delta\}$ is coarsely bounded in G and hence, for any $\delta > 0$, there is, as G has no proper open subgroup, some k so that $B_d(\Delta) \subseteq \left(B_d(\delta)\right)^k$. It follows that every $x \in B_d(\Delta)$ can be written as $x = z_1 z_2 \cdots z_k$ with $z_i \in B_d(\delta)$. So, letting $y_i = z_1 \cdots z_i$, we find that the above condition on d is verified.

Assume instead that d satisfies this condition and let $\Delta > 0$ be given. We must show that the ball $B_d(\Delta)$ is coarsely bounded in G. For this, note that, if V is any identity neighbourhood in G, then, because d is a compatible metric, there is $\delta > 0$ so that $B_d(\delta) \subseteq V$. Choosing k as in the assumption on d, we note that $B_d(\Delta) \subseteq \left(B_d(\delta)\right)^k \subseteq V^k$, thus verifying that $B_d(\Delta)$ is coarsely bounded in G. $\square$

Example 2.54 Consider the additive topological group $(X,+)$ of a Banach space $(X, \|\cdot\|)$. Because the norm metric is geodesic on X, by Lemma 2.53, we conclude that the latter is coarsely proper on $(X,+)$.

Example 2.55 (A topology for the coarse structure) Let G be a Hausdorff topological group. Then the ideal $\mathcal{CB}$ of coarsely bounded sets may be used to define a topology $\tau_{\mathcal{CB}}$ on the one-point extension $G \cup \{*\}$. Namely, for $U \subseteq G \cup \{*\}$, we set $U \in \tau_{\mathcal{CB}}$ if

(1) $U \cap G$ is open in G and
(2) $* \in U \Rightarrow G \setminus U \in \mathcal{CB}$.

Because, clearly, $\emptyset, G \cup \{*\} \in \tau_{\mathcal{CB}}$ and $\tau_{\mathcal{CB}}$ is stable under finite intersections and arbitrary unions, it is a topology on $G \cup \{*\}$. Moreover, when restricted to G, this is simply the usual topology on G and G is open in $G \cup \{*\}$. Also, because $\mathcal{CB}$ is stable under taking topological closures in G, we find that, for $A \subseteq G$,

$$ * \in \overline{A}^{\tau_{\mathcal{CB}}} \Leftrightarrow A \notin \mathcal{CB}. $$

We claim that

$$ \tau_{\mathcal{CB}} \text{ is Hausdorff} \Leftrightarrow G \text{ is locally bounded.} $$

Indeed, if $\tau_{\mathcal{CB}}$ is Hausdorff, then there is an identity neighbourhood U so that $* \notin \overline{U}^{\tau_{\mathcal{CB}}}$, i.e., $U \in \mathcal{CB}$, showing that G is locally bounded. Conversely, suppose G is locally bounded and that U is a coarsely bounded open identity neighbourhood. Then xU and $V = \{*\} \cup (G \setminus \overline{xU})$ are open sets separating any point $x \in G$ from $*$. Because G itself is Hausdorff and $\tau_{\mathcal{CB}}$-open, $\tau_{\mathcal{CB}}$ is a Hausdorff topology.

Now, suppose that G is a locally bounded Polish group and fix a coarsely proper continuous left-invariant écart d on G. Setting $U_n = \{x \in G \mid d(x,1) > n\} \cup \{*\}$, we see that the U_n form a neighbourhood basis at $*$ so that $\overline{U_{n+1}} \subseteq U_n$. Thus, for any open $W \ni *$, there is $U_n \ni *$ so that $\overline{U_n} \subseteq W$. Because G is regular also, this shows that $\tau_{\mathcal{CB}}$ is a regular topology. Being also Hausdorff and second countable, we conclude, by Urysohn's metrisation theorem, that $\tau_{\mathcal{CB}}$ is metrisable. Thus, if ∂ is a compatible metric on $G \cup \{*\}$, we see that, because G is Polish, G and thus also $G \cup \{*\}$ are G_δ in the completion $\overline{G \cup \{*\}}^{\partial}$. It follows that $G \cup \{*\}$ is a Polish space with respect to the topology $\tau_{\mathcal{CB}}$.

Example 2.56 (Two notions of divergence) If $\tau_{\mathcal{CB}}$ denotes the topology from Example 2.55, we see that, for $g_n \in G$, we have $g_n \xrightarrow[\tau_{\mathcal{CB}}]{} *$ if and only if the g_n eventually leave every coarsely bounded subset of G. As an alternative to this, we write $g_n \to \infty$ if there is a continuous left-invariant écart d on G so that $d(g_n, 1) \to \infty$. Then, because every coarsely bounded set must have finite d-diameter, we have

$$g_n \to \infty \;\Rightarrow\; g_n \xrightarrow[\tau_{CB}]{} *.$$

Conversely, if G admits a coarsely proper continuous left-invariant écart ∂, then $g_n \xrightarrow[\tau_{CB}]{} *$ implies that $\partial(g_n, 1) \to \infty$ and thus also that $g_n \to \infty$. However, as we shall see in Proposition 3.42, this implication fails without the existence of a coarsely proper écart.

Assume that $G \curvearrowright (X, d)$ is a continuous isometric action of a topological group G on a metric space (X, d). Then, for every $x \in X$, the orbit map

$$g \in G \mapsto g \cdot x \in X$$

is uniformly continuous and bornologous. To see that it is uniformly continuous, let $x \in X$ and $\epsilon > 0$ be given. Then, by continuity of the action, there is an identity neighbourhood $V \ni 1$ so that $d(x, vx) < \epsilon$ for all $v \in V$. It follows that

$$d(gx, fx) = d(x, g^{-1}fx) < \epsilon$$

whenever $(g, f) \in E_V$, i.e., $g^{-1}f \in V$, thus verifying uniform continuity. To verify that the map is bornologous, suppose E is a coarse entourage in G. We set $\partial(g, f) = d(gx, fx)$ and note that ∂ is a continuous left-invariant écart on G. By Lemma 2.11, we find that

$$\sup_{(g,f)\in E} d(gx, fx) = \sup_{(g,f)\in E} \partial(g, f) < \infty,$$

showing that $g \mapsto gx$ is bornologous.

We claim that, if the orbit map $g \mapsto gx$ is coarsely proper just for one point $x \in X$, then $g \mapsto gy$ is expanding for every $y \in X$. To see this, suppose that $x \in X$ is such that $g \mapsto gx$ is coarsely proper, that $y \in X$ and $K > 0$. Then, as $g \mapsto gx$ is coarsely proper, the set

$$B = \{g \in G \mid d(x, gx) \leqslant 2d(x, y) + K\}$$

is coarsely bounded. Therefore also

$$A = \{g \in G \mid d(y, gy) \leqslant K\},$$

which is a subset of B, must be coarsely bounded in G. So, if $(g, f) \notin E_A$, we have $d(gy, fy) = d(y, g^{-1}fy) > K$. As E_A is a coarse entourage in G, this shows that $g \mapsto gy$ is expanding.

We have thus verified the following statement.

Lemma 2.57 *Assume* $G \curvearrowright (X, d)$ *is a continuous isometric action of a topological group* G *on a metric space* (X, d). *Then the following are equivalent.*

(1) *For every* $x \in X$, *the orbit map* $g \mapsto gx$ *is a uniformly continuous coarse embedding of* G *into* X,
(2) *for some* $x \in X$, *the orbit map* $g \mapsto gx$ *is coarsely proper.*

As is easy to see, the orbit map $g \mapsto gy$ is cobounded for every $y \in X$ if and only if it is cobounded for some $y \in X$, which again is equivalent to there being an open set $U \subseteq X$ of finite diameter so that $X = G \cdot U$.

Definition 2.58 An isometric action $G \curvearrowright (X, d)$ of a topological group on a metric space is said to be *coarsely proper*, respectively *cobounded*, if the orbit map

$$g \in G \mapsto gx \in X$$

is coarsely proper, respectively cobounded, for some $x \in X$.

By Lemmas 2.45 and 2.57, we see that, if the action is both coarsely proper and cobounded, then the orbit map $g \in G \mapsto gx \in X$ will be a coarse equivalence between G and X.

Suppose d is a continuous left-invariant metric on G and consider the continuous isometric action $G \curvearrowright (G, d)$ given by left-translation. Setting x to be the identity element 1 in G, one sees that d is coarsely proper if and only if the left-multiplication action $G \curvearrowright (G, d)$ is coarsely proper.

Conversely, suppose $G \curvearrowright (X, d)$ is a coarsely proper continuous isometric action of a metrisable group G on a metric space (X, d). Let D be a compatible left-invariant metric on G, fix $x \in X$ and define $\partial(g, f) = d(gx, fx) + D(g, f)$. Then ∂ is a continuous left-invariant metric on G. Also, $\partial \geqslant D$, whereby ∂ is compatible with the topology on G. Moreover, as the action is coarsely proper, so is ∂.

To sum up, a metrisable topological group admits a coarsely proper compatible left-invariant metric if and only if it admits a coarsely proper continuous isometric action on a metric space.

2.8 Quasi-metric Spaces

We are now ready to introduce a refinement of coarse spaces, namely, quasi-metric spaces. The material presented here is all variations of well-known results.

Definition 2.59 Let (X, d_X) and (Y, d_Y) be pseudo-metric spaces. A map $X \xrightarrow{\phi} Y$ is said to be *Lipschitz for large distances* if there are constants K, C so that

$$d_Y(\phi x_1, \phi x_2) \leqslant K \cdot d_X(x_1, x_2) + C$$

for all $x_1, x_2 \in X$.

Similarly, a map $X \xrightarrow{\phi} Y$ is said to be a *quasi-isometric embedding* if there are constants K, C so that

$$\frac{1}{K} \cdot d_X(x_1, x_2) - C \leqslant d_Y(\phi x_1, \phi x_2) \leqslant K \cdot d_X(x_1, x_2) + C$$

for all $x_1, x_2 \in X$. Also, ϕ is a *quasi-isometry* if, moreover, $\phi[X]$ is cobounded in Y.

As is immediately obvious, a quasi-isometric embedding is also a coarse embedding, whereas a quasi-isometry is a coarse equivalence. Furthermore, every map that is Lipschitz for large distances is bornologous.

In the literature on Banach spaces (e.g., [9]), Lipschitz for large distances sometimes means something slightly weaker; namely, that, for all $\alpha > 0$, there is a constant K_α so that $d_Y(\phi x_1, \phi x_2) \leqslant K_\alpha \cdot d_X(x_1, x_2)$, whenever $d_X(x_1, x_2) \geqslant \alpha$. However, in most natural settings, it is equivalent to the above, which is more appropriate, because it is a strengthening of bornologous maps.

Remark 2.60 (Quasi-isometric spaces and écarts) We note that, if there is a quasi-isometry $X \xrightarrow{\phi} Y$ from X to Y, then one may construct a quasi-isometry $Y \xrightarrow{\psi} X$ so that $\psi \circ \phi$ and $\phi \circ \psi$ are close to the identities on X and Y respectively. This means that we can define two spaces (X, d_X) and (Y, d_Y) to be *quasi-isometric* if there is a quasi-isometry between them. Similarly, two écarts d and ∂ on the same set X will be called *quasi-isometric* if the identity map $(X, d) \xrightarrow{\text{id}} (X, \partial)$ is a quasi-isometry. This is easily seen to define equivalence relations on, respectively, pseudo-metric spaces and the set of écarts on a fixed set.

Definition 2.61 A *quasi-metric space* is a set X equipped with a quasi-isometric equivalence class $\mathcal{D}$ of écarts d on X.

Evidently, every quasi-metric space $(X, \mathcal{D})$ admits a canonical coarse structure, namely, the coarse structure $\mathcal{E}_d$ induced by some $d \in \mathcal{D}$. Indeed, any two d and ∂ in $\mathcal{D}$ are quasi-isometric and thus induce the same coarse structure on X. So $\mathcal{E}_d$ is independent of the specific choice of $d \in \mathcal{D}$. Similarly, every pseudo-metric space (X, d) admits a canonical quasi-metric structure, namely, the equivalence class of its écart d.

Also, suppose $X \xrightarrow{\phi} Y$ is a map that is Lipschitz for large distances between two pseudo-metric spaces (X, d_X) and (Y, d_Y). Then, if ∂_X and ∂_Y are écarts quasi-isometric to d_X and d_Y respectively, we see that ϕ will be Lipschitz for large distances for these écarts too (similarly, for quasi-isometric embeddings and quasi-isometries). This means that the notions of Lipschitz for large distances, quasi-isometric embeddings and quasi-isometries can unambiguously be applied to maps between quasi-metric spaces. Quasi-metric spaces may therefore be viewed as coarse spaces with an additional structure that allows us to talk about mappings between them as being, e.g., Lipschitz for large distances as opposed to just bornologous.

Definition 2.62 A pseudo-metric space (X, d) is said to be *large-scale geodesic* if there is some constant $K \geqslant 1$ so that, for all $x, y \in X$, there are $z_0 = x, z_1, z_2, \ldots, z_n = y$ for which $d(z_i, z_{i+1}) \leqslant K$ and

$$\sum_{i=0}^{n-1} d(z_i, z_{i+1}) \leqslant K \cdot d(x, y).$$

For example, if $\mathbb{X}$ is a connected graph, then the shortest-path metric ρ on $\mathbb{X}$ makes $(\mathbb{X}, \rho)$ large-scale geodesic with constant $K = 1$.

The following observation about large-scale geodicity will be employed at several points.

Lemma 2.63 *Suppose (X, d) is a large-scale geodesic pseudo-metric space with constant K. Then for any two points $x, y \in X$, there is a path $z_0 = x$, $z_1, z_2, \ldots, z_n = y$ so that $d(z_i, z_{i+1}) \leqslant K$ and*

$$\sum_{i=0}^{n-1} d(z_i, z_{i+1}) \leqslant K \cdot d(x, y),$$

where

$$n \leqslant 4d(x, y) + 1.$$

Proof Given $x, y \in X$, let $z_0 = x, z_1, z_2, \ldots, z_n = y$ be a path of shortest length so that $d(z_i, z_{i+1}) \leqslant K$ and

$$\sum_{i=0}^{n-1} d(z_i, z_{i+1}) \leqslant K \cdot d(x, y).$$

Observe that, if for some $i = 1, \ldots, n-1$ we had both $d(z_{i-1}, z_i) \leqslant \frac{K}{2}$ and $d(z_i, z_{i+1}) \leqslant \frac{K}{2}$, then $d(z_{i-1}, z_{i+1}) \leqslant K$ and therefore

$$z_0, \ldots, z_{i-1}, z_{i+1}, \ldots, z_n$$

would be a shorter such path, contradicting our assumption. This shows that, for at least $\frac{n-1}{2}$ of the $i = 0, \ldots, n - 1$, we have $d(z_i, z_{i+1}) \geqslant \frac{K}{2}$ and hence that

$$\frac{n-1}{2} \cdot \frac{K}{2} \leqslant \sum_{i=0}^{n-1} d(z_i, z_{i+1}) \leqslant K \cdot d(x, y).$$

We thus conclude that

$$n \leqslant 4 \cdot d(x, y) + 1$$

as claimed. □

This in turn has the following easy consequence.

Lemma 2.64 *Suppose* $X \xrightarrow{\phi} Y$ *is a quasi-isometry between pseudo-metric spaces* (X, d) *and* (Y, ∂). *Then, if* (X, d) *is large-scale geodesic, so is* (Y, ∂).

Proof Let $K \geqslant 1$ be a constant of large-scale geodecity for X and assume that we have chosen K large enough so that $\phi[X]$ is also K-cobounded in Y and that

$$\frac{1}{K} d(x_1, x_2) - K \leqslant \partial(\phi x_1, \phi x_2) \leqslant K \cdot d(x_1, x_2) + K$$

for all $x_1, x_2 \in X$.

Now, suppose that $y_1, y_2 \in Y$ are given and assume that $\partial(y_1, y_2) \geqslant K$. Pick $x_1, x_2 \in X$ so that $\partial(\phi x_i, y_i) \leqslant K$. Observe that

$$d(x_1, x_2) \leqslant K \cdot \partial(\phi x_1, \phi x_2) + K^2.$$

Pick a path $z_0 = x_1, z_1, z_2, \ldots, z_n = x_2$ in X with $n \leqslant 4 \cdot d(x_1, x_2) + 1$ so that $d(z_i, z_{i+1}) \leqslant K$ and

$$\sum_{i=0}^{n-1} d(z_i, z_{i+1}) \leqslant K \cdot d(x_1, x_2).$$

Then, for all i,

$$\partial(\phi z_i, \phi z_{i+1}) \leqslant K^2 + K$$

and

$$\sum_{i=0}^{n-1} \partial(\phi z_i, \phi z_{i+1}) \leqslant (4 \cdot d(x_1, x_2) + 1) \cdot (K^2 + K)$$

$$\leqslant (4K \cdot \partial(\phi x_1, \phi x_2) + 4K^2 + 1) \cdot 2K^2$$

$$\leqslant 8K^3 \cdot \partial(\phi x_1, \phi x_2) + 10K^4$$

$$\leqslant 8K^3 \cdot \partial(y_1, y_2) + 26K^4$$
$$\leqslant 34K^3 \cdot \partial(y_1, y_2).$$

It thus follows that

$$y_1, \; \phi z_0, \; \phi z_1, \ldots, \phi z_n, \; y_2$$

is a path from y_1 to y_2 in Y with steps of length $\leqslant K^2 + K$ and so that

$$\partial(y_1, \phi z_0) + \partial(\phi z_0, \phi z_1) + \partial(\phi z_1, \phi z_2) + \cdots + \partial(\phi z_{n-1}, \phi z_n) + \partial(\phi z_n, y_2)$$
$$\leqslant 34K^3 \cdot \partial(y_1, y_2).$$

On the other hand, if $\partial(y_1, y_2) < K$, then y_1, y_2 is itself a path from y_1 to y_2 with steps of length $\leqslant K^2 + K$ and total length $\leqslant 34K^3 \cdot \partial(y_1, y_2)$. This thus shows that Y is large-scale geodesic with constant $34K^3$. □

Remark 2.65 By Lemma 2.64 we find that large-scale geodecity is invariant under quasi-isometries between pseudo-metric spaces and thus, in particular, is a quasi-isometric invariant of écarts on a space X. Thus, we may define a quasi-metric space $(X, \mathcal{D})$ to be *large-scale geodesic* if some or, equivalently, all $d \in \mathcal{D}$ are large-scale geodesic.

Also, of central importance is the following well-known fact generalising the classical Corson–Klee Lemma [24].

Lemma 2.66 *Let* $X \xrightarrow{\phi} Y$ *be a bornologous map between quasi-metric spaces* $(X, \mathcal{D}_X)$ *and* $(Y, \mathcal{D}_Y)$ *and assume that* $(X, \mathcal{D}_X)$ *is large-scale geodesic. Then* ϕ *is Lipschitz for large distances.*

Proof Fix écarts $d_X \in \mathcal{D}_X$ and $d_Y \in \mathcal{D}_Y$ and let K be a constant of large-scale geodecity for (X, d_X). Let C be chosen large enough so that $d_Y(\phi x, \phi x') \leqslant C$ whenever $d_X(x, x') \leqslant K$.

Assume now that $x, x' \in X$ are given and let $z_0 = x, z_1, z_2, \ldots, z_n = x'$ be a path of minimal length in X so that $d_X(z_i, z_{i+1}) \leqslant K$ and

$$\sum_{i=0}^{n-1} d_X(z_i, z_{i+1}) \leqslant K \cdot d_X(x, x').$$

By Lemma 2.63, we have $n \leqslant 4d_X(x, x') + 1$. It thus follows that

$$d_Y(\phi x, \phi x') \leqslant \sum_{i=0}^{n-1} d_Y(\phi z_i, \phi z_{i+1}) \leqslant \sum_{i=0}^{n-1} C = nC \leqslant 4C \cdot d_X(x, x') + C,$$

which implies that ϕ is Lipschitz for large distances. □

J. Roe (Proposition 2.57 in [75]) showed that a connected coarse space $(X, \mathcal{E})$ is monogenic if and only if it is coarsely equivalent to a geodesic metric space, and a variation of this proof also works for large-scale geodesic quasi-metric spaces in place of geodesic metric spaces.

Lemma 2.67 *A connected coarse space $(X, \mathcal{E})$ is monogenic if and only if it is coarsely equivalent to a large-scale geodesic quasi-metric space.*

Proof Suppose first that $(X, \mathcal{E})$ is monogenic, i.e., that $\mathcal{E}$ is generated by a single entourage E. Without loss of generality, we may assume that $\Delta \subseteq E = E^\mathsf{T}$. Then $\{E^n\}_n$ is a basis for $\mathcal{E}$. Furthermore, as $\mathcal{E}$ is connected, (X, E) is a connected graph and the shortest-path distance ρ will induce the coarse structure $\mathcal{E}$.

Conversely, suppose $Y \xrightarrow{\phi} X$ is a coarse equivalence between a large-scale geodesic quasi-metric space $(Y, \mathcal{D})$ and $(X, \mathcal{E})$. Fix an écart $d \in \mathcal{D}$ and pick $K > 0$ and $E \in \mathcal{E}$ so that (Y, d) is large-scale geodesic with constant K, E is symmetric and contains the diagonal Δ, and $X = E[\phi[Y]]$.

We set

$$F = \{(y_1, y_2) \in Y \times Y \mid d(y_1, y_2) \leqslant K\}$$

and claim that the coarse entourage

$$E \circ (\phi \times \phi) F \circ E$$

generates $\mathcal{E}$. To see this, suppose $E' \in \mathcal{E}$ and let $F' = (\phi \times \phi)^{-1}(E \circ E' \circ E)$. As ϕ is a coarse equivalence, F' is a coarse entourage in Y and thus

$$C = \sup_{(y_1, y_2) \in F'} d(y_1, y_2) < \infty.$$

Suppose that $(y_1, y_2) \in F'$. Then, as d is large-scale geodesic, we may find a shortest path $z_0 = y_1, z_1, \ldots, z_p = y_2$ with $d(z_i, z_{i+1}) \leqslant K$ and

$$\sum_{i=0}^{p-1} d(z_i, z_{i+1}) \leqslant K \cdot d(y_1, y_2) \leqslant KC.$$

By Lemma 2.63, we have $p \leqslant 4C + 1$. It follows that $(y_1, y_2) \in F^{4C+1}$ and hence that

$$F' \subseteq F^{4C+1}.$$

Therefore,

$$(\phi \times \phi) F' \subseteq (\phi \times \phi)(F^{4C+1}) \subseteq ((\phi \times \phi) F)^{4C+1}.$$

Now assume that $(x_1, x_2) \in E'$ and find $y_1, y_2 \in Y$ so that $(x_i, \phi y_i) \in E$ and hence also $(\phi y_1, \phi y_2) \in E \circ E' \circ E$. Then $(y_1, y_2) \in F'$ and thus

$$(\phi y_1, \phi y_2) \in \big((\phi \times \phi)F\big)^{4C+1}.$$

It follows that $(x_1, x_2) \in E \circ \big((\phi \times \phi)F\big)^{4C+1} \circ E$, that is,

$$E' \subseteq E \circ \big((\phi \times \phi)F\big)^{4C+1} \circ E \subseteq \big(E \circ (\phi \times \phi)F \circ E\big)^{4C+1}.$$

This shows our claim that $E \circ (\phi \times \phi)F \circ E$ generates $\mathcal{E}$ and hence that $\mathcal{E}$ is monogenic.   □

2.9 Maximal Écarts

Observe that we may define an ordering on the space of écarts on a set X by letting $\partial \ll d$ if there are constants K and C so that $\partial \leqslant K \cdot d + C$, that is, if the identity map $(X, d) \xrightarrow{\text{id}} (X, \partial)$ is Lipschitz for large distances and thus, a fortiori, bornologous. Note also that $\ll$ is a directed quasi-ordering (or partial pre-ordering) of the écarts on X. For, if d and ∂ are two écarts on X, then so is $d + \partial$, and both $d \ll d + \partial$ and $\partial \ll d + \partial$. It follows that a maximal element in this quasi-ordering $\ll$ is automatically the maximum (or rather a representative for the maximum class).

Because, similarly, the continuous left-invariant écarts on a topological group form a directed quasi-ordering under $\ll$, these considerations motivate the following terminology.

Definition 2.68 A continuous left-invariant écart d on a topological group G is said to be *maximal* if, for every other continuous left-invariant écart ∂, there are constants K, C so that $\partial \leqslant K \cdot d + C$.

Note that, if d is a maximal écart on a metrisable group G and D is a compatible left-invariant metric, then $d + D$ is defines a compatible left-invariant maximal metric on G.

We remark also that, unless G is discrete, this is really the strongest notion of maximality possible for d. Indeed, if G is non-discrete and thus d takes arbitrarily small values, then

$$(G, d) \xrightarrow{\text{id}} (G, \sqrt{d})$$

fails to be Lipschitz for small distances, whereas $\sqrt{d}$ is a continuous left-invariant écart on G.

Note that, by Lemma 2.52, maximal écarts are automatically coarsely proper and thus the ordering $\ll$ provides a gradation within the class of coarsely proper écarts. Moreover, any two maximal écarts are clearly quasi-isometric. This latter observation shows the unambiguity of the following definition.

Definition 2.69 Let G be a topological group admitting a maximal écart. The *quasimetric structure* on G is the quasi-isometric equivalence class of its maximal écarts.

Let us now see how maximal metrics may be constructed. Recall first that, if Σ is a symmetric generating set for a topological group G, then we can define an associated *word metric* $\rho_\Sigma \colon G \to \mathbb{N} \cup \{0\}$ by

$$\rho_\Sigma(g,h) = \min\left(k \geqslant 0 \mid \exists s_1, \ldots, s_k \in \Sigma \ \ g = h s_1 \cdots s_k\right).$$

Thus, ρ_Σ is a left-invariant metric on G, but, because it takes only integral values, it will never be continuous unless of course G is discrete. However, in certain cases, this may be remedied via the following lemma.

Lemma 2.70 *Suppose d is a compatible left-invariant metric on a topological group G and V is a symmetric open identity neighbourhood generating G and having finite d-diameter. Define ∂ by*

$$\partial(f,h) = \inf\left(\sum_{i=1}^n d(g_i, 1) \,\middle|\, g_i \in V \text{ and } f = h g_1 \cdots g_n\right).$$

Then ∂ is a compatible left-invariant metric, quasi-isometric to the word metric ρ_V associated with the generating set V.

Proof We begin by observing that the definition of ∂ may be rewritten as

$$\partial(f,h) = \inf\left(\sum_{i=1}^n d(k_{i-1}, k_i) \,\middle|\, k_0 = f, \ k_n = h \ \text{ and } \ k_{i-1}^{-1}k_i \in V\right).$$

We note that ∂ is a left-invariant écart. Because V is open and d is continuous, ∂ is continuous also. Moreover, as $\partial \geqslant d$, it is a metric generating the topology on G.

To see that ∂ is quasi-isometric to ρ_V, note first that

$$\partial(f,h) \leqslant \mathsf{diam}_d(V) \cdot \rho_V(f,h).$$

For the other direction, pick some $\epsilon > 0$ so that V contains the identity neighbourhood $\{g \in G \mid d(g,1) < 2\epsilon\}$. Now, fix $f,h \in G$ and find a shortest sequence

$g_1, \ldots, g_n \in V$ so that $f = hg_1 \cdots g_n$ and $\sum_{i=1}^n d(g_i, 1) \leqslant \partial(f, h) + 1$. Note that, for all i, we have $g_i g_{i+1} \notin V$, because otherwise we could coalesce g_i and g_{i+1} into a single term $g_i g_{i+1}$ to get a shorter sequence where

$$d(g_i g_{i+1}, 1) \leqslant d(g_i g_{i+1}, g_i) + d(g_i, 1) = d(g_{i+1}, 1) + d(g_i, 1),$$

contradicting the minimal length of $g_1, \ldots, g_n$. It thus follows that $d(g_i g_{i+1}, 1) \geqslant 2\epsilon$ and hence either $d(g_i, 1) \geqslant \epsilon$ or $d(g_{i+1}, 1) \geqslant \epsilon$. As this holds for all i, we see that there are at least $\frac{n-1}{2}$ terms g_i so that $d(g_i, 1) \geqslant \epsilon$. In particular,

$$\frac{n-1}{2} \cdot \epsilon \leqslant \sum_{i=1}^n d(g_i, 1) \leqslant \partial(f, h) + 1$$

and so, as $\rho_V(f, h) \leqslant n$, we have

$$\frac{\epsilon}{2} \cdot \rho_V(f, h) - \left(1 + \frac{\epsilon}{2}\right) \leqslant \partial(f, h) \leqslant \mathrm{diam}_d(V) \cdot \rho_V(f, h).$$

As this inequality holds for all $f, h \in G$, this shows that ∂ and ρ_V are quasi-isometric. $\square$

Remark 2.71 The proof of Lemma 2.70 also applies to the case where d is only a continuous left-invariant écart on G and V is a symmetric open identity neighbourhood generating G, having finite d-diameter and containing an open d-ball centred at the identity. In that case, ∂ will be a continuous left-invariant écart quasi-isometric with ρ_V.

Proposition 2.72 *The following conditions are equivalent for a continuous left-invariant écart d on a topological group G:*

(1) *d is maximal,*
(2) *d is coarsely proper and (G, d) is large-scale geodesic,*
(3) *d is quasi-isometric to the word metric ρ_A given by a coarsely bounded symmetric generating set $A \subseteq G$.*

Proof (2)⇒(1) Assume that d is coarsely proper and (G, d) is large-scale geodesic with constant $K \geqslant 1$. Suppose ∂ is another continuous left-invariant écart on G. Because d is coarsely proper, Lemma 2.52 implies that the identity map

$$(G, d) \xrightarrow{\mathrm{id}} (G, \partial)$$

is bornologous. From Lemma 2.66 it follows that it is also Lipschitz for large distances, thus showing the maximality of d.

(1)$\Rightarrow$(3) Suppose d is maximal. We claim that G is generated by some closed ball $B_k = \{g \in G \mid d(g,1) \leqslant k\}$. Note that, if this fails, then G is the increasing union of the chain of proper open subgroups $V_n = \langle B_n \rangle, n \geqslant 1$. However, it is now easy, using Lemma 2.13, to construct an écart ∂ from the V_n contradicting the maximality of d. First, complementing with symmetric open sets $V_0 \supseteq V_{-1} \supseteq V_{-2} \supseteq \cdots \ni 1$ so that $V_{-n}^3 \subseteq V_{-n+1}$, and letting ∂ denote the continuous left-invariant écart obtained via Lemma 2.13 from $(V_n)_{n \in \mathbb{Z}}$, we see that, for all $g \in B_n \setminus V_{n-1} \subseteq V_n \setminus V_{n-1}$ with $n \geqslant 1$, we have

$$\partial(g,1) \geqslant 2^{n-1} \geqslant n \geqslant d(g,1).$$

Because $B_n \setminus V_{n-1} \neq \emptyset$ for infinitely many $n \geqslant 1$ and $\lim_n \frac{2^{n-1}}{n} = \infty$, this contradicts the maximality of d. We therefore conclude that $G = V_k = \langle B_k \rangle$ for some $k \geqslant 1$.

Let ∂ denote the écart obtained from $V = B_k$ and d via Remark 2.71. Then $d \leqslant \partial$ and, because d is maximal, we have $\partial \leqslant K \cdot d + C$ for some constants K, C, showing that d, ∂ and hence also the word metric ρ_{B_k} are all quasi-isometric. As d is maximal, the symmetric generating set B_k is coarsely bounded in G.

(3)$\Rightarrow$(2) We note that the word metric ρ_A is simply the shortest-path metric on the Cayley graph of G with respect to the symmetric generating set A, i.e., the graph whose vertex set is G and whose edges are $\{g, gs\}$, for $g \in G$ and $s \in A$. Thus, (G, ρ_A) is large-scale geodesic and, because d is quasi-isometric to ρ_A, so is (G, d). Again, as d and ρ_A are quasi-isometric, every d-bounded set is ρ_A-bounded and hence included in some power A^n. It follows that d-bounded sets are coarsely bounded, showing that d is coarsely proper. $\square$

Theorem 2.73 *Suppose G is a European topological group. Then the following conditions are equivalent.*

(1) *G admits a continuous left-invariant maximal écart d,*
(2) *G is generated by a coarsely bounded set,*
(3) *G is locally bounded and is not the union of a countable chain of proper open subgroups,*
(4) *G is monogenic.*

Proof The equivalence of (2), (3) and (4) has already been established in Theorem 2.40. Also, if these equivalent conditions hold, then, by Theorem 2.38, G admits a continuous left-invariant coarsely proper écart d. Furthermore, as G is generated by some coarsely bounded set, this set must be contained in an open ball $V = \{x \in G \mid d(1,x) < k\}$, which in turn generates G. Proposition 2.72 then implies that the écart ∂ defined from V and d by Lemma 2.70 is maximal.

Conversely, if d is a maximal écart on G, then, by Proposition 2.72, G is generated by a coarsely bounded set. □

Example 2.74 (Maximality of geodesic metrics) By Lemma 2.53 and Theorem 2.72, we see that every compatible left-invariant geodesic metric is maximal.

In general, if a topological group G is generated by a coarsely bounded set A, then G is also generated by the closed coarsely bounded set $\overline{A}$. Observe also that, by Theorem 2.73, if G is a European topological group generated by a coarsely bounded set, then G is also generated by an open coarsely bounded set, namely, a sufficiently large open ball of a maximal metric. In a Polish group, we can also show that, if A and B are analytic coarsely bounded symmetric generating sets containing 1, then $A \subseteq B^n$ and $B \subseteq A^n$ for some sufficiently large n and thus the word metrics ρ_A and ρ_B are quasi-isometric. We recall here that a set in a Polish space is *analytic* if it is the continuous image of some other Polish space. For example, Borel sets are analytic.

To verify the statement above, observe first that $G = \bigcup_{n \geqslant 1} A^n$. So, as the A^n are also analytic and thus have the Baire property, we have that by Baire's Theorem some A^n must be somewhere comeagre. From Pettis' Lemma [**71**] it follows that A^{2n} has non-empty interior, whereby $B \subseteq (A^{2n})^m = A^{2nm}$ for some m.

As the next example shows, this may fail if A and B are no longer assumed to be analytic. Nevertheless, there is a large number of Polish groups in which it is true for all symmetric generating sets A and B containing 1. For example, this holds in every group G having ample generics by Lemma 6.15 [**49**].

Before we state the next result, recall that the euclidean metric on $\mathbb{R}$ is geodesic and hence maximal by Example 2.74.

Example 2.75 (A word metric not quasi-isometric to a maximal metric) The abelian group $(\mathbb{R}, +)$ is generated by a coarsely bounded set D so that

$$\underbrace{D + \cdots + D}_{k \text{ times}}$$

has empty interior for all k. In particular, the word metric ρ_D is not quasi-isometric to the euclidean metric.

Proof Let H be a Hamel basis for $\mathbb{R}$, that is, a basis for $\mathbb{R}$ as a vector space over the field $\mathbb{Q}$. By rescaling each of the basis elements, we may assume that $H \subseteq \,]-1,1[$ and that H is dense in this interval. Let also

$$D = \{rh \mid r \in \mathbb{Q}, \ -1 < r < 1 \text{ and } h \in H\}.$$

Observe that, if $x \in \mathbb{R}$, it has a unique representation

$$x = q_1 h_1 + \cdots + q_n h_n$$

with $h_1, \ldots, h_n \in H$ distinct and $q_i \in \mathbb{Q}$. Furthermore, each q_i can be written as a sum of rationals $-1 < r < 1$ and so x belongs to some large enough finite sum

$$D + \cdots + D.$$

Also, suppose towards a contradiction that

$$\underbrace{D + \cdots + D}_{k \text{ times}}$$

contains some open interval I. Pick $m > k$ large enough so that $I \subseteq \,]-m,m[$. Then, because H is dense in $]-1,1[$, we can find some $h \in H$ so that $mh \in I$, whereby

$$mh = q_1 h_1 + \cdots + q_k h_k$$

for some $h_i \in H$ and rational numbers $-1 < q_i < 1$. However, because H is linearly independent over $\mathbb{Q}$, this means that the coefficient m equals a sum of a subsequence of the $q_1, \ldots, q_k$. Because $|q_1| + \cdots + |q_k| < k < m$, this is absurd.

It follows that

$$\sup_{0 \leqslant x \leqslant 1} \rho_D(x,0) = \infty$$

and so the word metric ρ_D fails to be quasi-isometric to the euclidean metric on $\mathbb{R}$. $\square$

Thus far, we have been able, on the one hand, to characterise the maximal écarts and, on the other, to characterise the groups admitting these. However, oftentimes it will be useful to have other criteria that guarantee existence. In the context of finitely generated groups, the main such criterion is the Milnor–Schwarz Lemma [**64, 83**] of which we will have a close analogue.

Theorem 2.76 (First Milnor–Schwarz Lemma) *Suppose $G \curvearrowright (X,d)$ is a coarsely proper, cobounded, continuous isometric action of a topological group on a connected metric space. Then G admits a maximal écart.*

Proof Because the action is cobounded, there is an open set $U \subseteq X$ of finite diameter so that $G \cdot U = X$. We let

$$V = \{g \in G \mid g \cdot U \cap U \neq \emptyset\}$$

and observe that V is an open identity neighbourhood in G. Because the action is coarsely proper, so is the orbit map $g \mapsto gx$ for any $x \in U$. But, as $\mathrm{diam}_d(V \cdot x) \leqslant 3 \cdot \mathrm{diam}_d(U) < \infty$, this shows that V is coarsely bounded in G.

To see that G admits a maximal écart, by Theorem 2.73, it now suffices to verify that G is generated by V. For this, observe that, if $g, f \in G$, then

$$\left(g\langle V \rangle \cdot U\right) \cap \left(f\langle V \rangle \cdot U\right) \neq \emptyset \Rightarrow \left(\langle V \rangle f^{-1} g\langle V \rangle \cdot U\right) \cap U \neq \emptyset$$

$$\Rightarrow \left(\langle V \rangle f^{-1} g\langle V \rangle\right) \cap V \neq \emptyset$$

$$\Rightarrow f^{-1} g \in \langle V \rangle$$

$$\Rightarrow g\langle V \rangle = f\langle V \rangle.$$

Thus, distinct left cosets $g\langle V \rangle$ and $f\langle V \rangle$ give rise to disjoint open subsets $g\langle V \rangle \cdot U$ and $f\langle V \rangle \cdot U$ of X. However, $X = \bigcup_{g \in G} g\langle V \rangle \cdot U$ and X is connected, which implies that there can only be a single left coset of $\langle V \rangle$, i.e., $G = \langle V \rangle$. $\qquad\square$

Theorem 2.77 (Second Milnor–Schwarz Lemma) *Suppose $G \curvearrowright (X, d)$ is a coarsely proper, cobounded, continuous isometric action of a topological group on a large-scale geodesic metric space. Then G admits a maximal écart.*

Moreover, for every $x \in X$, the map

$$g \in G \mapsto gx \in X$$

is a quasi-isometry between G and (X, d).

Proof Let $x \in X$ be given, and set $\partial(g, f) = d(gx, fx)$, which defines a continuous left-invariant écart on G. Moreover, because the action is coarsely proper, so is ∂. Also, $g \mapsto gx$ is a cobounded isometric embedding of (G, ∂) into (X, d), i.e., a quasi-isometry of (G, ∂) with (X, d). Thus, as (X, d) is large-scale geodesic, so is (G, ∂), whence ∂ is maximal by Proposition 2.72. $\qquad\square$

Example 2.78 (Power growth of group elements) One of the main imports of the existence of a canonical quasimetric structure on a topological group is that it allows for a formulation of the growth rate of elements and discrete subgroups. On the other hand, for a non-trivial definition of the growth rate of the topological group itself, one must require something more, namely, *bounded geometry*. We will return to this in Chapter 5.

Now, assume G is a topological group admitting a maximal écart d. To an element $g \in G$, we associate the function $p_{d,g}(n) = d(g^n, 1)$ defined for $n \in \mathbb{N}$. Observe that, if ∂ is a different maximal écart, then the associated power growth is equivalent to that of d, in the sense that

$$\frac{1}{K} p_{d,g}(n) - K \leqslant p_{\partial,g}(n) \leqslant K \cdot p_{d,g}(n) + K$$

for some K and all n. Observe also that, if $f \in G$ is another element, then for all n

$$
\begin{aligned}
p_{d, fgf^{-1}}(n) &= d\big((fgf^{-1})^n, 1\big) \\
&= d(fg^n f^{-1}, 1) \\
&\leqslant d(fg^n f^{-1}, fg^n) + d(fg^n, f) + d(f, 1) \\
&= d(f^{-1}, 1) + d(g^n, 1) + d(f, 1) \\
&= p_{d,g}(n) + 2d(f, 1).
\end{aligned}
$$

By symmetry we find that

$$\big| p_{d, fgf^{-1}}(n) - p_{d,g}(n) \big| \leqslant 2d(f, 1)$$

for all n.

Now, define an equivalence relation $\sim$ on maps $\mathbb{N} \to \mathbb{R}_+$ by setting

$$p \sim q \quad \Leftrightarrow \quad \exists K \; \forall n \quad \frac{1}{K} p(n) - K \leqslant q(n) \leqslant K \cdot p(n) + K.$$

Then we see that

$$g \in G \mapsto p_g = [p_{d,g}]_\sim$$

is a conjugacy and inversion invariant assignment that does not depend on the specific choice of maximal metric d. We view p_g as the *power growth function* of the element g in G.

Example 2.79 (Length functions) Just as Proposition 2.22 establishes a duality between left-invariant coarse structures on groups and special ideals of subsets, we have a natural duality between left-invariant écarts and special single-variable functions called length functions.

Definition 2.80 A *length function* on a group G is a mapping $\ell \colon G \to [0, \infty[$ satisfying the following conditions:

(1) $\ell(1) = 0$,
(2) $\ell(g^{-1}) = \ell(g)$,
(3) $\ell(gf) \leqslant \ell(g) + \ell(f)$.

Suppose G is a group. Then there is a bijective correspondence

$$\ell \rightsquigarrow d$$

between length functions ℓ on G and left-invariant écarts d on G given by

$$d(f,g) = \ell(f^{-1}g) \quad \text{and} \quad \ell(g) = d(1,g).$$

Now, suppose in addition that G is a topological group; then, as the map $(f,g) \mapsto f^{-1}g$ is continuous, we find that

$$\ell \text{ is continuous} \quad \Leftrightarrow \quad d\colon G \times G \to [0,\infty[\text{ is jointly continuous}$$

$$\Leftrightarrow \quad d\colon G \times G \to [0,\infty[\text{ is separately continuous.}$$

Observe also that d is a metric exactly when $\ell(g) > 0$ for all $g \neq 1$.

Because of this duality, it is of course entirely possible to express all facts about (continuous) left-invariant écarts in terms of the corresponding length functions. Indeed, in many cases, this can be preferable, but in order to have a unified presentation we have mostly eschewed length functions in this text.

For the record, let us reformulate a couple of the concepts encountered so far. Firstly, a continuous left-invariant écart d is coarsely proper if and only if the corresponding length function ℓ has the following property.

For every continuous length function L, there is a monotone function $\omega\colon [0,\infty[\to [0,\infty[$ so that

$$L(g) \leqslant \omega\big(\ell(g)\big)$$

for all $g \in G$.

Alternatively, ℓ is coarsely proper if and only if $\sup_{g \in A} \ell(g) = \infty$ for all coarsely unbounded A. Similarly, a set $A \subseteq G$ is coarsely bounded if and only if $\sup_{g \in A} \ell(g) < \infty$ for all continuous length functions ℓ on G.

3

Structure Theory

In Chapter 2, we introduced the coarse and quasi-metric structures along with characterisations of these and various criteria for the existence of the latter. It is now time to study these structures on concrete groups. In the following, we focus on Polish groups, which seem to be of most interest in practice. However, many of the tools presented here will probably be of use in a wider context.

3.1 The Roelcke Uniformity

A topological tool that will turn out to be fundamental in the study of the coarse structure of Polish groups is the so-called Roelcke uniformity.

Definition 3.1 The *Roelcke uniformity* on a topological group G is the meet $\mathcal{U}_L \wedge \mathcal{U}_R$ of the left- and right-uniformities on G. That is, it is the uniformity generated by the basic entourages

$$E_V = \{(x, y) \in G \times G \mid y \in VxV\},$$

where V ranges over identity neighbourhoods in G.

When G is metrisable, then by the Birkhoff–Kakutani Theorem, i.e., Theorem 2.14, one may pick a left-invariant compatible metric d on G, in which case, the metric

$$d_\wedge(x, y) = \inf_{z \in G} d(x, z) + d(z^{-1}, y^{-1})$$
$$= \inf_{\substack{v, w \in G \\ y = vxw}} d(v, 1) + d(w, 1)$$

is a compatible metric for the Roelcke uniformity. Note that, as opposed to d, the Roelcke metric $d_\wedge$ will not in general be left-invariant.

As with any uniform space, $(G, \mathcal{U}_L \wedge \mathcal{U}_R)$ has a unique completion, denoted $\widehat{G}$, which is obtained by considering the set of all $\mathcal{U}_L \wedge \mathcal{U}_R$-Cauchy nets in G. In the case where d is a compatible left-invariant metric on G, then $\widehat{G}$ is simply the completion of the metric space $(G, d_\wedge)$ and so, in particular, $d_\wedge$ extends to a compatible complete metric on $\widehat{G}$. We call $\widehat{G}$ the *Roelcke completion* of G.

Now, a subset A of G is said to be *Roelcke pre-compact* if it is relatively compact in $\widehat{G}$. This is equivalent to demanding that A is totally bounded in the uniformity; that is, for every identity neighbourhood V there is a finite set $F \subseteq G$ so that $A \subseteq VFV$. Thus, a Roelcke pre-compact set is automatically coarsely bounded. Observe that, if $A \subseteq VFV$, then also $\operatorname{cl} A \subseteq V^2 F V^2$. Therefore, the family of Roelcke pre-compact sets is a subideal of the coarsely bounded sets, stable under taking topological closures $A \mapsto \operatorname{cl} A$ and under inversion $A \mapsto A^{-1}$.

Roelcke pre-compact sets are more robust than coarsely bounded sets, because they are preserved by passing to open subgroups. More precisely, suppose H is an open subgroup of a topological group G equipped with the induced topology. Then the Roelcke uniformity on H is simply that induced from G, i.e., consists of entourages $E \cap (H \times H)$ where E varies over uniform entourages in G. It follows that a subset $A \subseteq H$ is Roelcke pre-compact in H if and only if it is Roelcke pre-compact in G.

Definition 3.2 A topological group G is *Roelcke pre-compact* if it is pre-compact in the Roelcke uniformity, and *locally Roelcke pre-compact* if it has a Roelcke pre-compact identity neighbourhood.

Within the class of Polish groups, there is a useful characterisation of the Roelcke pre-compact groups as the automorphism groups of ω-categorical metric structures [8, 78]. We can also describe the Roelcke pre-compact groups in less technical language. Namely, suppose (X, d) is a separable complete metric space and H is a closed subgroup of the isometry group $\operatorname{Isom}(X, d)$, where the latter is equipped with the topology of pointwise convergence. Then H acts diagonally on X^n for each $n \geqslant 1$, i.e.,

$$h \cdot (x_1, \ldots, x_n) = (hx_1, \ldots, hx_n)$$

and we can equip X^n with the the supremum metric

$$d_\infty\big((x_1, \ldots, x_n), (y_1, \ldots, y_n)\big) = \max_i d(x_i, y_i).$$

We say that H is *approximately oligomorphic* if, for every $n \geqslant 1$ and $\epsilon > 0$, there is a finite set $A \subseteq X^n$ so that $H \cdot A$ is ϵ-dense in X^n. Now, every such approximately oligomorphic group H is Roelcke pre-compact and, conversely, every Roelcke pre-compact group G is isomorphic to such an approximately oligomorphic group H.

From this description, one can verify that the class of Roelcke pre-compact Polish groups includes many familiar isometry groups of highly homogeneous metric structures, e.g., the infinite symmetric group S_∞, the unitary group $U(\mathcal{H})$ of separable Hilbert space with the strong operator topology and the group $\mathrm{Aut}([0,1],\lambda)$ of measure-preserving automorphisms of the unit interval with the weak topology. Furthermore, the homeomorphism group of the unit interval $\mathrm{Homeo}([0,1])$ equipped with the compact-open topology is also Roelcke precompact [76].

Obviously, every locally Roelcke pre-compact group is locally bounded and thus provides an important source of examples.

For the next few examples, we will need the following concept.

Definition 3.3 A metric space (X,d) is said to be *metrically homogeneous* if every surjective isometry $A \xrightarrow{\phi} B$ between two finite subsets of X extends to a surjective isometry of the entire space, $X \xrightarrow{\tilde{\phi}} X$.

Note that, because any two points x,y in a metric space X give isometric subspaces $\{x\}$ and $\{y\}$, being metrically homogeneous implies that the isometry group $\mathrm{Isom}(X,d)$ acts transitively on X.

Example 3.4 (Metrically homogeneous graphs) Every connected graph Γ is naturally a metric space when equipped with the shortest-path metric ρ, and automorphisms of Γ are then exactly the isometries of the metric space. We say that Γ is metrically homogeneous if the associated metric space is metrically homogeneous. As noted above, this is stronger than requiring Γ to be *vertex transitive*, i.e., that the automorphism group acts transitively on the set of vertices. A classification program of metrically homogeneous countable graphs is currently underway, see, e.g., the book by G. Cherlin [18].

When Γ is countable, we equip the automorphism group $\mathrm{Aut}(\Gamma)$ with the *permutation group topology*, which is obtained by declaring the pointwise stabilisers

$$V_A = \{g \in \mathrm{Aut}(\Gamma) \mid g(a) = a \text{ for } a \in A\}$$

of finite sets of vertices A to be open. This makes $\mathrm{Aut}(\Gamma)$ into a Polish group.

Theorem 3.5 *Let Γ be a countable metrically homogenous connected graph. Then the automorphism group $\mathrm{Aut}(\Gamma)$ is locally Roelcke pre-compact and, for any root $t \in \Gamma$, the mapping $g \in \mathrm{Aut}(\Gamma) \mapsto g(t) \in \Gamma$ is a quasi-isometry.*

Proof We claim that, for any root $a_0 \in \Gamma$ and any $k \geqslant 0$, the set

$$U = \{g \in \mathrm{Aut}(\Gamma) \mid \rho(g(a_0), a_0) \leqslant k\}$$

is Roelcke pre-compact. To see this, let V be any identity neighbourhood in $\mathrm{Aut}(\Gamma)$ and find vertices $a_1, \ldots, a_n \in \Gamma$ so that $V_{\{a_1, \ldots, a_n\}} \subseteq V$. Let

$$r = \mathrm{diam}_\rho(\{a_0, a_1, \ldots, a_n\})$$

and note that there are only finitely many isometry types of metric spaces of cardinality $\leqslant 2n + 2$ and whose distances are integral and at most $2r + k$. Observe that, if $g \in U$, then

$$\{a_0, a_1, \ldots, a_n, g(a_0), g(a_1), \ldots, g(a_n)\}$$

is such a space. That means that there is a finite set $F \subseteq \mathrm{Aut}(\Gamma)$ so that, for any $g \in U$, there is $f \in F$ for which

$$\rho(a_i, g(a_j)) = \rho(a_i, f(a_j))$$

for all i, j. Because Γ is metrically homogeneous, we find that for such g, f the isometry

$$\{a_0, \ldots, a_n, g(a_0), \ldots, g(a_n)\} \xrightarrow{\ \phi\ } \{a_0, \ldots, a_n, f(a_0), \ldots, f(a_n)\}$$

defined by $\phi(a_i) = a_i$ and $\phi(g(a_i)) = f(a_i)$ extends to an element

$$h \in V_{\{a_1, \ldots, a_n\}} \subseteq \mathrm{Aut}(\Gamma).$$

Then $f^{-1}hg(a_i) = a_i$ for all i and so $f^{-1}hg \in V_{\{a_1, \ldots, a_n\}}$, i.e.,

$$g \in h^{-1}f V_{\{a_1, \ldots, a_n\}} \subseteq V_{\{a_1, \ldots, a_n\}} F V_{\{a_1, \ldots, a_n\}}.$$

So

$$U \subseteq V_{\{a_1, \ldots, a_n\}} F V_{\{a_1, \ldots, a_n\}} \subseteq VFV,$$

showing that U is Roelcke pre-compact.

Now, observe that every such U is an identity neighbourhood, so $\mathrm{Aut}(\Gamma)$ is locally Roelcke pre-compact. Also, as Roelcke pre-compact sets are coarsely bounded, this shows that the transitive isometric action of $\mathrm{Aut}(\Gamma)$ on the geodesic metric space (Γ, ρ) is coarsely proper. Therefore, by the second Milnor–Schwarz Lemma, Theorem 2.77, the orbit mapping

$$g \in \mathrm{Aut}(\Gamma) \mapsto g(a_0) \in \Gamma$$

is a quasi-isometry between $\mathrm{Aut}(\Gamma)$ and Γ. $\square$

Example 3.6 (Regular trees) For a number $n = 2, 3, 4, \ldots, \aleph_0$, the *n-regular tree* is a connected, acyclic graph whose vertices all have valence n. We denote this by T_n for n finite and T_∞ for $n = \aleph_0$. Observe that, for example, T_{2k} with $k \geqslant 1$ can be realised as the Cayley graph of the non-abelian free group $\mathbb{F}_k$ on k generators with respect to its free generating set.

As is easy to see, any surjective isometry $A \xrightarrow{\phi} B$ between two finite subsets of T_n extends uniquely to an isometry between the two convex hulls of A and B in the graph T_n and further to a surjective isometry of T_n itself. So T_n is metrically homogeneous. It thus follows that, for any $t \in \mathsf{T}_n$, the mapping

$$g \in \mathrm{Aut}(\mathsf{T}_n) \mapsto g(t) \in \mathsf{T}_n$$

is a quasi-isometry between $\mathrm{Aut}(\mathsf{T}_n)$ and T_n itself.

Example 3.7 (The integral Urysohn metric space) The *integral Urysohn metric space* denoted $\mathbb{Z}\mathbb{U}$ is, up to isometry, the unique non-empty countable metric space with integral distances that satisfies the following extension property.

> For any finite metric space X with integral distances, any subspace $Y \subseteq X$ and any isometric embedding
>
> $$Y \xrightarrow{\phi} \mathbb{Z}\mathbb{U},$$
>
> there exists an extension $\tilde{\phi}$ of ϕ to an isometric embedding
>
> $$X \xrightarrow{\tilde{\phi}} \mathbb{Z}\mathbb{U}.$$

Thus, for example, any two points $x, y \in \mathbb{Z}\mathbb{U}$ with distance $d(x, y) = n$ will be connected by a path $z_0 = x, z_1, z_2, \ldots, z_n = y$ in $\mathbb{Z}\mathbb{U}$, where $d(z_i, z_{i+1}) = 1$. In particular, if we let two points x, y in $\mathbb{Z}\mathbb{U}$ be connected by an edge whenever $d(x, y) = 1$, then we see that d is simply the path metric of the resulting graph. Also, by a back-and-forth argument, the extension property ensures that $\mathbb{Z}\mathbb{U}$ is metrically homogeneous. Therefore, we conclude that, for any $x \in \mathbb{Z}\mathbb{U}$, the mapping

$$g \in \mathrm{Isom}(\mathbb{Z}\mathbb{U}) \mapsto g(x) \in \mathbb{Z}\mathbb{U}$$

is a quasi-isometry between $\mathrm{Isom}(\mathbb{Z}\mathbb{U})$ and $\mathbb{Z}\mathbb{U}$ itself. In order to avoid confusion, let us stress that, in this case, the isometry group $\mathrm{Isom}(\mathbb{Z}\mathbb{U})$ is equipped with the permutation group topology from Example 3.4, which, since $\mathbb{Z}\mathbb{U}$ is discrete, coincides with the topology of pointwise convergence on $\mathbb{Z}\mathbb{U}$.

Example 3.8 (The Urysohn space) The Urysohn space $\mathbb{U}$ is a separable complete metric space satisfying the following unrestricted form of the extension property from Example 3.7.

For any finite metric space X, any subspace $Y \subseteq X$ and any isometric embedding

$$Y \xrightarrow{\ \phi\ } \mathbb{U},$$

there exists an extension $\tilde{\phi}$ of ϕ to an isometric embedding

$$X \xrightarrow{\ \tilde{\phi}\ } \mathbb{U}.$$

By a result of P. S. Urysohn [90], these properties completely determine $\mathbb{U}$ up to isometry. Moreover, by a back-and-forth argument, one sees that $\mathbb{U}$ is metrically homogeneous.

We let $\mathrm{Isom}(\mathbb{U})$ be the group of all isometries of $\mathbb{U}$ equipped with the topology of pointwise convergence on $\mathbb{U}$, that is, $g_i \to g$ in $\mathrm{Isom}(\mathbb{U})$ if and only if $d(g_i x, g x) \to 0$ for all $x \in \mathbb{U}$. With this topology, $\mathrm{Isom}(\mathbb{U})$ is a Polish group.

Lemma 3.9 *Suppose $X = \{x_i\}_{i \in I}$ and $Y = \{y_i\}_{i \in I}$ are countable metric spaces indexed by a common set I. Then there are isometric embeddings $X \xrightarrow{\ \phi\ } \mathbb{U}$ and $Y \xrightarrow{\ \psi\ } \mathbb{U}$ so that*

$$\sup_i d_{\mathbb{U}}(\phi x_i, \psi y_i) \leqslant \frac{1}{2} \sup_{i,j} \big| d_X(x_i, x_j) - d_Y(y_i, y_j) \big|.$$

Proof By the extension property of $\mathbb{U}$, it suffices to produce an écart d on the disjoint union $X \sqcup Y$ that agrees with d_X and d_Y on each of X and Y respectively, and so that

$$d(x_k, y_k) \leqslant \frac{1}{2} \sup_{i,j} \big| d_X(x_i, x_j) - d_Y(y_i, y_j) \big|$$

for all $k \in I$. To do this, let $\Delta = \sup_{i,j} \big| d_X(x_i, x_j) - d_Y(y_i, y_j) \big|$ and set

$$d(x, y) = d(y, x) = \inf_{i \in I} \big(d_X(x, x_i) + \Delta/2 + d(y_i, y) \big)$$

for all $x \in X$ and $y \in Y$, while letting d extend d_X on X and d_Y on Y.

To verify the triangle inequality, note first that d restricts to a metric on each of X and Y. Also, the definition of d is clearly symmetric in X and Y, so it suffices to check that

(1) $d(x_i, x_j) \leqslant d(x_i, y_k) + d(y_k, x_j)$,

(2) $d(x_i, y_k) \leqslant d(x_i, x_j) + d(x_j, y_k)$

for all $i, j, k \in I$.

For (1), note that

$$
\begin{aligned}
d(x_i, x_j) &= d_X(x_i, x_j) \\
&\leqslant \inf_{n,m} d_X(x_i, x_n) + d_X(x_n, x_m) + d_X(x_m, x_j) \\
&\leqslant \inf_{n,m} d_X(x_i, x_n) + d_Y(y_n, y_m) + d_X(x_m, x_j) + \Delta \\
&\leqslant \inf_{n,m} d_X(x_i, x_n) + d_Y(y_n, y_k) + d_Y(y_k, y_m) + d_X(x_m, x_j) + \Delta \\
&= \inf_n \big(d_X(x_i, x_n) + \Delta/2 + d_Y(y_n, y_k) \big) \\
&\quad + \inf_m \big(d_Y(y_k, y_m) + \Delta/2 + d_X(x_m, x_j) \big) \\
&= d(x_i, y_y) + d(y_k, x_j),
\end{aligned}
$$

whereas for (2) we have

$$
\begin{aligned}
d(x_i, y_k) &= \inf_m \big(d_X(x_i, x_m) + \Delta/2 + d_Y(y_m, y_k) \big) \\
&\leqslant d_X(x_i, x_j) + \inf_m \big(d_X(x_j, x_m) + \Delta/2 + d_Y(y_m, y_k) \big) \\
&= d_X(x_i, x_j) + d(x_j, y_k) \\
&= d(x_i, x_j) + d(x_j, y_k),
\end{aligned}
$$

which finishes the proof.   □

Theorem 3.10 Isom($\mathbb{U}$) *is locally Roelcke pre-compact and, for any* $x \in \mathbb{U}$, *the map*

$$
g \in \mathrm{Isom}(\mathbb{U}) \mapsto gx \in \mathbb{U}
$$

is a quasi-isometry between Isom($\mathbb{U}$) *and* $\mathbb{U}$.

Proof Fix $x \in \mathbb{U}$ and $\alpha < \infty$. We claim that the set

$$
U = \{ g \in \mathrm{Isom}(\mathbb{U}) \mid d(g(x), x) < \alpha \}
$$

is Roelcke pre-compact in Isom($\mathbb{U}$).

To see this, suppose V is any identity neighbourhood. By shrinking V, we may suppose that

$$
V = \big\{ g \in \mathrm{Isom}(\mathbb{U}) \mid d(g(y), y) < \epsilon, \; \forall y \in A \big\},
$$

where $A \subseteq \mathbb{U}$ is finite, $x \in A$ and $\epsilon > 0$.

Now choose finitely many $f_1, \ldots, f_n \in U$ so that, for any $g \in U$, there is $k \leqslant n$ with

$$\left| d\big(y, g(z)\big) - d\big(y, f_k(z)\big) \right| < \epsilon$$

for all $y, z \in A$. To see that it is possible, just note that the set

$$\left\{ \big(d(y, g(z))\big)_{y,z \in A} \mid g \in U \right\}$$

is a bounded and hence totally bounded subset of $(\mathbb{R}^{A \times A}, \|\cdot\|_\infty)$. It thus suffices to choose the f_i so that $\left\{ \big(d(y, f_i(z))\big)_{y,z \in A} \mid i = 1, \ldots, n \right\}$ is ϵ-dense in $\left\{ \big(d(y, g(z))\big)_{y,z \in A} \mid g \in U \right\}$ for the supremum norm $\|\cdot\|_\infty$.

We claim that $U \subseteq V\{f_1, \ldots, f_n\}V$, showing that $\mathrm{Isom}(\mathbb{U})$ is locally Roelcke pre-compact. To see this, suppose $g \in U$ and pick f_k as above so that

$$\sup_{y,z \in A} \left| d\big(y, g(z)\big) - d\big(y, f_k(z)\big) \right| < \epsilon.$$

By Lemma 3.9, there are isometric embeddings $A \cup g[A] \xrightarrow{\phi} \mathbb{U}$ and $A \cup f_k[A] \xrightarrow{\psi} \mathbb{U}$ so that

$$\sup_{y \in A} d_\mathbb{U}(\phi(y), \psi(y)) < \epsilon/2, \qquad \sup_{y \in A} d_\mathbb{U}(\phi g(y), \psi f_k(y)) < \epsilon/2.$$

Because $A \cup g[A]$ and $A \cup f_k[A]$ are finite subsets of $\mathbb{U}$, the extension property of $\mathbb{U}$ implies that ϕ and ψ may be extended to surjective isometries from $\mathbb{U}$ to $\mathbb{U}$, which we will still denote by ϕ and ψ. Observe now that

$$\sup_{y \in A} d_\mathbb{U}(y, \phi^{-1}\psi(y)) = \sup_{y \in A} d_\mathbb{U}(\phi(y), \psi(y)) < \epsilon/2,$$

so $\phi^{-1}\psi \in V$. Finally,

$$\sup_{y \in A} d_\mathbb{U}(g(y), \phi^{-1}\psi f_k(y)) = \sup_{y \in A} d_\mathbb{U}(\phi g(y), \psi f_k(y)) < \epsilon/2,$$

whereby

$$g \in \phi^{-1}\psi f_k V \subseteq V f_k V$$

as claimed.

By the claim, we see that the orbit map $g \in \mathrm{Isom}(\mathbb{U}) \mapsto g(x) \in \mathbb{U}$ is coarsely proper, whence the tautological action of $\mathrm{Isom}(\mathbb{U})$ on $\mathbb{U}$ is coarsely proper. As the action is also transitive and $\mathbb{U}$ is geodesic, we conclude, by the Milnor–Schwarz Lemma, Theorem 2.77, that the orbit map is a quasi-isometry between $\mathrm{Isom}(\mathbb{U})$ and $\mathbb{U}$. $\qquad\square$

In the above four examples, one may observe that the orbit mappings are more than just coarsely proper. Namely, the inverse image of a set of bounded diameter is actually Roelcke pre-compact. This is not entirely by chance, as in locally Roelcke pre-compact Polish groups we always have a tight connection between the coarse geometry and the Roelcke uniformity. This is borne out by the following fundamental result by J. Zielinski that provides the framework for all work on locally Roelcke pre-compact groups.

Theorem 3.11 (J. Zielinski [99]) *Suppose G is a Polish group and $\widehat{G}$ is its completion in the Roelcke uniformity. Then the following are equivalent.*

(1) *G is locally Roelcke pre-compact,*
(2) *G is locally bounded and every coarsely bounded subset of G is Roelcke pre-compact,*
(3) *$\widehat{G}$ is locally compact.*

One specific consequence of Theorem 3.11 is that, in a locally Roelcke pre-compact Polish group, the product AB of two Roelcke pre-compact sets A and B is still Roelcke pre-compact. We note that this is far from true in general.

Nevertheless, if A, C are respectively Roelcke pre-compact and compact subsets of a Polish group G, then the product set AC is still Roelcke pre-compact. Indeed, given such sets A, C and an identity neghbourhood V in G, choose W to be a smaller identity neighbourhood so that $W^2 \subseteq V$. Then, being compact, C can be covered by finitely many left-translates of W, say $C \subseteq FW$ for some finite set $F \subseteq G$. Now choose an identity neighbourhood $U \subseteq W$ so that $UF \subseteq FW$ and pick $E \subseteq G$ finite so that $A \subseteq UEU$. Then

$$AC \subseteq UEU \cdot FW \subseteq UEFWW \subseteq VEFV,$$

showing that AC is Roelcke pre-compact.

Another very interesting example of a locally Roelcke pre-compact groups is the group $\mathrm{Aut}_{\mathbb{Z}}(\mathbb{Q})$ of order-preserving bijections of $\mathbb{Q}$ commuting with integral shifts. We will return to this in Chapter 5, but now tackle the closely related group $\mathrm{Homeo}_{\mathbb{Z}}(\mathbb{R})$.

Example 3.12 (Homeomorphisms of $\mathbb{R}$ commuting with integral shifts) Consider the groups

$$\mathrm{Homeo}_+(\mathbb{S}^1), \quad \mathrm{Homeo}_+([0,1]), \quad \mathrm{Homeo}_{\mathbb{Z}}(\mathbb{R})$$

of, respectively, orientation-preserving homeomorphisms of the circle $\mathbb{S}^1$, orientation- preserving homeomorphisms of the interval $[0, 1]$ and homeomorphisms of the real line commuting with integral shifts, i.e., commuting with the maps $\tau_n(x) = x + n$ for $n \in \mathbb{Z}$. Alternatively, $\mathrm{Homeo}_{\mathbb{Z}}(\mathbb{R})$ may be described as the group of orientation preserving homeomorphisms of $\mathbb{R}$ preserving the relation $|x - y| = 1$.

Now, $\mathrm{Homeo}_+([0, 1])$ can be seen as the isotropy subgroup in $\mathrm{Homeo}_+(\mathbb{S}^1)$ of any point on the circle and we can therefore write $\mathrm{Homeo}_+(\mathbb{S}^1)$ as a *Zappa–Szép product*

$$\mathrm{Homeo}_+(\mathbb{S}^1) = \mathrm{SO}(2) \cdot \mathrm{Homeo}_+([0, 1]),$$

where $\mathrm{SO}(2)$ or $\mathbb{T}$ is the group of rotations of $\mathbb{S}^1$. In other words, every element of $\mathrm{Homeo}_+(\mathbb{S}^1)$ can be written uniquely as a product gf, where $g \in \mathrm{SO}(2)$ and $f \in \mathrm{Homeo}_+([0, 1])$.

On the other hand, $\mathrm{Homeo}_\mathbb{Z}(\mathbb{R})$ can be seen as the group of lifts of homeomorphisms of the circle $\mathbb{S}^1 = \mathbb{R}/\mathbb{Z}$ to its universal cover $\mathbb{R}$ and thus as a central extension of $\mathrm{Homeo}_+(\mathbb{S}^1)$ by $\mathbb{Z}$,

$$0 \to \mathbb{Z} \to \mathrm{Homeo}_\mathbb{Z}(\mathbb{R}) \to \mathrm{Homeo}_+(\mathbb{S}^1) \to 1.$$

Also, if H denotes the isotropy subgroup of 0 inside $\mathrm{Homeo}_\mathbb{Z}(\mathbb{R})$, then every element of H fixes all the points of $\mathbb{Z}$ and is thus just a homeomorphism of $[0, 1]$ replicated on each interval $[n, n + 1]$. So H is isomorphic to $\mathrm{Homeo}_+([0, 1])$ and $\mathrm{Homeo}_\mathbb{Z}(\mathbb{R})$ factors as the Zappa–Szép product

$$\mathrm{Homeo}_\mathbb{Z}(\mathbb{R}) = \mathbb{R} \cdot H,$$

where we view $\mathbb{R}$ as the group of translations $\tau_r(x) = x + r$ by real numbers r.

Nevertheless, as none of the factors $\mathrm{SO}(2)$ and $\mathrm{Homeo}_+([0, 1])$ in $\mathrm{Homeo}_+(\mathbb{S}^1)$ or $\mathbb{R}$ and H in $\mathrm{Homeo}_\mathbb{Z}(\mathbb{R})$ is normal, the above factorisations of $\mathrm{Homeo}_+(\mathbb{S}^1)$ and of $\mathrm{Homeo}_\mathbb{Z}(\mathbb{R})$ are not semi-direct products. We similarly point out that the central extension $\mathbb{Z} \to \mathrm{Homeo}_\mathbb{Z}(\mathbb{R}) \to \mathrm{Homeo}_+(\mathbb{S}^1)$ does not split, so $\mathrm{Homeo}_\mathbb{Z}(\mathbb{R})$ is not a semi-direct product of $\mathbb{Z}$ and $\mathrm{Homeo}_+(\mathbb{S}^1)$ either.

Now, as shown by W. Roelcke and S. Dierolf [76], $\mathrm{Homeo}_+([0, 1])$ is Roelcke pre-compact. Thus, as $\mathrm{Homeo}_+(\mathbb{S}^1)$ is the product of the compact set $\mathrm{SO}(2)$ and the Roelcke pre-compact set $\mathrm{Homeo}_+([0, 1])$, we find that $\mathrm{Homeo}_+(\mathbb{S}^1)$ is itself Roelcke pre-compact.

Observe also that the identity neighbourhood

$$V = \{f \in \mathrm{Homeo}_\mathbb{Z}(\mathbb{R}) \mid -2 < f(0) < 2\}$$

in $\mathrm{Homeo}_\mathbb{Z}(\mathbb{R})$ is contained in the product of a compact set and a Roelcke pre-compact set, namely,

$$\{\tau_r \mid -2 \leqslant r \leqslant 2\} \cdot H.$$

So V is a Roelcke pre-compact identity neighbourhood in $\mathrm{Homeo}_{\mathbb{Z}}(\mathbb{R})$ and the latter group is locally Roelcke pre-compact.

3.2 Examples of Polish Groups

In the listing of the geometry of various groups, the first to be mentioned are the geometrically trivial examples, i.e., those quasi-isometric to a single-point space.

Definition 3.13 A topological group G is said to be *coarsely bounded*[1] if it is coarsely bounded in itself, i.e., if every continuous left-invariant écart on G is bounded.

Whereas coarsely bounded topological groups may not be very small in a topological sense, they may be viewed as those that are 'geometrically compact' and indeed contain the compact groups as a subclass. Note also that a European group G is coarsely bounded if and only if, for every identity neighbourhood V, there is a finite set F and a k so that $G = (FV)^k$.

Any compact group is Roelcke pre-compact and any Roelcke pre-compact group is coarsely bounded, but there are many sources of coarsely bounded groups beyond Roelcke pre-compactness.

Example 3.14 (Homeomorphism groups of spheres) As shown by M. Culler and the present author in [**78**], homeomorphism groups of compact manifolds of dimension $\geqslant 2$ are never locally Roelcke pre-compact. This is essentially because Dehn twists of different orders can be shown to be well separated in the Roelcke uniformity.[2] Similarly, the homeomorphism group $\mathrm{Homeo}([0,1]^{\mathbb{N}})$ of the Hilbert cube fails to be Roelcke pre-compact [**78**]. However, as shown in [**77**], for all dimensions $n \geqslant 1$, the homeomorphism group $\mathrm{Homeo}(\mathbb{S}^n)$ of the n-sphere is coarsely bounded and the same holds for $\mathrm{Homeo}([0,1]^{\mathbb{N}})$.

Beyond the geometrically trivial groups, we can identify the coarse or quasimetric geometry of many other groups. First of all, the left-coarse structure $\mathcal{E}_L$ of a countable discrete group is that given by any left-invariant proper metric, i.e., whose balls are finite. In the case of a finitely generated group Γ, we see by Proposition 2.72 that the word metric ρ_S induced by a finite symmetric generating set $S \subseteq \Gamma$ is maximal, whence the quasi-isometry

---

[1] We note that these are exactly the groups that have *property (OB)* in the language of [**77**].
[2] The result in [**78**] states only that the groups are not Roelcke pre-compact, but the proof implicitly treats local Roelcke pre-compactness too.

type of Γ is the usual one. The same holds true for a compactly generated locally compact Hausdorff group, i.e., its quasi-isometry type is given by the word metric of a compact generating set.

Example 3.15 (The additive group of a Banach space) Consider again the additive topological group $(X, +)$ of a Banach space $(X, \|\cdot\|)$. Because the norm metric on X is geodesic, it follows from Example 2.74 that it is maximal. In other words, the quasi-isometry type of the topological group $(X, +)$ is none other than the quasi-isometry type of $(X, \|\cdot\|)$ itself. In particular, we need not worry too much about the distinction between the additive topological group $(X, +)$ and the Banach space $(X, \|\cdot\|)$.

Example 3.16 (Groups of affine isometries of Banach spaces) Assume $(X, \|\cdot\|)$ is a Banach space and let $\mathrm{Aff}(X)$ be the group of all *affine* surjective isometries $X \xrightarrow{f} X$, which, by the Mazur–Ulam Theorem, coincides with the group of all surjective isometries of X. Thus, every element $X \xrightarrow{f} X$ splits into a linear isometry $X \xrightarrow{\pi(f)} X$ and a vector $b(f) \in X$ so that

$$f(y) = \pi(f)(y) + b(f)$$

for all $y \in X$.

We equip $\mathrm{Aff}(X)$ with the topology of pointwise convergence on X, i.e., $g_i \to g$ if and only if $\|g_i(x) - g(x)\| \to 0$ for all $x \in X$, and let $\mathrm{Isom}(X)$ be the closed subgroup consisting of linear isometries. So the induced topology on $\mathrm{Isom}(X)$ is simply the strong operator topology. Also, $(X, +)$ may be identified with the closed group of translations in $\mathrm{Aff}(X)$. The discussion above shows that $\mathrm{Aff}(X)$ may be written as a topological semi-direct product

$$\mathrm{Aff}(X) = \mathrm{Isom}(X) \ltimes (X, +)$$

for the natural action of $\mathrm{Isom}(X)$ on X, and so is Polish provided X is separable. In particular, the topology on $\mathrm{Aff}(X)$ is simply the product of the topologies on $\mathrm{Isom}(X)$ and $(X, +)$.

Proposition 3.17 *Let* $(X, \|\cdot\|)$ *be a Banach space. Then the transitive isometric action* $\mathrm{Aff}(X) \curvearrowright X$ *is coarsely proper if and only if* $\mathrm{Isom}(X)$ *is a coarsely bounded group. In that case, the cocycle*

$$\mathrm{Aff}(X) \xrightarrow{b} X$$

is a quasi-isometry between $\mathrm{Aff}(X)$ *and* $(X, \|\cdot\|)$.

Proof Observe first that the cocycle $\mathrm{Aff}(X) \xrightarrow{b} X$ is simply the orbit map

$$f \in \mathrm{Aff}(X) \mapsto f(0) \in X.$$

So the action of $\mathrm{Aff}(X)$ on X is coarsely proper if and only if b is coarsely proper.

Now, $b(f) = 0$ for all $f \in \mathrm{Isom}(X)$. So, if b is coarsely proper, then $\mathrm{Isom}(X)$ must be coarsely bounded in $\mathrm{Aff}(X)$. However, because $\mathrm{Isom}(X)$ is the quotient of $\mathrm{Aff}(X)$ by the normal subgroup $(X, +)$, we find that, if $\mathrm{Isom}(X)$ is coarsely bounded in $\mathrm{Aff}(X)$, then also the quotient group $\mathrm{Isom}(X)$ is coarsely bounded in itself.

Conversely, assume that $\mathrm{Isom}(X)$ is a coarsely bounded group. We must show that, for any α, the set

$$U = \{f \in \mathrm{Aff}(X) \mid \|b(f)\| \leqslant \alpha\}$$

is coarsely bounded in $\mathrm{Aff}(X)$. But, for any $f \in \mathrm{Aff}(X)$, $f = \tau_{b(f)} \circ \pi(f)$, where $\tau_{b(f)} \in \mathrm{Aff}(X)$ denotes the translation by $b(f)$. Therefore $U \subseteq B_\alpha \cdot \mathrm{Isom}(X)$, where B_α is the ball of radius α in X, which is coarsely bounded in $(X, +)$ and thus in $\mathrm{Aff}(X)$.

That b is a quasi-isometry follows immediately from the second Milnor–Schwarz Lemma, Theorem 2.77. $\square$

Because, as noted in Section 3.1, the unitary group $\mathrm{U}(\mathcal{H})$ of separable infinite-dimensional Hilbert space with the strong operator topology is Roelcke pre-compact and thus coarsely bounded, we see that the group of affine isometries of $\mathcal{H}$ is quasi-isometric to $\mathcal{H}$ itself.

Also, S. Banach described the linear isometry groups of ℓ^p, $1 < p < \infty$, $p \neq 2$, as consisting entirely of sign changes and permutations of the basis elements. Thus, the isometry group is the semi-direct product $S_\infty \ltimes \{-1, 1\}^{\mathbb{N}}$ of the Roelcke pre-compact group S_∞ and a compact group, and hence is Roelcke pre-compact itself. Therefore, the affine isometry group $\mathrm{Aff}(\ell^p)$ is quasi-isometric to ℓ^p.

By results due to C. W. Henson [6], the L^p-lattice $L^p([0, 1], \lambda)$, with λ being the Lebesgue measure and $1 < p < \infty$, is ω-categorical in the sense of model theory for metric structures. This also implies that the Banach space reduct $L^p([0, 1], \lambda)$ is ω-categorical and hence that the action by its isometry group on the unit ball is approximately oligomorphic. By Theorem 5.2 [77], it follows that the isometry group $\mathrm{Isom}(L^p)$ is coarsely bounded and thus, as before, that the affine isometry group $\mathrm{Aff}(L^p)$ is quasi-isometric to L^p.

Now, by results of W. B. Johnson, J. Lindenstrauss and G. Schechtman [**42**] (see also Theorem 10.21 in [**9**]), any Banach space quasi-isometric to ℓ^p for $1 < p < \infty$ is, in fact, linearly isomorphic to ℓ^p. Also, for $1 < p < q < \infty$, the spaces L^p and L^q are not coarsely equivalent because they then would be quasi-isometric (being geodesic spaces) and, by taking ultrapowers, would be Lipschitz equivalent, contradicting Corollary 7.8 in [**9**].

Thus, it follows that all of $\mathrm{Aff}(\ell^p)$ and $\mathrm{Aff}(L^p)$ for $1 < p < \infty$, $p \neq 2$, have distinct quasi-isometry types and, in particular, cannot be isomorphic as topological groups.

Example 3.18 Consider again the infinite symmetric group S_∞ of all permutations of $\mathbb{N}$ with the permutation group topology and let $F \leqslant S_\infty$ be the normal subgroup of finitely supported permutations. Viewing F as a countable discrete group, we may define a continous action by automorphisms

$$S_\infty \curvearrowright F$$

simply by setting $\alpha.f = \alpha f \alpha^{-1}$ for $f \in F$ and $\alpha \in S_\infty$. Let $S_\infty \ltimes F$ be the corresponding topological semi-direct product, which is a Polish group. Thus, elements of $S_\infty \ltimes F$ may be represented uniquely as products $f \cdot \alpha$, where $f \in F$ and $\alpha \in S_\infty$. Moreover, $\alpha \cdot f = f^\alpha \cdot \alpha$, where $f^\alpha \in F$ is the conjugation of f by α.

Letting (nm) denote the transposition switching n and m, and noting that S_∞ is coarsely bounded, we see that $B = \{(12) \cdot \alpha \mid \alpha \in S_\infty\}$ is coarsely bounded in $S_\infty \ltimes F$. Noting then that

$$(12) \cdot (12)\alpha \cdot (12)\alpha^{-1}\beta = \big(\alpha(1)\ \alpha(2)\big)\beta,$$

we see that B^3 contains the set A of all products $(nm)\beta$ for $n \neq m$ and $\beta \in S_\infty$. Moreover, because every finitely supported permutation may be written as a product of transpositions, we find that the coarsely bounded set A generates $S_\infty \ltimes F$.

We claim that $S_\infty \ltimes F$ is quasi-isometric to F with the metric

$$d(f,g) = \min(k \mid g^{-1}f \text{ can be written as a product of } k \text{ transpositions}).$$

Indeed, because $S_\infty \ltimes F$ may be written as the product FS_∞, we find that F is cobounded in $S_\infty \ltimes F$. Moreover, for $f \in F$ and $\alpha_i \in S_\infty$,

$$f_1\alpha_1 f_2\alpha_2 \cdots f_n\alpha_n = f_1 f_2^{\alpha_1} f_3^{\alpha_1\alpha_2} \cdots f_n^{\alpha_1\alpha_2\cdots\alpha_{n-1}} \alpha_1\alpha_2\cdots\alpha_n.$$

So, if the f_i are all transpositions, then $f_1\alpha_1 f_2\alpha_2 \ldots f_n\alpha_n = g\beta$, where $g \in F$ is a product of n transpositions and $\beta \in S_\infty$. It follows that, up to an additive error of 1, we have that the A-word length of a product $g\beta$, $g \in F$ and $\beta \in S_\infty$, equals

$$\min(k \mid g \text{ can be written as a product of } k \text{ transpositions}).$$

Now, for $f, g \in F$ and $\alpha, \beta \in S_\infty$,

$$\rho_A(f\alpha, g\beta) = \rho_A(\beta^{-1}g^{-1}f\alpha, 1) = \rho_A((g^{-1}f)^{\beta^{-1}}\beta^{-1}\alpha, 1).$$

So because the minimal number of transpositions with product $(g^{-1}f)^{\beta^{-1}}$ equals that for $g^{-1}f$, we find that

$$\left|\rho_A(f\alpha, g\beta) - d(f, g)\right| \leqslant 1.$$

In other words, $S_\infty \ltimes F$ is quasi-isometric to (F, d).

Example 3.19 (The fragmentation norm on homeomorphism groups) Let M be a closed manifold and $\mathrm{Homeo}_0(M)$ the identity component of its homeomorphism group $\mathrm{Homeo}(M)$ with the compact-open topology. That is, $\mathrm{Homeo}_0(M)$ is the group of isotopically trivial homeomorphisms of M. In [57], it is shown that $\mathrm{Homeo}_0(M)$ is a Polish group generated by a coarsely bounded set and thus has a well-defined quasi-isometry type. Moreover, in contradistinction to the case of spheres, it is also shown that this quasi-isometry type is highly non-trivial once the fundamental group $\pi_1(M)$ has an element of infinite order.

The fact that $\mathrm{Homeo}_0(M)$ is generated by a coarsely bounded set relies on the Fragmentation Lemma of G. M. Fisher [31] and R. D. Edwards and R. C. Kirby [26], which states that, if $\mathcal{U} = \{U_1, \ldots, U_n\}$ is an open cover of M, there is an identity neighbourhood V in $\mathrm{Homeo}_0(M)$ so that every $g \in V$ can be factored into $g = h_1 \cdots h_n$ with $\mathrm{supp}(h_i) \subseteq U_i$. It follows that we may define a fragmentation norm on $\mathrm{Homeo}_0(M)$ by letting

$$\|g\|_{\mathcal{U}} = \min(k \mid g = h_1 \cdots h_k \text{ and } \forall i \; \exists j \; \mathrm{supp}(h_i) \subseteq U_j).$$

As is observed in [57], provided the cover $\mathcal{U}$ is sufficiently fine, the identity neighbourhood V is coarsely bounded and thus the maximal metric on $\mathrm{Homeo}_0(M)$ is quasi-isometric to the metric induced by the fragmentation norm $\|\cdot\|_{\mathcal{U}}$.

In previous work, E. Militon [62] was able to take this even further for the case of compact surfaces by identifying the fragmentation norm $\|\cdot\|_{\mathcal{U}}$ with a metric of maximal displacement on the universal cover of M.

Example 3.20 (Diffeomorphism groups) In [19], M. Cohen considers the diffeomorphism groups $\mathrm{Diff}_+^k(M)$ for $1 \leqslant k \leqslant \infty$ of the one-dimensional manifolds $M = \mathbb{S}^1$ and $M = [0, 1]$. In particular, he shows that a subset $A \subseteq \mathrm{Diff}^k_+(M)$ is coarsely bounded if and only if

$$\sup_{f \in A} \sup_{x \in M} \left| \log f'(x) \right| < \infty \quad \text{and} \quad \sup_{f \in A} \sup_{x \in M} \left| f^{(i)}(x) \right| < \infty$$

for all integers $2 \leqslant i \leqslant k$. If follows that, for $1 \leqslant k < \infty$, the group $\mathrm{Diff}^k_+(M)$ is generated by a coarsely bounded set and, in fact, is quasi-isometric to the Banach space $C([0, 1])$.

Example 3.21 P. J. Cameron and A. M. Vershik [16] have shown that there is an invariant metric d on the group $\mathbb{Z}$ for which the metric space $(\mathbb{Z}, d)$ is isometric to the rational Urysohn metric space $\mathbb{QU}$. Because d is two-sided invariant, the topology τ it induces on $\mathbb{Z}$ is necessarily a group topology, i.e., the group operations are continuous. Thus, $(\mathbb{Z}, \tau)$ is a metrisable topological group and we claim that $(\mathbb{Z}, \tau)$ has a well-defined quasi-isometry type, namely, the Urysohn metric space $\mathbb{U}$ or, equivalently, $\mathbb{QU}$.

To see this, we first verify that d is coarsely proper on $(\mathbb{Z}, \tau)$. For this, note that, because $(\mathbb{Z}, \tau)$ is isometric to $\mathbb{QU}$, we have that, for all $n, m \in \mathbb{Z}$ and $\epsilon > 0$, if $r = \lceil \frac{d(n,m)}{\epsilon} \rceil$, then there are $k_0 = n, k_1, k_2, \ldots, k_r = m \in \mathbb{Z}$ so that $d(k_{i-1}, k_i) \leqslant \epsilon$. Thus, as r is a function only of ϵ and of the distance $d(n, m)$, we see that d satisfies the criteria in Example 2.53 and hence is coarsely proper on $(\mathbb{Z}, \tau)$. Also, as $\mathbb{QU}$ is large-scale geodesic, so is $(\mathbb{Z}, d)$. It follows that the shift action of the topological group $(\mathbb{Z}, \tau)$ on $(\mathbb{Z}, d)$ is a coarsely proper transitive action on a large scale geodesic space. So, by the Milnor–Schwarz Lemma, Theorem 2.77, the identity map is a quasi-isometry between the topological group $(\mathbb{Z}, \tau)$ and the metric space $(\mathbb{Z}, d)$. As the latter is quasi-isometric to $\mathbb{QU}$, so is $(\mathbb{Z}, \tau)$.

By taking the completion of $(\mathbb{Z}, \tau)$, this also provides us with monothetic Polish groups quasi-isometric to the Urysohn space $\mathbb{U}$.

3.3 Rigidity of Categories

In our study we have considered topological groups in three different categories, namely, as uniform and coarse spaces and then as quasi-metric spaces. Each of these come with appropriate notions of morphisms, i.e., uniformly continuous and bornologous maps, whereas the latter also allow for the finer concept of Lipschitz for large-distance maps.

We have already seen some relation between the uniform and coarse structures in that a European group has a metrisable coarse structure if and only if it is locally bounded or, equivalently, if $\mathcal{E}_L \cap \mathcal{U}_L \neq \emptyset$. Not surprisingly, there is also a connection at the level of morphisms. We remind the reader that, in the following, we will exclusively be referring to the left-uniformity $\mathcal{U}_L$ on topological groups.

Proposition 3.22 *Suppose* $G \xrightarrow{\phi} (X,d)$ *is a uniformly continuous map from a topological group* G *to a pseudo-metric space* (X,d) *and assume that* G *has no proper open subgroups. Then* ϕ *is bornologous.*

Proof Because ϕ is uniformly continuous, there is an identity neighbourhood $V \subseteq G$ so that $d(\phi x, \phi y) < 1$, whenever $x^{-1}y \in V$. So, if $x^{-1}y \in V^n$, write $y = xv_1 \cdots v_n$, for $v_i \in V$, and note that

$$d(\phi x, \phi y) \leqslant d\big(\phi x, \phi(xv_1)\big) + d\big(\phi(xv_1), \phi(xv_1v_2)\big)$$
$$+ \cdots + d\big(\phi(xv_1 \cdots v_{n-1}), \phi y\big) < n.$$

On the other hand, because G has no proper open subgroups, then, if E is a coarse entourage on G, there is an n so that $x^{-1}y \in V^n$ for all $(x,y) \in E$. It follows that $d(\phi x, \phi y) < n$ for all $(x,y) \in E$, showing that ϕ is bornologous. $\qquad\square$

Corollary 3.23 *Suppose* $G \xrightarrow{\phi} H$ *is a uniformly continuous map from a topological group* G *without proper open subgroups to a topological group* H. *Then* ϕ *is bornologous.*

Proof Suppose that d is a continuous left-invariant écart on H. Then $G \xrightarrow{\phi} (H,d)$ remains uniformly continuous and therefore is bornologous by Proposition 3.22. This means that, if E is a coarse entourage on G, then $(\phi \times \phi)E \in \mathcal{E}_d$. Therefore, by the definition of the left-coarse structure $\mathcal{E}_L$ on H, we have

$$(\phi \times \phi)E \in \bigcap_\partial \mathcal{E}_\partial = \mathcal{E}_L,$$

where ∂ ranges over continuous left-invariant écarts on H. So ϕ is bornologous. $\qquad\square$

Proposition 3.24 *Suppose* $G \xrightarrow{\phi} H$ *is a bornologous cobounded map between European topological groups. Then, if* G *is locally bounded, so is* H. *Similarly, if* G *is monogenic, so is* H.

Proof As ϕ is cobounded, there is a coarsely bounded set $B \subseteq H$ for which $H = \phi[G] \cdot B$.

Assume first that G is locally bounded. Then G admits a countable covering by open sets U_n coarsely bounded in G, whence $H = \phi[G] \cdot B = \bigcup_n \overline{\phi[U_n] \cdot B}$. Now, as ϕ is bornologous, the $\phi[U_n]$ are coarsely bounded, whence the same is true of the $\overline{\phi[U_n] \cdot B}$. By the Baire Category Theorem, it follows that some $\overline{\phi[U_n] \cdot B}$ has non-empty interior, showing that H is locally bounded.

Assume now that G is monogenic and fix a coarsely bounded symmetric generating set $A \subseteq G$ containing 1. Thus, $G = \bigcup_n A^n$. Because ϕ is bornologous, there is a coarsely bounded set $C \subseteq H$ so that $(\phi \times \phi)E_A \subseteq E_C$ and thus that $(\phi \times \phi)E_{A^n} \subseteq E_{C^n}$ for all n. As $\{1\} \times A^n \subseteq E_{A^n}$, we have that $\{\phi(1)\} \times \phi[A^n] \subseteq E_{C^n}$, whereby $\phi[A^n] \subseteq \phi(1)C^n$. Thus, because $G = \bigcup_n A^n$, we find that

$$H = \phi[G] \cdot B = \bigcup_n \phi[A^n] \cdot B = \bigcup_n \phi(1)C^n B,$$

showing that the coarsely bounded set $\{\phi(1)\} \cup C \cup B$ generates H. $\square$

A complication in the above proposition is that even a dense subgroup of a Polish group need not in general be cobounded (cf. Example 2.49).

Proposition 3.25 *Let $G \xrightarrow{\phi} H$ be a bornologous map between monogenic European groups. Then ϕ is Lipschitz for large distances.*

Proof By Theorem 2.73 and Proposition 2.72, a monogenic European group admits a maximal metric, which furthermore is large-scale geodesic. The proposition therefore follows from an application of Lemma 2.66. $\square$

Corollary 3.26 *Among European groups, the properties of being locally bounded and of being monogenic are both invariant under coarse equivalence. Moreover, every coarse equivalence between monogenic European groups is automatically a quasi-isometry.*

It is natural to wonder whether the coarse properties of a map can be matched by topological properties. Clearly, a bornologous map cannot, in general, be approximated by a continuous bornologous map, e.g., the integral part map $\lfloor \cdot \rfloor \colon \mathbb{R} \to \mathbb{Z}$ is a coarse equivalence, whereas every continuous map $\mathbb{R} \to \mathbb{Z}$ is constant. Nevertheless, measurability can be attained. See also Theorem 5.52 for more a more substantial result in this direction.

Recall first that a subset A of a topological space X is said to have the *Baire property* in X if there is an open set $V \subseteq X$ so that $A \triangle V$ is meagre, i.e., $A \triangle V$ is the union of countable many nowhere dense sets. Similarly, A is a *C-set* if it belongs to the smallest σ-algebra in $\mathcal{P}(X)$ closed under the Souslin operation $\mathcal{A}$ (see Section 29.D in [47]). Every Borel set is a C-set and, by a theorem of O. M. Nikodým (see (29.14) in [47]), every C-set has the Baire property.

Moreover, a map $X \xrightarrow{\phi} Y$ between topological spaces is said to be *Baire measurable* if the inverse image $\phi^{-1}(U)$ of every open set $U \subseteq Y$ has the property of Baire in X. Similarly for C and Borel measurable. By the above, we have the following implications among functions between topological spaces:

$$\text{Borel measurable} \Rightarrow C\text{-measurable} \Rightarrow \text{Baire measurable.}$$

However, in contradistinction to Baire measurable functions, the composition of two C-measurable functions is again C-measurable, which makes it an appropriate class of functions to work with.

In the next lemma, by $\mathcal{G}\phi$, we denote the graph of a function ϕ.

Lemma 3.27 *Suppose $\phi, \psi \colon H \to G$ are maps between topological groups so that $\mathcal{G}\psi \subseteq \overline{\mathcal{G}\phi}$. Assume also that H is metrisable. Then the following hold.*

(1) *If ϕ is modest, so is ψ.*
(2) *If ϕ is bornologous, then ψ is close to ϕ and thus is bornologous too.*

Proof Note first that $\mathcal{G}\psi \subseteq \overline{\mathcal{G}\phi}$ is equivalent to the condition

$$\psi(x) \in \bigcap_{\substack{U \ni x \\ \text{open}}} \overline{\phi[U]} \quad \text{for all } x \in H.$$

To see this, note that $\mathcal{G}\psi \not\subseteq \overline{\mathcal{G}\phi}$ if and only if there is some $x \in H$ and open sets U, V with $(x, \psi(x)) \in U \times V$, while $(U \times V) \cap \mathcal{G}\phi = \emptyset$. However, $(U \times V) \cap \mathcal{G}\phi = \emptyset$ is, in turn, equivalent to $V \cap \phi[U] = \emptyset$. Thus, $\mathcal{G}\psi \not\subseteq \overline{\mathcal{G}\phi}$ if and only if there are $x \in U \subseteq H$ with U open so that $\psi(x) \notin \overline{\phi[U]}$.

For (1), suppose for a contradiction that $A \subseteq H$ is coarsely bounded and that $\psi[A]$ fails to be coarsely bounded in G. Then there is a continuous left-invariant écart d on G and a sequence $x_n \in A$ so that $d(\psi(x_n), \psi(x_1)) \xrightarrow[n \to \infty]{} \infty$. Because H is metrisable and thus first countable, the condition above implies that we can find $z_n \in H$ with $z_n \xrightarrow[n \to \infty]{} 1$ so that also

$$d(\psi(x_n), \phi(x_n z_n)) \xrightarrow[n \to \infty]{} 0,$$

whereby

$$d(\phi(x_n z_n), \phi(x_1 z_1)) \xrightarrow[n \to \infty]{} \infty.$$

However, $C = \{z_n\}_n \cup \{1\}$ is compact and thus coarsely bounded, whereby AC and the subset $\{x_n z_n\}_n$ are also coarsely bounded in H. However, then $\{\phi(x_n z_n)\}_n$ will be coarsely bounded in G, contradicting the above.

For (2), assume instead for a contradiction that ψ is not close to ϕ, i.e., that

$$\{(\psi x, \phi x) \mid x \in H\}$$

is not a coarse entourage in G. Then there is a continuous left-invariant écart d on G and a sequence $x_n \in H$ so that $d(\psi(x_n), \phi(x_n)) \xrightarrow[n \to \infty]{} \infty$. As before, we find $z_n \in H$ with $z_n \xrightarrow[n \to \infty]{} 1$ so that

$$d\big(\psi(x_n), \phi(x_n z_n)\big) \xrightarrow[n \to \infty]{} 0,$$

whence

$$d\big(\phi(x_n), \phi(x_n z_n)\big) \xrightarrow[n \to \infty]{} \infty,$$

showing that $\big\{\big(\phi(x_n), \phi(x_n z_n)\big)\big\}_n$ is not a coarse entourage in G. However, $C = \{z_n\}_n \cup \{1\}$ is compact and thus coarsely bounded, while

$$\{(x_n, x_n z_n)\}_n \subseteq E_C.$$

So this contradicts that ϕ is bornologous. □

Proposition 3.28 *Every bornologous map* $H \xrightarrow{\phi} G$ *between Polish groups is close to a C-measurable bornologous map* ψ *with* $\mathcal{G}\psi \subseteq \overline{\mathcal{G}\phi}$.

Similarly, if instead $H \xrightarrow{\phi} G$ *is modest, then there is a C-measurable modest map* ψ *with* $\mathcal{G}\psi \subseteq \overline{\mathcal{G}\phi}$.

Proof Observe that $\overline{\mathcal{G}\phi}$ is a closed subset of the Polish space $H \times G$ all of whose vertical sections

$$\left(\overline{\mathcal{G}\phi}\right)_x = \{g \in G \mid (x,g) \in \overline{\mathcal{G}\phi}\}$$

are non-empty. So, by the Jankov–von Neumann selection theorem (see (18.1) in [**47**]), there is a C-measurable selector for $\overline{\mathcal{G}\phi}$, i.e., a map $H \xrightarrow{\psi} G$ with $\mathcal{G}\psi \subseteq \overline{\mathcal{G}\phi}$. The properties of ψ now follow from Lemma 3.27. □

The assumption that $\mathcal{G}\psi \subseteq \overline{\mathcal{G}\phi}$ can be useful in different contexts. Here is a particular example.

Corollary 3.29 *Suppose* $G \xrightarrow{\pi} H$ *is a continuous epimorphism between Polish groups. If there is a bornologous section* $H \xrightarrow{\phi} G$ *for* π, *i.e.,* $\pi \circ \phi = \mathsf{id}$, *then there is a C-measurable bornologous section for* π.

Proof Since π is continuous, its graph is closed. Also, by Proposition 3.28, we may pick a C-measurable bornologous map $H \xrightarrow{\psi} G$ with

$$\mathcal{G}\psi \subseteq \overline{\mathcal{G}\phi} \subseteq (\mathcal{G}\pi)^{\mathsf{T}},$$

i.e., ψ is a section for π.   □

3.4 Comparison of Left- and Right-Coarse Structures

Whereas hitherto we have only studied the left-invariant coarse structure $\mathcal{E}_L$ generated by the ideal $\mathcal{CB}$ of coarsely bounded sets in a topological group G, one may equally well consider the *right-coarse structure* $\mathcal{E}_R$ generated by the entourages

$$F_A = \{(x, y) \in G \times G \mid xy^{-1} \in A\},$$

where A varies over coarsely bounded subsets of G. Of course, because the inversion map $\mathsf{inv}\colon x \mapsto x^{-1}$ is seen to be a coarse equivalence between $(G, \mathcal{E}_L)$ and $(G, \mathcal{E}_R)$, the coarse spaces are very much alike and we are instead interested in when they outright coincide.

Proposition 3.30 *Let G be a topological group and $\mathcal{E}_L$ and $\mathcal{E}_R$ be its left- and right-coarse structures. Then the following conditions are equivalent.*

(1) *The left- and right-coarse structures coincide, $\mathcal{E}_L = \mathcal{E}_R$,*
(2) *the inversion map $(G, \mathcal{E}_L) \xrightarrow{\mathsf{inv}} (G, \mathcal{E}_L)$ is bornologous,*
(3) *if A is coarsely bounded is G, then so is $A^G = \{gag^{-1} \mid a \in A$ and $g \in G\}$.*

Proof The equivalence between (1) and (2) follows easily from the fact that the inversion map is a coarse equivalence between $(G, \mathcal{E}_L)$ and $(G, \mathcal{E}_R)$.

(2)$\Rightarrow$(3) Assume that $(G, \mathcal{E}_L) \xrightarrow{\mathsf{inv}} (G, \mathcal{E}_L)$ is bornologous and that $A \subseteq G$ is coarsely bounded in G. Replacing A by $A \cup A^{-1}$, we may suppose that A is symmetric. Then $(\mathsf{inv} \times \mathsf{inv})E_A \in \mathcal{E}_L$ and hence is contained in some basic entourage E_B, with B coarsely bounded in G. In other words, $x^{-1}y \in A$ implies $xy^{-1} \in B$ for all $x, y \in G$. In particular, if $a \in A$ and $g \in G$, then $g^{-1} \cdot ga^{-1} = a^{-1} \in A$, whence $g \cdot ag^{-1} \in B$, showing that $A^G \subseteq B$ and thus that A^G is coarsely bounded in G.

(3)$\Rightarrow$(2) Assume that (3) holds and that E_A is a basic coarse entourage with $A \subseteq G$ symmetric. Then

$$(\mathsf{inv} \times \mathsf{inv})E_A = \{(x^{-1}, y^{-1}) \mid x^{-1}y \in A\} = \{(x, y) \mid xy^{-1} \in A\}.$$

But clearly, if $xy^{-1} \in A$, then $yx^{-1} \in A^{-1} = A$ and thus $x^{-1}y = x^{-1} \cdot yx^{-1} \cdot x \in A^G$, showing that $(\mathsf{inv} \times \mathsf{inv})E_A \subseteq E_{A^G}$. As A^G is coarsely bounded in G, this shows that $\mathsf{inv} \colon (G, \mathcal{E}_L) \to (G, \mathcal{E}_L)$ is bornologous. $\qquad\square$

Observe that condition (3) may equivalently be stated as the ideal $\mathcal{CB}$ having a basis consisting of conjugacy invariant sets.

Example 3.31 If G is a countable discrete group, the coarsely bounded sets in G are simply the finite sets, so $\mathcal{E}_L = \mathcal{E}_R$ if and only if every conjugacy class is finite, i.e., if G is an FCC group (for *finite conjugacy classes*).

Observe that, if d_L is a coarsely proper continuous left-invariant écart on G, i.e., inducing the left-invariant coarse structure $\mathcal{E}_L$, then the écart d_R given by $d_R(g,h) = d_L(g^{-1}, h^{-1})$ is right-invariant. Moreover, because $d_L(g,1) = d_L(1, g^{-1}) = d_R(g,1)$, we see that sets of finite d_R-diameter are coarsely bounded in G and hence that d_R induces the right-coarse structure $\mathcal{E}_R$. Thus, by Proposition 3.30, the identity mapping

$$(G, d_L) \xrightarrow{\ \mathsf{id}\ } (G, d_R)$$

is bornologous if and only if A^G is coarsely bounded in G for every coarsely bounded set $A \subseteq G$.

Example 3.32 It is worth pointing out that the equivalent properties of Proposition 3.30 are not coarse invariants of topological groups, that is, are not preserved under coarse equivalence. Indeed, let D_∞ denote the infinite dihedral group, i.e., the group of all isometries of $\mathbb{Z}$ with the euclidean metric. Then every element g of D_∞ can be written uniquely as $g = \tau_n$ or $g = \tau_n \cdot \rho$, where τ_n is a translation of amplitude n and ρ is the reflection around 0. It follows that the group $\mathbb{Z}$ of translations is an index-2 subgroup of D_∞ and hence is quasi-isometric to D_∞. On the other hand, as $\tau_m \rho \tau_{-m} = \tau_{2m}\rho$, we see that ρ has an infinite conjugacy class in D_∞. So $\mathbb{Z}$ and D_∞ are quasi-isometric, but only $\mathbb{Z}$ is FCC.

Proposition 3.33 *Suppose* $G \curvearrowright X$ *is a coarsely proper continuous isometric action of a topological group* G *on a metric space* (X, d) *commuting with a cobounded isometric group action* $H \curvearrowright X$. *Then the left- and right-coarse structures coincide on* G.

Proof Fix some $x \in X$ and let $\alpha = \sup_{y \in X} \inf_{h \in H} d(y, h \cdot x)$. That the H-action is cobounded simply means that $\alpha < \infty$. Let us begin by noticing that, for all $g \in G$ and $h \in H$, we have

$$d(gx, hx) = d(h^{-1}gx, x) = d(gh^{-1}x, x) = d(h^{-1}x, g^{-1}x).$$

We use this to show that the inversion map $\mathsf{inv}\colon (G,\mathcal{E}_L) \to (G,\mathcal{E}_L)$ is bornologous.

So let E be any coarse entourage on G and let

$$\beta = \sup_{(g,f)\in E} d(gx, fx).$$

Because the G-action is coarsely proper,

$$F = \{(g,f) \in G \times G \mid d(gx, fx) \leqslant \beta + 2\alpha\}$$

is a coarse entourage on G. Now, suppose $(g,f) \in E$ and find some $h \in H$ so that $d(gx, hx) \leqslant \alpha$. Then

$$
\begin{aligned}
d(g^{-1}x, f^{-1}x) &\leqslant d(g^{-1}x, h^{-1}x) + d(h^{-1}x, f^{-1}x) \\
&= d(gx, hx) + d(hx, fx) \\
&\leqslant d(gx, hx) + d(hx, gx) + d(gx, fx) \\
&\leqslant 2\alpha + \beta,
\end{aligned}
$$

whereby $(g^{-1}, f^{-1}) \in F$. This shows that inv is bornologous. $\qquad\square$

Similar to the considerations above is the question of when the left- and right-uniformities $\mathcal{U}_L$ and $\mathcal{U}_R$ coincide. As before one sees that this is equivalent to the inversion map on G being left-uniformly continuous, which again is equivalent to there being a neighbourhood basis at the identity consisting of conjugacy-invariant sets. Groups with this property are called SIN for *small invariant neighbourhoods*. Moreover, by a result of V. Klee [50], metrisable SIN groups are exactly those admitting a compatible bi-invariant metric.

Proposition 3.34 *The following are equivalent for a Polish group G.*

(1) *G admits a coarsely proper bi-invariant compatible metric d,*
(2) *G is SIN, locally bounded and every conjugacy class is coarsely bounded.*

Proof The implication $(1) \Rightarrow (2)$ is trivial because d-balls are conjugacy invariant.

$(2) \Rightarrow (1)$ Assume that (2) holds and let U be a coarsely bounded identity neighbourhood. Since G is SIN, there is a symmetric open conjugacy-invariant identity neighbourhood $V \subseteq U$. Suppose that $A \subseteq G$ is a coarsely bounded set and find a finite $F \subseteq G$ and $k \geqslant 1$ so that $A \subseteq (FV)^k$, whereby also $A^G \subseteq (F^G V^G)^k = (F^G V)^k$. As, F^G is a union of finitely many conjugacy classes, it is coarsely bounded in G, showing that also $(F^G V)^k$ and thus A^G are coarsely bounded in G. In other words, if A is coarsely bounded, so is A^G. Since G

is separable, we may therefore find symmetric, open, coarsely bounded and conjugacy-invariant sets $V = V_0 \subseteq V_1 \subseteq V_2 \subseteq \cdots$ so that

$$V_n^3 \subseteq V_{n+1}$$

for all $n \geqslant 0$ and with $G = \bigcup_k V_k$.

Using that G is SIN, there is a neighbourhood basis at the identity $V_0 \supseteq V_{-1} \supseteq V_{-2} \supseteq \cdots \ni 1$ consisting of conjugacy-invariant, coarsely bounded, symmetric open sets so that now $V_n^3 \subseteq V_{n+1}$ for all $n \in \mathbb{Z}$. It follows that the metric defined from $(V_n)_{n \in \mathbb{Z}}$ via Lemma 2.13 is bi-invariant, coarsely proper and compatible with the topology on G.   □

If G is a locally bounded Polish group whose left- and right-coarse structures coincide, but which fails to be SIN, one may still hope that this is reflected by continuous bi-invariant écarts on the group. However, this turns out to be false, as seen from the following example.

Example 3.35 (Bi-invariant continuous écarts on $\mathrm{Homeo}_{\mathbb{Z}}(\mathbb{R})$ are trivial) Recall from Example 3.12 that $\mathrm{Homeo}_{\mathbb{Z}}(\mathbb{R})$ is the group of homeomorphisms of the real line commuting with integral shifts. Also,

$$d_\infty(g,h) = \sup_{x \in \mathbb{R}} |g^{-1}(x) - f^{-1}(x)|$$

defines a compatible left-invariant metric on $\mathrm{Homeo}_{\mathbb{Z}}(\mathbb{R})$, which clearly is unbounded. In particular, $\mathrm{Homeo}_{\mathbb{Z}}(\mathbb{R})$ is not coarsely bounded. As will be shown in Example 5.16, the left- and right-coarse structures coincide on $\mathrm{Homeo}_{\mathbb{Z}}(\mathbb{R})$. Nevertheless, we claim that every bi-invariant continuous écart d on $\mathrm{Homeo}_{\mathbb{Z}}(\mathbb{R})$ will be constantly 0.

To see this, because $\mathrm{Homeo}_{\mathbb{Z}}(\mathbb{R})$ is connected, it suffices to show that there is an identity neighbourhood V so that $d(v,1) = 0$ for all $v \in V$. First let

$$A = \left\{ g \in \mathrm{Homeo}_{\mathbb{Z}}(\mathbb{R}) \;\middle|\; \mathrm{supp}(g) \subseteq \bigcup_{n \in \mathbb{Z}} \left[n, n + \frac{3}{4}\right] \right\},$$

$$B = \left\{ g \in \mathrm{Homeo}_{\mathbb{Z}}(\mathbb{R}) \;\middle|\; \mathrm{supp}(g) \subseteq \bigcup_{n \in \mathbb{Z}} \left[n - \frac{1}{2}, n + \frac{1}{4}\right] \right\}$$

and note that $V = AB$ is an identity neighbourhood in $\mathrm{Homeo}_{\mathbb{Z}}(\mathbb{R})$. Also, if $\tau_{\frac{1}{2}}$ is the translation $\tau_{\frac{1}{2}}(x) = x + \frac{1}{2}$, then $B = \tau_{\frac{1}{2}} A \tau_{\frac{1}{2}}^{-1}$. Now, because d is continuous, for every $\epsilon > 0$, the open ball

$$B_d(\epsilon) = \{ g \in \mathrm{Homeo}_{\mathbb{Z}}(\mathbb{R}) \mid d(g,1) < \epsilon \}$$

is an identity neighbourhood and thus contains some set

$$D = \left\{ g \in \mathrm{Homeo}_{\mathbb{Z}}(\mathbb{R}) \; \middle| \; \mathrm{supp}(g) \subseteq \bigcup_{n \in \mathbb{Z}} [n, n + \delta] \right\},$$

where $0 < \delta < 1$ depends on ϵ. However, we can then find some $\sigma \in \mathrm{Homeo}_{\mathbb{Z}}(\mathbb{R})$ satisfying $\sigma(n) = n$ and $\sigma(n + \delta) = n + \frac{3}{4}$ for all $n \in \mathbb{Z}$. It then follows that $\sigma D \sigma^{-1} = A$. By the bi-invariance of d, it thus follows that $A, \tau_{\frac{1}{2}} A \tau_{\frac{1}{2}}^{-1} \subseteq B_d(\epsilon)$ and thus that

$$V = A \cdot \tau_{\frac{1}{2}} A \tau_{\frac{1}{2}}^{-1} \subseteq B_d(\epsilon) \cdot B_d(\epsilon) \subseteq B_d(2\epsilon).$$

As $\epsilon > 0$ was arbitrary, we conclude that $V \subseteq \bigcap_{\epsilon > 0} B_d(2\epsilon)$ and so $d(v, 1) = 0$ for all $v \in V$.

3.5 Coarse Geometry of Product Groups

Note that, if $H = \prod_{i \in I} H_i$ is a product of topological groups H_i equipped with the Tychonoff topology, then a basis for the uniformity on H is given by entourages

$$E = \left\{ \big((x_i), (y_i) \big) \in H \mid x_i^{-1} y_i \in V_i, \; \forall i \in J \right\},$$

where J is a *finite* subset of I and $V_i \subseteq H_i$ are identity neighbourhoods. This means that the entourages depend only on a finite set of coordinates, which, as we shall see, is opposed to the coarse structure.

Lemma 3.36 *Let $H = \prod_{i \in I} H_i$ be a product of topological groups H_i. Then a subset $A \subseteq H$ is coarsely bounded in H if and only if there are coarsely bounded sets $A_i \subseteq H_i$ with $A \subseteq \prod_{i \in I} A_i$.*

Proof Suppose that $A \subseteq H$ and set $A_i = \mathrm{proj}_i(A)$. Then, if some A_j is not coarsely bounded in H_j, we may find a continuous left-invariant écart d on H_j with $\mathrm{diam}_d(A_j) = \infty$. So let ∂ be defined on H by

$$\partial \big((x_i), (y_i) \big) = d(x_j, y_j)$$

and observe that ∂ is a continuous left-invariant écart with respect to which A has infinite diameter, i.e., A is not coarsely bounded in H.

Conversely, if each A_i is coarsely bounded in H_i and d is a continuous left-invariant écart on H, then, by continuity of d, there is a finite set $J \subseteq I$ so that

$$\mathrm{diam}_d\left(\prod_{i\in J}\{1\}\times\prod_{i\notin J}H_i\right)<1.$$

Moreover, because each A_j is coarsely bounded in H_j, also $A_j\times\prod_{i\neq j}\{1\}$ is coarsely bounded in H and hence has finite d-diameter. Writing $J=\{j_1,\dots,j_n\}$, this implies that the finite product with respect to the group multiplication in H,

$$\prod_{i\in J}A_i\times\prod_{i\notin J}H_i=\left(A_{j_1}\times\prod_{i\neq j_1}\{1\}\right)\cdots\left(A_{j_n}\times\prod_{i\neq j_n}\{1\}\right)\cdot\left(\prod_{i\in J}\{1\}\times\prod_{i\notin J}H_i\right),$$

has finite d-diameter. As $A\subseteq\prod_{i\in J}A_i\times\prod_{i\notin J}H_i$, this shows that A has finite diameter with respect to any continuous left-invariant écart on H, i.e., that A is coarsely bounded in H. $\qquad\square$

By Lemma 3.36, product sets $\prod_{i\in I}A_i$ with A_i coarsely bounded in H_i thus form a basis for $\mathcal{CB}$, which means that the entourages

$$E_{\prod_{i\in I}A_i}=\left\{\bigl((x_i),(y_i)\bigr)\mid x_i^{-1}y_i\in A_i,\ \forall i\in I\right\}=\prod_{i\in I}E_{A_i}$$

also form a basis for the coarse structure on H. Therefore, a map $X\xrightarrow{\phi}H$ from a coarse space $(X,\mathcal{E})$ is bornologous if and only if all of the compositions $X\xrightarrow{\mathrm{proj}_i\circ\phi}H_i$ are bornologous.

The above considerations also motivate the following definition.

Definition 3.37 Let $\{(X_i,\mathcal{E}_i)\}_{i\in I}$ be a family of coarse spaces. Then the coarse structure on the product $\prod_{i\in I}X_i$ is that generated by entourages of the form

$$\prod_{i\in I}E_i$$

with $E_i\in\mathcal{E}_i$.

Thus, with this definition, the coarse structure on a product group $\prod_{i\in I}H_i$ is simply the product of the coarse structures on the H_i.

Example 3.38 (Inverse limits of Polish groups) If $(G_n)_{n=1}^{\infty}$ is a sequence of Polish groups with bonding continuous homomorphisms

$$\cdots\xrightarrow{p_4}G_4\xrightarrow{p_3}G_3\xrightarrow{p_2}G_2\xrightarrow{p_1}G_1,$$

then the *inverse limit* is the closed subgroup of $\prod_{n=1}^{\infty}G_n$ given by

$$\varprojlim G_n=\left\{(x_1,x_2,\dots)\in\prod_{n=1}^{\infty}G_n\mid p_n(x_{n+1})=x_n,\ \forall n\right\}.$$

For every k, the coordinate projection $\prod_{n=1}^{\infty} G_n \xrightarrow{\mathrm{proj}_k} G_k$ restricts to a mapping

$$\varprojlim G_n \xrightarrow{\mathrm{proj}_k} G_k$$

and, furthermore, the sets

$$\mathrm{proj}_k^{-1}(U),$$

where U ranges over open identity neighbourhoods in G_k and $k \in \mathbb{N}$, form a neighbourhood basis at the identity in $\varprojlim G_n$.

Assume moreover that, for all k, the image $p_k[G_{k+1}]$ is dense in G_k. Then we can show that also

$$\mathrm{proj}_k\left[\varprojlim G_n\right]$$

is dense in G_k. Indeed, note first that, for any non-empty open set $U \subseteq G_k$, the pre-image $p_k^{-1}(U)$ is non-empty open in G_{k+1} and hence, by induction, that

$$(p_k \circ p_{k+1} \circ \cdots \circ p_n)^{-1}(U)$$

is non-empty open in G_{n+1} for all $n \geqslant k$. This means that, if U is any non-empty open set in some G_k, we can inductively choose

$$x_k \in U \subseteq G_k, \quad x_{k+1} \in p_k^{-1}(U) \subseteq G_{k+1},$$

$$x_{k+2} \in (p_k \circ p_{k+1})^{-1}(U) \subseteq G_{k+2}, \ldots$$

so that, for every $m \geqslant k$, the sequence

$$x_m, \quad p_m(x_{m+1}), \quad p_m \circ p_{m+1}(x_{m+2}), \ldots$$

is Cauchy with respect to some fixed complete metric in G_m. Letting $y_m \in G_m$ be the limit, we find that $p_m(y_{m+1}) = y_m$ for all m and so the $y_k, y_{k+1}, \ldots$ correspond to a point $g \in \varprojlim G_m$ so that $\mathrm{proj}_k(g) = y_k \in \overline{U}$. Because U was arbitrary, this ensures density of $\mathrm{proj}_k\left[\varprojlim G_n\right]$ in G_k.

Lemma 3.39 *Suppose $\varprojlim G_n$ is an inverse limit of Polish groups with bonding maps*

$$\cdots \xrightarrow{p_3} G_3 \xrightarrow{p_2} G_2 \xrightarrow{p_1} G_1$$

so that $p_k[G_{k+1}]$ is dense in G_k for all k. Then a subset $A \subseteq \varprojlim G_n$ is coarsely bounded if and only if $\mathrm{proj}_k[A]$ is coarsely bounded in G_k for all k.

Proof Because $p_k[G_{k+1}]$ is dense in G_k for all k, also $\mathrm{proj}_k\left[\varprojlim G_n\right]$ is dense in G_k for all k. Now, as the proj_k are continuous homomorphisms and thus are bornologous, the forward implication of the lemma is clear.

So assume instead that $\mathsf{proj}_k[A]$ is coarsely bounded in G_k for every k and let us see that A is coarsely bounded. Thus, let V be a given identity neighbourhood in $\varprojlim G_n$ and pick some k and identity neighbourhood $U \subseteq G_k$ so that $\mathsf{proj}_k^{-1}(U) \subseteq V$. Since $\mathsf{proj}_k[A]$ is coarsely bounded in G_k and the subgroup $H = \mathsf{proj}_k\big[\varprojlim G_n\big]$ is dense in G_k, we can find a finite set $F \subseteq H$ and some m so that

$$\mathsf{proj}_k[A] \subseteq (FU)^m.$$

Let $E \subseteq \varprojlim G_k$ be a finite set with $\mathsf{proj}_k[E] = F$. Then, using again that H is dense in G_k, that FU is open in G_k and that $F \subseteq H$, we have

$$\mathsf{proj}_k[A] \subseteq (FU)^m \cap H$$
$$= \big((FU) \cap H\big)^m$$
$$= \big(F(U \cap H)\big)^m$$
$$\subseteq \big(\mathsf{proj}_k[E] \cdot \mathsf{proj}_k[V]\big)^m$$
$$= \mathsf{proj}_k\big[(EV)^m\big].$$

Therefore, as $\mathsf{ker}(\mathsf{proj}_k) \subseteq V$, we find that $A \subseteq (EV)^m V$. This shows that A is coarsely bounded in $\varprojlim G_m$. □

We now return to the Urysohn space $\mathbb{U}$ from Example 3.8. Whereas the Urysohn space is universal for all separable metric spaces, i.e., embeds any separable metric space, its isometry group $\mathsf{Isom}(\mathbb{U})$ is universal for all Polish groups as shown by V. V. Uspenskiĭ [91]. We show that similar results are valid in the coarse category.

Theorem 3.40 *Let G be a locally bounded Polish group. Then G admits a simultaneously coarse and isomorphic embedding $G \xrightarrow{\phi} \mathsf{Isom}(\mathbb{U})$. Moreover, if G is monogenic, then ϕ can be made a simultaneously quasi-isometric and isomorphic embedding.*

Proof If G is locally bounded, we let ∂ be a coarsely proper compatible left-invariant metric on G, whereas if G is, furthermore, monogenic, we choose ∂ to be maximal.

We now reprise a construction due to M. Katětov [46] and used by Uspenskiĭ [91] to prove universality of $\mathsf{Isom}(\mathbb{U})$. The construction associates to every separable metric space X an isometric embedding $X \xrightarrow{\iota} \mathbb{U}$, which is functorial in the sense that there is a corresponding isomorphic embedding $\mathrm{Isom}(X) \xrightarrow{\theta} \mathrm{Isom}(\mathbb{U})$ so that

$$\theta(g)(\iota x) = \iota\big(g(x)\big)$$

for all $x \in X$ and $g \in \mathrm{Isom}(X)$.

Taking $X = (G, \partial)$ and embedding G into $\mathrm{Isom}(G, \partial)$ via the left-regular representation $g \mapsto \lambda_g$, we obtain an isometric embedding

$$(G, d) \xrightarrow{\iota} \mathbb{U}$$

and an isomorphic embedding $G \xrightarrow{\phi} \mathrm{Isom}(\mathbb{U})$, $\phi(g) = \theta(\lambda_g)$, so that

$$\phi(g)(\iota f) = \iota\big(\lambda_g(f)\big) = \iota(gf)$$

for all $g, f \in G$. Thus, if $x = \iota 1$ is the image of the identity element in G, we have

$$d\big(\phi(g)x, \phi(f)x\big) = d\big(\iota(g1), \iota(f1)\big) = d(\iota g, \iota f) = \partial(g, f),$$

i.e., the orbit map $g \in G \mapsto \phi(g)x \in \mathbb{U}$ is an isometric embedding of (G, ∂) into $\mathbb{U}$. As, on the other hand, the orbit map $a \in \mathrm{Isom}(\mathbb{U}) \mapsto ax \in \mathbb{U}$ is a quasi-isometry between $\mathrm{Isom}(\mathbb{U})$ and $\mathbb{U}$ by Theorem 3.10, we find that $g \mapsto \phi(g)$ is a quasi-isometric embedding of (G, ∂) into $\mathrm{Isom}(\mathbb{U})$.

Thus, when ∂ is coarsely proper, we get a coarse embedding $G \to \mathrm{Isom}(\mathbb{U})$, whereas when ∂ is maximal the embedding is quasi-isometric. □

Theorem 3.41 *Let G be Polish group. Then G admits a simultaneously coarse and isomorphic embedding into $\prod_{n \in \mathbb{N}} \mathrm{Isom}(\mathbb{U})$.*

Proof Let $\{U_n\}_{n \in \mathbb{N}}$ be a neighbourhood basis at the identity consisting of open symmetric sets. Then, by Lemma 2.35, there are continuous left-invariant écarts d_n so that a subset $A \subseteq G$ is d_n-bounded if and only if there is a finite set $D \subseteq G$ and an m with $A \subseteq (DU_n)^m$. Fixing a compatible left-invariant metric $\partial \leqslant 1$ on G and replacing d_n with $d_n + \partial$, we may suppose that the d_n are compatible left-invariant metrics.

By the proof of Theorem 3.40, there are isomorphic embeddings

$$G \xrightarrow{\phi_n} \mathrm{Isom}(\mathbb{U}),$$

which are quasi-isometric embeddings with respect to the metric d_n on G. So let $G \xrightarrow{\phi} \prod_{n \in \mathbb{N}} \mathrm{Isom}(\mathbb{U})$ be the product $\phi = \bigotimes_{n \in \mathbb{N}} \phi_n$. Evidently, ϕ is an isomorphic embedding and hence also bornologous. So to see that it is a coarse embedding, by Lemma 2.46, it remains to verify that it is coarsely proper. To this aim, assume A is a coarsely unbounded subset of G. Then A has infinite d_n-diameter for some n. But ϕ_n is a quasi-isometric embedding of (G, d_n) into $\mathrm{Isom}(\mathbb{U})$, so $\phi_n[A]$ is coarsely unbounded in $\mathrm{Isom}(\mathbb{U})$. Because

$$\mathrm{proj}_n \circ \phi[A] = \phi_n[A],$$

it follows that $\phi[A]$ is coarsely unbounded in $\prod_{k \in \mathbb{N}} \mathrm{Isom}(\mathbb{U})$ and thus that ϕ is coarsely proper.   □

For the next proposition, recall that, as in Examples 2.55 and 2.56, for elements h_m in a topological group G, we write $g_n \xrightarrow[\tau_{CB}]{} *$ if the g_n eventually leave every coarsely bounded set and $g_n \to \infty$ if there is a continuous left-invariant écart d so that $d(g_n, 1) \to \infty$.

Proposition 3.42 *The following are equivalent for a Polish group G.*

(1) *G is locally bounded,*
(2) *for all sequences (g_n), we have $g_n \xrightarrow[\tau_{CB}]{} *$ if and only if $g_n \to \infty$.*

Proof The implication from (1) to (2) has already been established in Example 2.56. So suppose instead that G fails to be locally bounded. We will construct a sequence (g_n) in G that eventually leaves every coarsely bounded set, but so that nevertheless $g_n \nrightarrow \infty$. As in the proof of Theorem 3.41, let $\{U_n\}_{n \in \mathbb{N}}$ be a neighbourhood basis at the identity consisting of open symmetric sets and find compatible left-invariant metrics d_n so that a subset $A \subseteq G$ is d_n-bounded if and only if there is a finite set $D \subseteq G$ and an m with $A \subseteq (DU_n)^m$. Set also $\partial_n = d_1 + \cdots + d_n$. Then a subset A of G is coarsely bounded if and only if $\mathrm{diam}_{\partial_n}(A) < \infty$ for all n. Moreover, if d is any continuous left-invariant écart on G, then there is some n so that

$$(G, \partial_n) \xrightarrow{\mathrm{id}} (G, d)$$

is bornologous and therefore $d \leqslant \rho \circ \partial_n$ for some increasing function $\rho \colon \mathbb{R}_+ \to \mathbb{R}_+$.

Note that, because G fails to be locally bounded, no ball $B_{\partial_n}(1_G, 1)$ is coarsely bounded. Thus, by passing to a subsequence, we may suppose that $\mathrm{diam}_{\partial_1}(G) = \infty$ and that

$$\mathrm{diam}_{\partial_{n+1}}\big(B_{\partial_n}(1_G, 1)\big) = \infty$$

for all n. So pick $f_n \in G$ with $\partial_1(f_n, 1) \geqslant n$ and find $h_{n,k} \in B_{\partial_n}(1_G, 1)$ so that $\partial_{n+1}(h_{n,k}, 1) \geqslant k$. We let $g_{n,k} = f_n h_{n,k}$ and consider the sequence $(g_{n,k})_{(n,k) \in \mathbb{N}^2}$ re-enumerated in ordertype $\mathbb{N}$.

Observe first that $(g_{n,k})_{(n,k)\in\mathbb{N}^2} \not\to \infty$. Indeed, if d is any continuous left-invariant écart on G, pick n and a monotone function $\rho\colon \mathbb{R}_+ \to \mathbb{R}_+$ for which $d \leqslant \rho \circ \partial_n$. Then, for all k,

$$
\begin{aligned}
d(1, g_{n,k}) &\leqslant \rho\big(\partial_n(1, g_{n,k})\big) \\
&\leqslant \rho\big(\partial_n(1, f_n) + \partial_n(f_n, f_n h_{n,k})\big) \\
&= \rho\big(\partial_n(1, f_n) + \partial_n(1, h_{n,k})\big) \\
&\leqslant \rho\big(\partial_n(1, f_n) + 1\big),
\end{aligned}
$$

showing that the infinite subsequence $(g_{n,k})_{k\in\mathbb{N}}$ is d-bounded.

On the other hand, no coarsely bounded set A contains more than finitely many terms of $(g_{n,k})_{(n,k)\in\mathbb{N}^2}$. If $g_{n,k} = f_n h_{n,k} \in A$, then

$$
\begin{aligned}
n &\leqslant \partial_1(f_n, 1) \\
&\leqslant \partial_1(f_n, f_n h_{n,k}) + \partial_1(f_n h_{n,k}, 1) \\
&\leqslant \partial_1(1, h_{n,k}) + \partial_1(g_{n,k}, 1) \\
&\leqslant \partial_n(1, h_{n,k}) + \mathsf{diam}_{\partial_1}(A) \\
&\leqslant 1 + \mathsf{diam}_{\partial_1}(A),
\end{aligned}
$$

whereas

$$
\begin{aligned}
k &\leqslant \partial_{n+1}(h_{n,k}, 1) \\
&\leqslant \partial_{n+1}(h_{n,k}, f_n^{-1}) + \partial_{n+1}(f_n^{-1}, 1) \\
&\leqslant \partial_{n+1}(f_n h_{n,k}, 1) + \partial_{n+1}(f_n^{-1}, 1) \\
&\leqslant \mathsf{diam}_{\partial_{n+1}}(A) + \partial_{n+1}(f_n^{-1}, 1).
\end{aligned}
$$

Thus, on the one hand, n is bounded independently of k, whereas k is bounded as a function of n, whence only finitely many terms $g_{n,k}$ belong to A. In other words, the sequence $(g_{n,k})_{(n,k)\in\mathbb{N}^2}$ eventually leaves any coarsely bounded set. □

Our discussion here also provides a framework for investigating a problem concerning coarse embeddability of groups. So let us first recall that, if H is a closed subgroup of a Polish group G, then H is coarsely embedded in G if the coarse structure on H, when H is viewed as a topological group in its own right, coincides with the coarse structure on G restricted to H. In other words, H is coarsely embedded in G when every subset $A \subseteq H$ is coarsely bounded in H exactly when it is coarsely bounded in G. In particular, the latter reformulation shows that being coarsely embedded is independent of whether we talk of the left- or the right-coarse structure on H and G (as long as we make the same choice for H and G).

Assume now that, in addition to H being coarsely embedded in G, the supergroup G is also locally bounded. Then the restriction $d|_H$ to H of a coarsely proper metric d on G will also be coarsely proper on H and so also H is locally bounded. However, one can have compatible left-invariant metrics d on G whose restrictions $d|_H$ to H are coarsely proper without d itself being coarsely proper.

Definition 3.43 Let H be a closed subgroup of a Polish group G. We say that H is *well embedded* in G if there is a compatible left-invariant metric d on G so that the restriction $d|_H$ is coarsely proper on H.

For future reference, let us note the following implications.

G is locally bounded and H is coarsely embedded in G

$\Rightarrow H$ is well embedded in G

$\Rightarrow H$ is locally bounded and is coarsely embedded in G.

To understand the link with product groups, let us first prove the following basic result.

Proposition 3.44 *Let H be a closed subgroup of a product $\prod_{n=1}^{\infty} G_n$ of locally bounded Polish groups G_n. Then H is well embedded in $\prod_{n=1}^{\infty} G_n$ if and only if there is some k so that the projection*

$$\prod_{n=1}^{\infty} G_n \xrightarrow{\ \pi_k\ } \prod_{n=1}^{k} G_n$$

is coarsely proper when restricted to H.

Proof Assume first that some π_k is coarsely proper on H, that is,

$$H \xrightarrow{\ \pi_k\ } \prod_{n=1}^{k} G_n$$

is a coarse embedding. Because each of the G_n are locally bounded, so is the finite product $\prod_{n=1}^{k} G_n$ and hence this latter admits a coarsely proper compatible left-invariant metric d_1. Let also d_2 be a compatible left-invariant metric on the infinite product $\prod_{n=1}^{\infty} G_n$. We may then define a compatible left-invariant metric d on $\prod_{n=1}^{\infty} G_n$ by setting

$$d(g, f) = d_1\big(\pi_k(g), \pi_k(f)\big) + d_2(g, f).$$

Then $d|_H$ is coarsely proper. Indeed, if $A \subseteq H$ is coarsely unbounded, then $\pi_k[A]$ is coarsely unbounded in $\prod_{n=1}^{k} G_n$ and hence has infinite d_2 diameter, whence A has infinite d-diameter. So H is well embedded in $\prod_{n=1}^{\infty} G_n$.

For the converse, suppose that H is well embedded in the product $\prod_{n=1}^{\infty} G_n$. Pick a compatible left-invariant metric d on $\prod_{n=1}^{\infty} G_n$ so that the restriction $d|_H$ is coarsely proper. Because d is continuous at the point $(1, 1, 1, \ldots)$, we can find some large enough k so that

$$\underbrace{\{1\} \times \cdots \times \{1\}}_{k \text{ times}} \times G_{k+1} \times G_{k+2} \times \cdots$$

has d-diameter < 1. Let also ι denote the canonical embedding of $\prod_{n=1}^{k} G_n$ into $\prod_{n=1}^{\infty} G_n$,

$$\iota(g_1, \ldots, g_k) = (g_1, \ldots, g_k, 1, 1, \ldots).$$

Then we see that

$$d(x, \iota \circ \pi_k(x)) < 1$$

for all $x \in \prod_{n=1}^{\infty} G_n$. In particular, if A is a coarsely unbounded subset of H, then A has infinite d-diameter, whereby $\iota \circ \pi_k[A]$ has infinite d-diameter and thus is coarsely unbounded in $\prod_{n=1}^{\infty} G_n$. It follows that the composition $\iota \circ \pi_k$

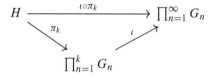

is coarsely proper and therefore that the first factor map

$$H \xrightarrow{\pi_k} \prod_{n=1}^{k} G_n$$

is coarsely proper too. $\qquad\qquad\qquad\qquad\qquad\qquad\qquad\qquad\qquad\qquad\square$

So are coarsely embedded, locally bounded subgroups automatically well embedded? As seen in the next example, the answer is in general no, even considering countable discrete groups H.

Example 3.45 (Coarsely, but not well-embedded subgroups) Consider the free abelian group $\mathbb{A}_X \cong \bigoplus_n \mathbb{Z}$ on a denumerable set of generators X. Let also $w_0 \colon X \to \mathbb{N}$ be a function so that every fibre $w_0^{-1}(n)$ is infinite and choose, for every $n \geqslant 1$, some function $w_n \colon X \to \mathbb{N}$ agreeing with w_0 on $X \setminus w_0^{-1}(n)$, while being injective on $w_0^{-1}(n)$. For every $n \geqslant 0$, let $d_n \colon \mathbb{A}_X \times \mathbb{A}_X \to \mathbb{N} \cup \{0\}$ be the invariant metric with weight w_n, i.e., for distinct $g, f \in \mathbb{A}_X$,

$$d_n(g, f) = \min \left(w_n(x_1) + \cdots + w_n(x_k) \mid g = f x_1^{\pm} \cdots x_k^{\pm} \text{ and } x_i \in X \right).$$

We claim that, if $A \subseteq \mathbb{A}_X$ is infinite, then A is d_n-unbounded for some $n \geqslant 0$. Indeed, if the reduced form of elements of A include generators from infinitely many distinct fibres $w_0^{-1}(n)$, then A is already d_0-unbounded. Similarly, if the elements of A include infinitely many distinct generators from some single fibre $w_0^{-1}(n)$, then A is d_n-unbounded. And finally, the last remaining option is that only finitely many generators appear in elements of A, whereby the word length must be unbounded on A and so A is d_0-unbounded.

On the other hand, for every k, the fibre $A = w_0^{-1}(k+1)$ is an infinite set bounded in each of the metrics $d_0, d_1, \ldots, d_k$.

Now, as in the proof of Theorem 3.40, the discrete metric group $(\mathbb{A}_X, d_n)$ admits a quasi-isometric isomorphic embedding into $\mathrm{Isom}(\mathbb{U})$,

$$\mathbb{A}_X \xrightarrow{\phi_n} \mathrm{Isom}(\mathbb{U}).$$

Then $\phi = \bigotimes_{n=0}^{\infty} \phi_n$ defines an isomorphic embedding

$$\prod_{n=0}^{\infty} \mathbb{A}_X \xrightarrow{\phi} \prod_{n=0}^{\infty} \mathrm{Isom}(\mathbb{U}).$$

We claim that the discrete diagonal subgroup

$$\Delta = \{(f, f, f, \ldots) \mid f \in \mathbb{A}_X\}$$

of $\prod_{n=0}^{\infty} \mathbb{A}_X$ is coarsely embedded into $\prod_{n=0}^{\infty} \mathrm{Isom}(\mathbb{U})$ via ϕ. To see this, it suffices to verify that an infinite subset $B \subseteq \Delta$ has coarsely unbounded image $\phi[B]$. Now, since B is infinite, also $A = \mathrm{proj}_0[B]$ is an infinite subset of $\mathbb{A}_X$ and thus is d_n-unbounded for some $n \geqslant 0$. It thus follows that

$$\mathrm{proj}_n \circ \phi[B] = \phi_n[A]$$

is coarsely unbounded in the nth factor $\mathrm{Isom}(\mathbb{U})$ of $\prod_{n=0}^{\infty} \mathrm{Isom}(\mathbb{U})$ and hence, by Lemma 3.36, that $\phi[B]$ is coarsely unbounded in $\prod_{n=0}^{\infty} \mathrm{Isom}(\mathbb{U})$.

On the other hand, the projections

$$\prod_{n=0}^{\infty} \mathrm{Isom}(\mathbb{U}) \xrightarrow{\pi_k} \prod_{n=0}^{k} \mathrm{Isom}(\mathbb{U})$$

onto the first $k+1$ coordinates all fail to be coarsely proper when restricted to $\phi[\Delta]$. Indeed, let $A = w_0^{-1}(k+1)$ and set $B = \{(f, f, \ldots) \mid f \in A\}$. Then B is an infinite subset of Δ, but, because A is bounded in each of the metrics $d_0, \ldots, d_k$, also

$$\phi_0[A], \phi_1[A], \ldots, \phi_k[A]$$

are coarsely bounded in $\mathrm{Isom}(\mathbb{U})$. It therefore follows that

$$\pi_k \circ \phi[B] \;\subseteq\; \phi_0[A] \times \phi_1[A] \times \cdots \times \phi_k[A]$$

are coarsely bounded.

Thus, by Proposition 3.44, we see that $\phi[\Delta] \cong \bigoplus_n \mathbb{Z}$ is a coarsely embedded countable discrete subgroup of $\prod_{n=0}^{\infty} \mathrm{Isom}(\mathbb{U})$, but fails to be well embedded.

As it turns out, we have positive results when considering coarsely embedded subgroups of products $\prod_n G_n$ of a very important restricted class of groups that we shall return to in Chapter 5.

Definition 3.46 A Polish group G is said to have *bounded geometry* if there is a coarsely bounded set $A \subseteq G$ covering every coarsely bounded set $B \subseteq G$ by finitely many left-translates, i.e., $B \subseteq FA$ for some finite $F \subseteq G$.

Evidently, every locally compact second countable group has bounded geometry, but also more complex groups such as the group $\mathrm{Homeo}_{\mathbb{Z}}(\mathbb{R})$ of homeomorphisms of $\mathbb{R}$ commuting with integral translations have bounded geometry.

Theorem 3.47 *Suppose H is a locally bounded Polish group and $H \xrightarrow{\phi_n} G_n$ is a sequence of continuous homomorphisms into Polish groups G_n of bounded geometry. Assume also that, for every coarsely unbounded set $A \subseteq H$, there is some n so that $\phi_n[A]$ is coarsely unbounded in G_n. Then there is some k so that, for every coarsely unbounded set $A \subseteq H$, there is some $n \leqslant k$ so that $\phi_n[A]$ is coarsely unbounded in G_n.*

Observe that an equivalent formulation of the theorem is that, if

$$H \xrightarrow{\phi} \prod_{n=1}^{\infty} G_n$$

is a coarsely proper continuous homomorphism, then there is some k so that already

$$H \xrightarrow{\pi_k \circ \phi} \prod_{n=1}^{k} G_n$$

is coarsely proper.

Proof Since the G_n have bounded geometry, we may find coarsely bounded symmetric subsets $A_n \subseteq G_n$ that cover every other coarsely bounded set in G_n by finitely many left-translates.

We claim that there is some k so that

$$B_k = \bigcap_{n=1}^{k} \phi_n^{-1}(A_n^2)$$

is coarsely bounded in H. To see this, fix a coarsely proper metric d on H and assume towards a contradiction that no B_k is coarsely bounded. We may then choose $x_k \in B_k$ so that $d(x_k, 1) > k$. In particular, $\{x_k\}_{k=1}^{\infty}$ is coarsely unbounded and so $\phi_n[\{x_k\}_{k=1}^{\infty}]$ is coarsely unbounded in G_n for some n. However, $\phi_n(x_k) \in A_n^2$ whenever $k \geqslant n$ and so $\phi_n[\{x_k\}_{k=1}^{\infty}]$ is contained in

$$A_n^2 \cup \{\phi_n(x_1), \dots, \phi_n(x_{n-1})\},$$

contradicting that this latter set is coarsely bounded.

So, fix k so that B_k is coarsely bounded. Suppose also that $C \subseteq H$ is coarsely unbounded. Then C cannot be covered by finitely many left-translates of the coarsely bounded symmetric set B_k and we can therefore find an infinite subset $D \subseteq C$ so that $y \notin x B_k$, whenever x and y are distinct points in D. Therefore, for distinct $x, y \in D$, there is some $n \leqslant k$ so that

$$\phi_n(x^{-1})\phi_n(y) \notin A_n^2.$$

By Ramsey's theorem for pairs, we find some infinite subset $E \subseteq D$ so that the same $n \leqslant k$ works for all pairs of distinct elements $x, y \in E$. However, this shows that $\phi_n[E]$ cannot be covered by finitely many left-translates of A_n in G_n and therefore, by the choice of A_n, that $\phi_n[E]$ and a fortiori $\phi_n[C]$ is coarsely unbounded in G_n. $\qquad\square$

Corollary 3.48 *Suppose $\varprojlim G_n$ is the inverse limit of a sequence of Polish groups of bounded geometry with continuous bonding maps*

$$\cdots \xrightarrow{p_3} G_3 \xrightarrow{p_2} G_2 \xrightarrow{p_1} G_1,$$

so that $p_k[G_{k+1}]$ is dense in G_k for all k. Suppose also that $H \xrightarrow{\phi} \varprojlim G_n$ is a coarsely proper continuous homomorphism from a locally bounded group H. Then there is some k so that also

$$H \xrightarrow[\mathrm{proj}_k \circ \phi]{} G_k$$

is coarsely proper.

Proof Suppose A is a coarsely unbounded subset of H. Then $\phi[A]$ is coarsely unbounded in $\varprojlim G_n$ and hence, by Lemma 3.39, there is some m so that

$\text{proj}_m \circ \phi[A]$ is coarsely unbounded in G_m. Thus, by Theorem 3.47 applied to the mappings $\phi_n = \text{proj}_k \circ \phi$, we can find some k, so that, if A is any coarsely unbounded subset of H, then $\text{proj}_m \circ \phi[A]$ is coarsely unbounded in G_m for some $m \leqslant k$. Because for $m < k$, $\text{proj}_m \circ \phi[A]$ is the image of $\text{proj}_k \circ \phi[A]$ by $p_m \circ \cdots \circ p_{k-1}$, it follows that $\text{proj}_k \circ \phi[A]$ is coarsely unbounded. It follows that the map $\text{proj}_k \circ \phi$ is coarsely proper. □

The following corollary is a direct consequence of Proposition 3.44, Theorem 3.47 and the fact that Polish groups of bounded geometry are locally bounded (see Lemma 5.6).

Corollary 3.49 *Suppose G_n is a sequence of Polish groups with bounded geometry. Then every coarsely embedded, closed, locally bounded subgroup of $\prod_n G_n$ is well embedded.*

When the G_n are not just of bounded geometry, but locally compact, an even stronger statement is true.

Proposition 3.50 *Suppose G_n is a sequence of locally compact Polish groups. Then every closed, locally bounded subgroup of $\prod_n G_n$ is well embedded.*

Proof Observe that, by Lemma 3.36, the coarsely bounded sets in $\prod_n G_n$ are exactly the relatively compact sets. Thus, if H is a closed subgroup of $\prod_n G_n$ and $A \subseteq H$ is coarsely bounded in $\prod_n G_n$, then $\overline{A}$ is compact and thus A must be coarsely bounded in H too. So H is coarsely embedded. If, moreover, H is locally bounded, i.e., locally compact, it follows from Corollary 3.49, that H is also well embedded. □

3.6 Groups of Finite Asymptotic Dimension

Mathematical literature is rife with competing dimension concepts and, not surprisingly, coarse geometry is no exception to this. One enduring notion is M. Gromov's *asymptotic dimension* [**36**], which for general coarse spaces, is defined as follows.

Definition 3.51 A coarse space $(X, \mathcal{E})$ has *asymptotic dimension* $\leqslant n$ if, for every entourage $E \in \mathcal{E}$, there is an entourage $F \in \mathcal{E}$ and families $\mathcal{B}_0, \ldots, \mathcal{B}_n$ of subsets of X so that

(1) $\bigcup_{i=0}^n \mathcal{B}_i$ is a covering of X,
(2) $B \times B \subseteq F$ for all $B \in \bigcup_{i=0}^n \mathcal{B}_i$,
(3) $(A \times B) \cap E = \emptyset$ for distinct A and B belonging to the same $\mathcal{B}_i$.

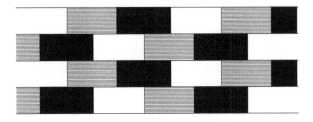

Figure 3.1 Tiling of $\mathbb{R}^2$ by rectangles.

Observe that E here indicates a discreteness factor of the covering, whereas F is a uniform bound on the sizes of sets in the covering. Thus, for a metric space (X, d), having asymptotic dimension $\leqslant n$, means that, for every α, there is a covering $\mathcal{B}$ of X into sets of uniformly bounded diameter so that $\mathcal{B}$ can be partitioned into $\mathcal{B} = \mathcal{B}_0 \cup \cdots \cup \mathcal{B}_n$, where any two distinct sets in the same $\mathcal{B}_i$ are at least α-apart.

The basic example is, of course, the plane $\mathbb{R}^2$, which can be seen to have asymptotic dimension 2 via the tiling by rectangles as in Figure 3.1.

The asymptotic dimension can be seen to be a coarse invariant, which will eventually provide us with examples of non-locally compact Polish groups with finite asymptotic dimension. For example, the group $\mathrm{Homeo}_{\mathbb{Z}}(\mathbb{R})$ of homeomorphisms of $\mathbb{R}$ commuting with integral translations is coarsely equivalent to $\mathbb{R}$ and thus has asymptotic dimension 1, cf. Chapter 5. However, for a more interesting example, consider the n-regular tree T_n for $n = 2, 3, \ldots, \infty$. As is fairly straightforward to see, irrespectively of the value of n, T_n has finite asymptotic dimension 1 (see, e.g., Proposition 2.3.1 in [68]). So again, $\mathrm{Aut}(\mathsf{T}_\infty)$ is a non-locally compact Polish group of finite asymptotic dimension 1.

In any case, these examples show that the notion of finite asymptotic dimension is integral to the geometric study of Polish groups and should be investigated carefully. In particular, we suggest a relation to the metrisability of the coarse structure.

Problem 3.52 Let G be a Polish group of finite asymptotic dimension. Must G be locally bounded?

To explain the motivation for this problem, note that, by Theorem 3.41, every Polish group G can be seen as a coarsely embedded closed subgroup of the infinite product

$$\prod_{n \in \mathbb{N}} \mathrm{Isom}(\mathbb{U}).$$

Moreover, if G is not locally bounded, all of the projections to an initial segment $\prod_{n=1}^{N} \mathrm{Isom}(\mathbb{U})$ will collapse the coarse structure on G. Thus, non-locally bounded Polish groups will have some infinite product structure associated with them via this embedding and, for this reason, seem unlikely to be of finite asymptotic dimension. Observe also that this strategy is not invalidated by Example 3.45, because the latter concerns the group $\bigoplus_{n=1}^{\infty} \mathbb{Z}$ that has infinite asymptotic dimension.

3.7 Distorted Elements and Subgroups

Suppose G is a Polish group with a maximal metric d_G. Recall that a closed subgroup H of G is coarsely embedded if the coarse structure of H agrees with that induced from G. However, if H itself admits a maximal metric d_H, one may also directly compare the metrics d_G and d_H.

Definition 3.53 Suppose G is a Polish group with a maximal metric d_G. Assume also that H is a subgroup of G that is Polish in a finer group topology and admits a maximal metric d_H. We say that H is *undistorted* in G if the inclusion map

$$(H, d_H) \to (G, d_G)$$

is a quasi-isometric embedding, i.e., when $d_H \ll d_G$.

A case of special interest is when H is a finitely generated subgroup of G. Then, for H to be undistorted in G, it must, in particular, be discrete in G. Although finite subgroups are always undistorted in G, it will be convenient to consider an element $g \in G$ to be distorted even if it has finite order.

Definition 3.54 Let G be a Polish group with a maximal metric d. An element $g \in G$ is said to be *distorted* or to be a *distortion element* if

$$\lim_{n \to \infty} \frac{d(g^n, 1)}{n} = 0.$$

Let us observe that, because $d(g^{n+m}, 1) \leqslant d(g^n, 1) + d(g^m, 1)$ for all n, m, the limit $\lim_{n \to \infty} \frac{d(g^n, 1)}{n}$ exists for all elements of G. Indeed, let $\epsilon > 0$ be given and pick n so that

$$\frac{d(g^n, 1)}{n} < \liminf_k \frac{d(g^k, 1)}{k} + \epsilon.$$

Then, for all $m > \frac{n \cdot d(g,1)}{\epsilon}$, write $m = pn + j$ with $0 \leqslant j < n$ and $p \geqslant 0$, whereby

$$\frac{d(g^m,1)}{m} \leqslant \frac{p \cdot d(g^n,1) + j \cdot d(g,1)}{m} < \frac{d(g^n,1)}{n} + \frac{n \cdot d(g,1)}{m}$$

$$< \liminf_k \frac{d(g^k,1)}{k} + 2\epsilon.$$

Thus $\limsup_m \frac{d(g^m,1)}{m} \leqslant \liminf_k \frac{d(g^k,1)}{k} + 2\epsilon$ and hence $\lim_{n\to\infty} \frac{d(g^n,1)}{n}$ exists.

Note also that, in the language of Example 2.78, the distorted elements of G are exactly those that have sublinear associated power growth function $\mathfrak{p}_g$.

We see that the set of distortion elements

$$\left\{ g \in G \ \middle| \ \lim_n \frac{d(g^n,1)}{n} = 0 \right\} = \bigcap_k \bigcup_{n \geqslant k} \left\{ g \in G \ \middle| \ \frac{d(g^n,1)}{n} < \frac{1}{k} \right\}$$

is G_δ in G.

Also, although it is easy to see that the set of distortion elements is invariant under conjugacy, a bit more can be said. For this, we let g^G denote the conjugacy class of g and $\overline{g^G}$ its closure.

Lemma 3.55 *Let G be a Polish group with a maximal metric d and let $g \in G$. If $\overline{g^G}$ contains a distortion element, then also g is distorted. In particular, if g is undistorted, then $\inf_{h \in G} d(hgh^{-1}, 1) > 0$.*

Proof As we are dealing with powers of individual elements, it will be useful to work with the length function ℓ associated to d, namely, $\ell(h) = d(h,1)$. Recall that $\ell(h_1^{-1}) = \ell(h_1)$ and $\ell(h_1 h_2) \leqslant \ell(h_1) + \ell(h_2)$ for all $h_1, h_2 \in G$.

Suppose $f \in \overline{g^G}$ is distorted. Fix $\epsilon > 0$ and pick n so that $\frac{\ell(f^n)}{n} < \epsilon$. Pick $h \in G$ so that $d(hg^n h^{-1}, f^n) = d\big((hgh^{-1})^n, f^n\big) < \epsilon$. Then, for $k \geqslant 1$, we have

$$\ell(g^{kn}) \leqslant \ell(hg^{kn}h^{-1}) + 2\ell(h)$$
$$= \ell\big((hg^n h^{-1})^k\big) + 2\ell(h)$$
$$\leqslant k \cdot \ell(hg^n h^{-1}) + 2\ell(h)$$
$$\leqslant k \cdot \big(\ell(f^n) + d(hg^n h^{-1}, f^n)\big) + 2\ell(h)$$
$$\leqslant kn\epsilon + k\epsilon + 2\ell(h)$$

and so

$$\lim_{k\to\infty} \frac{\ell(g^k)}{k} = \lim_{k\to\infty} \frac{\ell(g^{kn})}{kn} \leqslant \liminf_{k\to\infty} \left(\epsilon + \frac{\epsilon}{n} + \frac{2\ell(h)}{kn} \right) \leqslant 2\epsilon.$$

As $\epsilon > 0$ is arbitrary, this shows that g is distorted in G.

As 1 is distorted in G, we have as an immediate consequence that, if $g \in G$ is undistorted, then $\inf_{h \in G} d(hgh^{-1}, 1) > 0$. □

From Lemma 3.55, we see that

$$C = \{g \in G \mid \inf_{h \in G} d(hgh^{-1}, 1) = 0\} = \{g \in G \mid 1 \in \overline{g^G}\},$$

is a conjugacy invariant G_δ subset of the distortion elements. Observe also that, similarly to Lemma 3.55, if $\overline{g^G} \cap C \neq \emptyset$, then $g \in C$ too. Indeed, if $f \in \overline{g^G} \cap C$, then $\overline{f^G} \subseteq \overline{g^G}$ and so $1 \in \overline{f^G} \subseteq \overline{g^G}$, i.e., $g \in C$.

Example 3.56 (Distortion elements in affine isometry groups) In the following, let X be a separable reflexive real Banach space and A a surjective isometry of X. Then, by the Mazur–Ulam Theorem, A is affine and hence can be written as $A(x) = T(x) + b$ for some linear isometry $T \in \mathrm{Isom}(X)$ and a vector $b \in X$. Let

$$X = X^T \oplus X_T$$

be the Yosida decomposition [96] associated with the linear isometry T. That is, $X^T = \ker(T - \mathrm{id})$ and $X_T = \overline{\mathrm{rg}}(T - \mathrm{id})$ are closed linear subspaces decomposing X into a direct sum and

$$\frac{1}{n}\left(T^{n-1} + T^{n-2} + \cdots + \mathrm{id}\right)(x) \xrightarrow[n \to \infty]{} P(x)$$

for all $x \in X$, where P is the projection onto X^T with $\ker P = X_T$.

Observe that, for $x \in X$, we have

$$\frac{1}{n}A^n x = \frac{1}{n}\left(T^n x + T^{n-1}b + T^{n-2}b + \cdots + Tb + b\right) \xrightarrow[n \to \infty]{} 0 + Pb = Pb$$

and

$$\begin{aligned}
\|A^n x - x\| &\leqslant \|A^n x - A^{n-1}x\| + \|A^{n-1}x - A^{n-2}x\| + \cdots + \|Ax - x\| \\
&= \|Ax - x\| + \|Ax - x\| + \cdots + \|Ax - x\| \\
&= n\|Ax - x\|.
\end{aligned}$$

So, for any $x \in X$,

$$\|Pb\| = \lim_{n \to \infty}\frac{1}{n}\|A^n x\| = \lim_{n \to \infty}\frac{1}{n}\|A^n x - x\| \leqslant \|Ax - x\|.$$

On the other hand, $Pb - b \in \ker P = \overline{\mathrm{rg}}(T - \mathrm{id})$ and hence

$$Pb - b = \lim_{n \to \infty} Tx_n - x_n$$

for some sequence $x_n \in X$. Thus, for these x_n, we have

$$\lim_{n \to \infty} \|Ax_n - x_n\| = \lim_{n \to \infty} \|Tx_n + b - x_n\| = \|Pb - b + b\| = \|Pb\|.$$

All in all, this shows that

$$\|Pb\| = \inf_{x \in X} \|Ax - x\| = \lim_{n \to \infty} \frac{1}{n} \|A^n y\|$$

for all $y \in X$.

Proposition 3.57 *Let X be a separable reflexive Banach space so that the group $\mathrm{Isom}(X)$ of linear isometries is coarsely bounded in the strong operator topology. Let A be an affine isometry with linear part T and translation vector b.*

(1) *Then A is distorted in the group $\mathrm{Aff}(X)$ of affine isometries if and only if*

$$\inf_{x \in X} \|Ax - x\| = 0,$$

which happens if and only if $Pb = \lim_n \frac{1}{n}(T^{n-1} + \cdots + T + \mathrm{id})b = 0$.
(2) *Moreover, id is a limit of conjugates of A in $\mathrm{Aff}(X)$ if and only if id is a limit of conjugates of T in $\mathrm{Isom}(X)$ and $\inf_{x \in X} \|Ax - x\| = 0$.*

Proof Because $\mathrm{Isom}(X)$ is coarsely bounded, by Proposition 3.17, the orbit map evaluating an affine isometry at the point $0 \in X$ is a quasi-isometry between $\mathrm{Aff}(X)$ and X. Therefore, A is distorted in $\mathrm{Aff}(X)$ if and only if

$$\lim_{n \to \infty} \frac{1}{n} \|A^n(0)\| = \lim_{n \to \infty} \frac{1}{n} \|A^n(0) - \mathrm{id}(0)\| = 0.$$

However, as explained above, for all $y \in X$,

$$\|Pb\| = \inf_{x \in X} \|Ax - x\| = \lim_{n \to \infty} \frac{1}{n} \|A^n y\|$$

and hence (1) follows by taking $y = 0$.

For (2), suppose B_n are affine isometries of X with linear part R_n and translation vector c_n. Then, for all $x \in X$,

$$\begin{aligned}
B_n A B_n^{-1}(x) &= B_n A(R_n^{-1}x - R_n^{-1}c_n) \\
&= B_n(TR_n^{-1}x - TR_n^{-1}c_n + b) \\
&= R_n TR_n^{-1}x - R_n TR_n^{-1}c_n + R_n b + c_n
\end{aligned}$$

and so $\lim_n B_n A B_n^{-1} = \mathrm{id}$ in $\mathrm{Aff}(X)$ if and only if $\lim_n R_n TR_n^{-1} = \mathrm{id}$ in $\mathrm{Isom}(X)$ and

$$\lim_n \left\| b - (T - \mathrm{id})R_n^{-1}c_n \right\| = \lim_n \left\| - R_n TR_n^{-1}c_n + R_n b + c_n \right\| = 0.$$

In particular, if $\lim_n B_n A B_n^{-1} = \mathrm{id}$ in $\mathrm{Aff}(X)$, then id is a limit of conjugates of T in $\mathrm{Isom}(X)$ and $b \in \overline{\mathrm{rg}}(T - \mathrm{id}) = \ker P$, which by (1) happens if A is distorted in $\mathrm{Aff}(X)$.

Conversely, if A is distorted in $\mathrm{Aff}(X)$ and $\lim_n R_n T R_n^{-1} = \mathrm{id}$ in $\mathrm{Isom}(X)$ for some sequence R_n in $\mathrm{Isom}(X)$, then $b \in \overline{\mathrm{rg}}(T - \mathrm{id})$ and hence $b = \lim_n (T - \mathrm{id}) b_n$ for some sequence of $b_n \in X$. Setting B_n to be the affine isometry with linear part R_n and translation vector $c_n = R_n b_n$, we see that $\lim_n B_n A B_n^{-1} = \mathrm{id}$ in $\mathrm{Aff}(X)$, which finishes the proof of (2). $\qquad\square$

Example 3.58 (Distortion elements in homeomorphism groups of surfaces) Suppose S is a compact surface different from the disc, the closed annulus and the Möbius strip. Assume also that S has non-empty boundary ∂S. Let $\tilde{S} \xrightarrow{p} S$ be the universal cover, and equip $\tilde{S}$ with a compatible proper metric d that is invariant under the group $\pi_1(S)$ of deck-transformations. Fix also a fundamental domain D for $\tilde{S}$, i.e., a compact connected subset mapping onto S via the covering map and so that $\mathrm{int}(D) \cap \mathrm{int}(\gamma(D)) = \emptyset$ for any non-trivial deck-transformation γ.

Suppose f is an isotopically trivial homeomorphism of S. Then if $\tilde{f}_1$ and $\tilde{f}_2$ are any two lifts of f to $\tilde{S}$ and $n \geqslant 1$, both $\tilde{f}_1^n$ and $\tilde{f}_2^n$ are lifts of f^n and therefore differ by a deck-transformation γ, i.e., $\tilde{f}_1^n = \gamma \tilde{f}_2^n$. As d is invariant under deck-transformations, it thus follows that

$$\mathrm{diam}_d(\tilde{f}_1^n[D]) = \mathrm{diam}_d(\gamma \tilde{f}_2^n[D]) = \mathrm{diam}_d(\tilde{f}_2^n[D]).$$

Therefore,

$$\lim_n \frac{\mathrm{diam}_d(\tilde{f}^n[D])}{n} = 0$$

holds for *some* lift $\tilde{f}$ of f if and only if it holds for *all* lifts $\tilde{f}$ of f. A homeomorphism f with this property is said to be *non-spreading* in E. Militon's paper [63].

Theorem 1.11 of [63] states that, if f is a non-spreading homeomorphism of S, then the identity homeomorphism id_S is a limit of conjugates of f, that is

$$\mathrm{id}_S \in \overline{f^{\mathrm{Homeo}_0(S)}}.$$

Now, by Theorem 1 of [57], the group $\mathrm{Homeo}_0(S)$ of isotopically trivial homeomorphisms of S admits a maximal metric. Coupling this with Militon's theorem, we obtain the following.

Proposition 3.59 *The following are equivalent for an isotopically trivial homeomorphism f of S:*

(1) *f distorted in $\mathrm{Homeo}_0(S)$,*
(2) id_S *is a limit of conjugates of f,*
(3) *f is non-spreading.*

Proof The implication from (3) to (2) is just the above cited result of Militon, whereas the implication from (2) to (1) follows immediately from Lemma 3.55.

For the implication of (1) to (3), let $\mathcal{U} = \{U_1, \ldots, U_n\}$ be an open covering of S so that each $p^{-1}(U_i)$ is a disjoint union of d-bounded sets $W_{i,j}$ that are all mapped homeomorphically onto U_i via p. Note that every homeomorphism h with $\mathsf{supp}(h) \subseteq U_i$ admits a lift $\tilde{h}$ that leaves each $W_{i,j}$ invariant and hence satisfies

$$\sup_{x \in \tilde{S}} d(\tilde{h}(x), x) \leqslant K,$$

where K is an upper bound for the diameters of the sets $W_{i,j}$.

Now, suppose $f \in \mathrm{Homeo}_0(S)$ can be written as a product $f = h_1 \cdots h_m$ of homeomorphisms h_i each supported in some $U_{j(i)}$. Then we can choose lifts $\tilde{h}_i$ as above and find that

$$\sup_{x \in \tilde{S}} d(\tilde{h}_1 \cdots \tilde{h}_m(x), x) \leqslant \sup_{y \in \tilde{S}} d(\tilde{h}_1(y), y) + \sup_{x \in \tilde{S}} d(\tilde{h}_2 \cdots \tilde{h}_k(x), x)$$

$$\leqslant K + \sup_{x \in \tilde{S}} d(\tilde{h}_2 \cdots \tilde{h}_m(x), x)$$

$$\leqslant \cdots$$

$$\leqslant m \cdot K.$$

In particular, for $\tilde{f} = \tilde{h}_1 \cdots \tilde{h}_m$, we have

$$\mathsf{diam}_d(\tilde{f}[D]) \leqslant 2K \cdot m + \mathsf{diam}_d(D).$$

As discussed in Example 3.19, it follows from the Fragmentation Lemma of R. D. Edwards and R. C. Kirby [26] that there is an identity neighbourhood V in $\mathrm{Homeo}_0(S)$ so that every $g \in V$ can be factored into $g = h_1 \cdots h_n$ with $\mathsf{supp}(h_i) \subseteq U_i$. Thus, from the above it follows that

$$f \in V^m \implies \mathsf{diam}_d(\tilde{f}[D]) \leqslant 2Kn \cdot m + \mathsf{diam}_d(D).$$

As $\mathrm{Homeo}_0(S)$ is connected, it is generated by V and we let ρ_V designate the word metric induced by generating set V. If also ∂ denotes the maximal metric on $\text{Homeo}_0(S)$, we have $\rho_V \leqslant C \cdot \partial + C$ for some C. Now, suppose f is distorted, whence $\lim_k \frac{\partial(f^k,1)}{k} = 0$. Then, if $\tilde{f}$ is a lift of f, we have

$$\limsup_k \frac{\text{diam}_d(\tilde{f}^k[D])}{k} \leqslant 2Kn \cdot \limsup_k \frac{\rho_V(f^k,\text{id})}{k}$$

$$= 2KCn \cdot \lim_k \frac{\partial(f^k,1)}{k} = 0,$$

i.e., f is non-spreading. □

3.8 Groups with Property (PL)

We shall now investigate a very interesting rigidity property of topological groups introduced by Y. de Cornulier [20] in the setting of locally compact groups. Because the basic definition admits a direct generalisation to general topological groups, we shall keep the terminology the same. For another recent study of this property, one can consult the paper [88].

Definition 3.60 A topological group G is said to have *property (PL)* if every continuous left-invariant écart on G is either coarsely proper or bounded.

Note that, if G has property (PL), then either every continuous left-invariant écart on G will be bounded and thus G itself is coarsely bounded or G has a coarsely proper écart. In either case, we see that G has a coarsely proper continuous left-invariant écart and thus must be locally bounded.

Lemma 3.61 *If G is a topological group with property (PL), then G admits a coarsely proper continuous left-invariant écart.*

Typical applications of property (PL) concern dichotomies for isometric actions on metric spaces with various geometric qualities. For example, if G is a group with property (PL), then every continuous affine isometric action $G \curvearrowright \mathcal{H}$ on a real Hilbert space will either be coarsely proper or have bounded orbits. In the latter case, G will fix a point in the closed convex hull of any orbit. It follows that every group with property (PL) will either have the *Haagerup property*, meaning that it has a coarsely proper continuous affine isometric action on a Hilbert space, or it will have *property (FH)*, that is, every continuous affine isometric action on a Hilbert space fixes a point. Using C. Ryll-Nardczewski's fixed-point theorem, the same dichotomy holds for affine isometric actions on reflexive Banach spaces.

Property (PL) is evidently a rigidity property for the coarse structure on a group G. For example, it implies that every continuous isometric action of G on a metric space is either coarsely proper or has bounded orbits. Just how strong this property is can be seen from the following characterisation.

Theorem 3.62 *The following conditions are equivalent for a topological group G.*

(1) *G has property (PL),*
(2) *whenever A is a coarsely unbounded symmetric subset of G, there is a coarsely bounded set B and some n so that*

$$G = (AB)^n,$$

(3) *whenever A is a coarsely unbounded symmetric subset of G and V is an identity neighbourhood, there is a coarsely bounded set B and some n so that*

$$G = (ABV)^n,$$

(4) *the only left-invariant coarse structures containing $\mathcal{E}_L$ are the trivial coarse structure $\mathcal{P}(G \times G)$ and $\mathcal{E}_L$ itself.*

Proof (1)$\Rightarrow$(2) Suppose G has property (PL) and that A is a coarsely unbounded symmetric subset of G. As noted above, G has a coarsely proper continuous left-invariant écart, so we can let B_n denote the corresponding open ball of radius n centred at the identity. Define

$$V_n = \left(B_n A^2 B_n \right)^{3^n}$$

and note that the V_n are symmetric open identity neighbourhoods satisfying $V_n^3 \subseteq V_{n+1}$. Supplementing with appropriate $V_{-1}, V_{-2}, \ldots$, we thus obtain a bi-infinite sequence $(V_n)_{n \in \mathbb{Z}}$ of symmetric open identity neighbourhoods so that $V_n^3 \subseteq V_{n+1}$ for all $n \in \mathbb{Z}$. Let d be the associated continuous left-invariant écart given by Lemma 2.13. By property (PL), d is either coarsely proper or bounded. Now, clearly A is d-bounded, so as A is not coarsely bounded, d cannot be coarsely proper. But then d must be bounded, which means that G equals some V_n, thus establishing (2).

  (2)$\Rightarrow$(3) This is trivial.

  (2)$\Rightarrow$(4) Assume now that (2) holds and the $\mathcal{E}$ is a left-invariant coarse structure on G containing $\mathcal{E}_L$. If $\mathcal{E} \neq \mathcal{E}_L$, then some coarsely unbounded symmetric subset A of G must be $\mathcal{E}$-bounded and also $G = (AB)^n$ for some coarsely bounded B and some n. But, in this case, both A and B are $\mathcal{E}$-bounded and therefore, by left-invariance of $\mathcal{E}$, also $G = (AB)^n$ is $\mathcal{E}$-bounded and thus $\mathcal{E} = \mathcal{P}(G \times G)$.

(4)$\Rightarrow$(1) This is trivial, because every continuous left-invariant écart d induces a left-invariant coarse structure $\mathcal{E}_d$ containing $\mathcal{E}_L$.

(3)$\Rightarrow$(1) Suppose d is a continuous left-invariant écart on G that fails to be coarsely proper. Then there is some coarsely unbounded set A with $\mathrm{diam}_d(A) < \infty$. Let also V denote the open unit ball for d centred at 1. Then, by (3), there are a coarsely bounded set B and some n so that $G = (ABV)^n$. As A, B and V are d-bounded, we see that d is a bounded écart. $\square$

Note that the utility of (3) lies in cases where G is not a priori known to be locally bounded.

Corollary 3.63 *Suppose G is a Polish group with property (PL). Then either*

(1) *G is monogenic or*
(2) *G has asymptotic dimension 0, has the Haagerup property and is the union $G = \bigcup_{n=1}^{\infty} H_n$ of a chain $H_1 \leqslant H_2 \leqslant \ldots$ of open subgroups that are coarsely bounded in G.*

Proof Suppose that A is a coarsely bounded subset of G. Observe that, if the subgroup $\langle A \rangle$ generated by A is coarsely unbounded in G, we may find some coarsely bounded subset B and some n so that $G = (\langle A \rangle \cdot B)^n$. But then $A \cup B$ generates G and so G is monogenic.

Therefore, if G is not monogenic, then every subgroup $\langle A \rangle$ generated by a coarsely bounded subset $A \subseteq G$ will be coarsely bounded. Now, as noted above, G is locally bounded and hence can be written as the union of a chain $V_1 \subseteq V_2 \subseteq \cdots \subseteq G$ of coarsely bounded open sets. Set $H_n = \langle V_n \rangle$. Then each H_n is open coarsely bounded in G and $G = \bigcup_{n=1}^{\infty} H_n$.

To see that G has asymptotic dimension 0, we must show that, for every coarsely bounded set A, there is a covering $G = \bigcup_{i \in I} C_i$ by sets so that $C_i^{-1} C_j \cap A = \emptyset$ for all $i \neq j$ and so that the C_i are uniformly coarsely bounded; that is, for some coarsely bounded B we have $C_i^{-1} C_i \subseteq B$ for all $i \in I$. So let A be given. Then as the H_n exhaust G, we have $A \subseteq H_n$ for some n. So let $\{C_i\}_{i \in I}$ list the distinct left cosets of H_n in G. Then $C_i^{-1} C_j \cap H_n = \emptyset$ for $i \neq j$ whereas $C_i^{-1} C_i = H_n$ for all i, thus establishing asymptotic dimension 0.

Finally, to see that G has the Haagerup property, that is, has a coarsely proper continuous affine isometric action on a Hilbert space, it suffices to show that G has a coarsely proper isometric action on a tree (see Section 2.3 in [5]). In fact, because G has property (PL), it suffices to show that it acts isometrically on a tree without fixing any vertex and any edge. For, in that case, the action will have unbounded orbits and thus must be coarsely proper. The construction of such an action is standard from the sequence $H_1 \leqslant H_2 \leqslant \cdots$. Namely, consider the tree X with vertex set

$$\text{Vert } X = \{g H_n \mid n \in \mathbb{N} \text{ and } g \in G\}$$

and edges

$$\{g H_n, g H_{n+1}\}$$

for $g \in G$ and $n \in \mathbb{N}$. Clearly, the left-multiplication action of G on the vertices preserves the edge relation and thus G acts by isometries on X. Moreover, because all of the H_n are proper subgroups, G fixes neither a vertex nor an edge.   □

Corollary 3.64 *Suppose G is a Polish group with property (PL) and $A \ni 1$ is a symmetric closed generating set for G. Then either A is coarsely bounded or $G = A^n$ for some n. In particular, the word metric ρ_A is either bounded or induces the quasi-metric structure on G.*

Proof Assume that A is coarsely unbounded. Because A is closed and generates G, some power A^m must have non-empty interior. Thus, if B is a coarsely bounded set so that $(AB)^k = G$, we may find some high enough power n so that $B \subseteq A^n$, whereby $G = (AB)^k = A^{kn+k}$.   □

Corollary 3.65 *Suppose $G \xrightarrow{\phi} \prod_{n=1}^{\infty} H_n$ is a continuous homomorphism from a topological group with property (PL) into the product of a sequence of topological groups. Then either $\phi[G]$ is coarsely bounded in $\prod_{n=1}^{\infty} H_n$ or there is a k so that*

$$G \xrightarrow{\text{proj}_k \circ \phi} H_k$$

is coarsely proper.

Proof Suppose that $\phi[G]$ is coarsely unbounded in $\prod_{n=1}^{\infty} H_n$. Then there is some k so that $\text{proj}_k \circ \phi[G]$ is coarsely unbounded in H_k. In particular, there is some continuous left-invariant écart d on H_k so that $\text{diam}_d(\text{proj}_k \circ \phi[G]) = \infty$. It thus follows that

$$D(g, h) = d(\text{proj}_k \circ \phi(g), \text{proj}_k \circ \phi(f))$$

defines an unbounded continuous left-invariant écart on G that therefore must be coarsely proper. This shows that $G \xrightarrow{\text{proj}_k \circ \phi} H_k$ is coarsely proper.   □

Corollary 3.66 *Suppose G is a Polish group and H is a closed subgroup with property (PL). Then H is either coarsely bounded in G or is well embedded.*

Proof Suppose H is coarsely unbounded as a subset of G. Then there is a continuous left-invariant écart d on G so that $\mathrm{diam}_d(H) = \infty$, whereby $d|_H$ is an unbounded écart on H and hence must be coarsely proper. So H is well embedded.   □

Owing to the perceived strength of property (PL), one might be led to believe that examples should be rare. This, in fact, is not the case and we shall now proceed to identify many instances. The main source of examples is isometry groups.

For the following, we say that an isometric action $G \curvearrowright X$ is *almost transitive on pairs* if, for all $x, y, z, u \in X$ with $d(x, y) = d(z, u)$ and all $\epsilon > 0$, there is some $g \in G$ so that either

$$d(gx, z) < \epsilon \quad \text{and} \quad d(gy, u) < \epsilon$$

or

$$d(gx, u) < \epsilon \quad \text{and} \quad d(gy, z) < \epsilon.$$

In other words, the set $\{x, y\}$ can be brought as close to $\{z, u\}$ in Hausdorff distance as needed.

Proposition 3.67 *Suppose X is a metric space so that, for all distances $\alpha < \beta$ in X, there is an isosceles triangle $\{x, y, z\}$ in X with side lengths β, β, α. Assume also that $G \curvearrowright X$ is a coarsely proper continuous isometric action of a topological group which is almost transitive on pairs. Then G has property (PL).*

Proof Suppose that A is a coarsely unbounded subset of G and fix some $x \in X$. Because the action $G \curvearrowright X$ is coarsely proper, the partial orbit $A \cdot x$ is unbounded in X. Now, suppose $g \in G$ is given and pick any $f \in A$ so that $d(gx, x) < d(fx, x)$. Let also $\epsilon = \frac{d(gx, x)}{8}$ and set $V = \{h \in G \mid d(hx, x) < 4\epsilon\}$, which is coarsely bounded.

Choose an isosceles triangle in X with side lengths $d(gx, x), d(fx, x)$ and $d(fx, x)$, say $y, z, u \in X$ with $d(y, z) = d(gx, x), d(u, y) = d(u, z) = d(fx, x)$. As the action is almost transitive on pairs, by moving the triangle and possibly renaming the vertices, we may assume that $d(y, x) < \epsilon$ and $d(z, gx) < \epsilon$. Now choose $h, k \in G$ so that

$$d_H\big(\{kx, kfx\}, \{y, u\}\big) < \epsilon \qquad \text{and} \qquad d_H\big(\{hx, hfx\}, \{z, u\}\big) < \epsilon$$

and thus

$$d_H(\{kx, kfx\}, \{x, u\}) < 2\epsilon \qquad \text{and} \qquad d_H(\{hx, hfx\}, \{gx, u\}) < 2\epsilon.$$

There are now four cases to consider. Because they are all very similar, we shall just verify one and leave the others to the reader.

So suppose, for example, that $d(kx, u)$, $d(kfx, x)$, $d(hx, gx)$ and $d(hfx, u)$ are all less than 2ϵ. Then also $d(kx, hfx) < 4\epsilon$ and so kf, $h^{-1}g$, $k^{-1}hf \in V$, whereby

$$g \in h \cdot V \subseteq kVf^{-1} \cdot V \subseteq Vf^{-1} \cdot Vf^{-1} \cdot V.$$

With the three remaining cases, we find that $g \in Vf^{\pm}Vf^{\pm}V$.

As g is an arbitrary element of G, this shows that $G = VA^{\pm}VA^{\pm}V$, which implies that G has property (PL). $\qquad\qquad\square$

For the next result, we recall that a Banach space $(X, \|\cdot\|)$ is said to be *almost transitive* if the group of linear isometries of X induces dense orbits on the unit sphere $S_X = \{x \in X \mid \|x\| = 1\}$.

Corollary 3.68 *Let $(X, \|\cdot\|)$ be a separable almost transitive Banach space of dimension $\geqslant 2$ and suppose that its group of linear isometries $\mathrm{Isom}(X, \|\cdot\|)$ is coarsely bounded. Then the group $\mathrm{Aff}(X, \|\cdot\|)$ of affine isometries has property (PL).*

Proof Because $\mathrm{Isom}(X, \|\cdot\|)$ is coarsely bounded, by Proposition 3.17, we see that the tautological action $\mathrm{Aff}(X, \|\cdot\|) \curvearrowright X$ is coarsely proper. Note also that, as X has dimension at least 2, for any two $\alpha < \beta$ there is an isosceles triangle in X with side lengths β, β and α. Furthermore, if $x, y, z, u \in X$ satisfy $\|x - y\| = \|z - u\|$ and $\epsilon > 0$, then by almost transitivity of X we can find some linear isometry T so that $\|T(y - x) - (u - z)\| < \epsilon$, whereby the affine isometry $T(\cdot - x) + z$ maps x to z and maps y to a point within distance ϵ of u. Thus $\mathrm{Aff}(X, \|\cdot\|) \curvearrowright X$ is almost transitive on pairs and $\mathrm{Aff}(X, \|\cdot\|)$ has property (PL). $\qquad\qquad\square$

Example 3.69 Of examples of almost transitive Banach spaces, we can mention $L^p([0, 1])$ for all $1 < p < \infty$. Also, as noted in Example 3.16, the isometry group $\mathrm{Isom}(L^p)$ is coarsely bounded. Therefore, the group $\mathrm{Aff}(L^p)$ of affine isometries of L^p, $1 < p < \infty$, has property (PL).

A separable Banach space X is said to be a *Gurarij space* if, for all $\epsilon > 0$ and finite-dimensional Banach spaces $E \subseteq F$, any linear isometry from E into X can be extended to a $(1+\epsilon)$-embedding of F into X. By a result of W. Lusky [55], all separable Gurarij spaces are linearly isometric, which shows that the theory of Gurarij spaces is ω-categorical in the sense of model theory from metric structures. In particular, by the Engeler–Ryll-Nardzewski–Svenonius theorem for continuous logic, the linear isometry group of the Gurarij space X acts almost oligomorphically on the sphere S_X, which (by Theorem 5.2 in [77]) implies that $\mathrm{Isom}(X)$ is coarsely bounded (see also [13] for an explicit proof of this). Moreover, the separable Gurarij space is almost transitive. It thus follows that the group of affine isometries of the Gurarij space has property (PL).

Example 3.70 (Isometry groups of Urysohn spaces) All of the isometry groups of the Urysohn metric spaces $\mathbb{U}$, $\mathbb{QU}$ and $\mathbb{ZQ}$ act coarsely properly on the Urysohn metric spaces themselves. Furthermore, their actions are (almost) transitive on pairs and the spaces themselves contain triangles of any possible configuration. It thus follows that the isometry groups

$$\mathrm{Isom}(\mathbb{U}), \ \mathrm{Isom}(\mathbb{QU}) \ \text{and} \ \mathrm{Isom}(\mathbb{ZU})$$

all have property (PL).

Example 3.71 (Automorphism groups of regular trees T_n, $n \geqslant 3$) The regular trees T_n are are metrically transitive and thus the automorphism group $\mathrm{Aut}(\mathsf{T}_n)$ acts (almost) transitively on pairs. Furthermore, by Example 3.6, the tautological action on T_n is coarsely proper. Nevertheless, if k is an odd natural number and $m > k$ any other natural number, then there is no triangle in T_n with side lengths m, m and k. On the other hand, provided that $n \geqslant 3$ and k is even, then such an isosceles triangle always exists. In this latter setting, the proof of Proposition 3.67 then immediately shows that, for a fixed vertex $x \in \mathsf{T}_n$ and any $g, f \in \mathrm{Aut}(\mathsf{T}_n)$ with $d(gx, x)$ even and $d(gx, x) < d(fx, x)$, we have

$$g \in Vf^{\pm}Vf^{\pm}V,$$

where V denotes the pointwise stabiliser of x. Therefore, if we let

$$U = \{h \in \mathrm{Aut}(\mathsf{T}_n) \mid d(hx, x) \leqslant 1\},$$

then for any coarsely unbounded subset A of $\mathrm{Aut}(\mathsf{T}_n)$ we have

$$\mathrm{Aut}(\mathsf{T}_n) = VA^{\pm}VA^{\pm}VU.$$

So $\mathrm{Aut}(\mathsf{T}_n)$ has property (PL) for all $3 \leqslant n \leqslant \aleph_0$. In the locally compact range $3 \leqslant n < \infty$, a stronger result by C. Ciobotaru is included in [88].

Example 3.72 (The general linear group $\mathrm{GL}(E)$ of a Banach space E) Suppose E is a Banach space and let $\mathrm{GL}(E)$ denote the group of all invertible bounded linear operators $E \xrightarrow{T} E$. When equipped with the topology induced by the operator norm

$$\|T\| = \sup_{\|x\|=1} \|Tx\|,$$

GL(E) is a metrisable topological group. Indeed, that the composition is continuous follows directly from the inequality

$$\|ST - MN\| \leqslant \|ST - SN\| + \|SN - MN\|$$

$$\leqslant \|S\|\|T - N\| + \|S - M\|\|N\|.$$

To see that the inversion operation is continuous at some element $S \in$ GL(E), note that, if $T \in$ GL(E) with $\|S - T\| < \frac{1}{\|S^{-1}\|}$, also $S^{-1}T = I - S^{-1}(S - T)$ is invertible with inverse

$$(S^{-1}T)^{-1} = \left(I - S^{-1}(S - T)\right)^{-1} = \sum_{n=0}^{\infty} \left(S^{-1}(S - T)\right)^n.$$

As $T = S \cdot S^{-1}T$, we have $T^{-1} = (S^{-1}T)^{-1} \cdot S^{-1}$ and thus

$$\|T^{-1} - S^{-1}\| \leqslant \left\|(S^{-1}T)^{-1} - I\right\| \cdot \|S^{-1}\|$$

$$\leqslant \sum_{n=1}^{\infty} \|S^{-1}\|^{n+1}\|S - T\|^n$$

$$= \frac{\|S^{-1}\|^2\|S - T\|}{1 - \|S^{-1}\|\|S - T\|}.$$

The latter fraction tends to 0 as $\|S - T\|$ goes to 0, so inversion is continuous at S.

Because we are dealing with invertible operators and not isometries, the norm metric $d_{\|\cdot\|}(S, T) = \|S - T\|$ will be neither left- nor right-invariant. However, by the Birkhoff–Kakutani metrisation theorem, as GL(E) is metrisable, it does admit a compatible left-invariant metric. Still, there is another interesting continuous length function on GL(E) given by

$$\ell(S) = \left|\log\|S\|\right| + \left|\log\|S^{-1}\|\right|.$$

We note that S is an isometry exactly when $\ell(S) = 0$ and so $\ell(S)$ measures how far S is from being an isometry. Also, that ℓ is continuous follows immediately from the inequality $\left|\|S\| - \|T\|\right| \leqslant \|S - T\|$ and continuity of inversion. Thus, the expression

$$\rho(S, T) = \left|\log\|S^{-1}T\|\right| + \left|\log\|T^{-1}S\|\right|$$

is a continuous left-invariant écart on GL(E).

Example 3.73 (Continuous linear actions on Banach spaces) Suppose $G \overset{\pi}{\curvearrowright} E$ is a linear action by a topological group G on a Banach space E. There are two natural, but distinct, notions of continuity of π. On the one hand, we can demand that π defines a continuous homomorphism

$$G \longrightarrow \mathrm{GL}(E),$$

where $\mathrm{GL}(E)$ is given the operator norm topology, and, on the other, we can ask that the action map

$$G \times E \xrightarrow{\ \pi\ } E$$

is jointly continuous.

We claim that the action map $G \times E \xrightarrow{\ \pi\ } E$ is jointly continuous exactly when the following three conditions hold:

(1) each $\pi(g)$ is a bounded operator, i.e., $\|\pi(g)\| < \infty$,
(2) the action is *strongly continuous*, i.e., the map $g \in G \mapsto \pi(g)x \in E$ is continuous for every $x \in E$,
(3) the function $\|\pi(\cdot)\|$ is *locally bounded* on G, i.e., is bounded on some identity neighbourhood.

Indeed, observe first that conditions (1) and (2) are simply the separate continuity of the action map in, respectively, the second and first variable, so these follow immediately from joint continuity. Also, if the action map $G \times E \xrightarrow{\ \pi\ } E$ is jointly continuous at $(1,0)$, then there is some identity neighbourhood $V \subseteq G$ and closed ball ϵB_E so that $\|\pi(g)x\| = \|\pi(g)x - \pi(1)0\| < 1$ for all $g \in V$ and $x \in \epsilon B_E$. In particular, $\|\pi(g)\| = \sup_{\|x\|=1}\|\pi(g)x\| \leqslant \frac{1}{\epsilon}$ for all $g \in V$ and thus $\|\pi(\cdot)\|$ is locally bounded.

Conversely, to see that these conditions suffice for joint continuity, suppose (g_i, x_i) is a net in $G \times E$ converging to some (g, x). Then

$$\|\pi(g_i)x_i - \pi(g)x\| \leqslant \|\pi(g_i)\|\|x_i - x\| + \|\pi(g_i)x - \pi(g)x\|.$$

So, if $\|\pi(\cdot)\|$ is locally bounded on G, then $\|\pi(g_i)\|$ is eventually uniformly bounded and thus, by strong continuity, $\lim_i \|\pi(g_i)x_i - \pi(g)x\| = 0$, which shows joint continuity.

We also note that, as shown in Lemma 21 in [35], if G is first countable (i.e., metrisable) or is Baire, then joint continuity follows from separate continuity. Indeed, suppose G is metrisable and $G \stackrel{\pi}{\curvearrowright} E$ is a strongly continuous action by bounded linear operators. Then, if (g_n, x_n) is a sequence converging to (g, x), we find that $\|\pi(g_n)x\|$ is bounded for all $x \in E$ and thus that $\|\pi(g_n)\|$ is bounded by the Banach–Steinhaus uniform boundedness principle. Again one sees that this implies joint continuity of the action map.

To avoid confusion, let us stress the terminology.

Definition 3.74 A linear action $G \overset{\pi}{\curvearrowright} E$ of a topological group G on a Banach space E is *continuous* if the action map $G \times E \overset{\pi}{\longrightarrow} E$ is jointly continuous.

From the conditions above, it is obvious that, if π defines a continuous homomorphism $G \longrightarrow \mathrm{GL}(E)$, then the action $G \overset{\pi}{\curvearrowright} E$ is continuous. On the other hand, continuity of the action $G \overset{\pi}{\curvearrowright} E$ does not imply continuity of the associated homomorphism $G \longrightarrow \mathrm{GL}(E)$, where $\mathrm{GL}(E)$ is given the operator norm topology. For a simple example of this, consider the left-regular unitary representation

$$\mathbb{T} \overset{\lambda}{\longrightarrow} \mathcal{U}(L^2(\mathbb{T})),$$

which is clearly strongly continuous and thus corresponds to a continuous action $\mathbb{T} \curvearrowright L^2(\mathbb{T})$. On the other hand, $\|\lambda(s) - \lambda(t)\| \geqslant \sqrt{2}$ for all $s \neq t$ in $\mathbb{T}$.

In fact, if the linear action $G \overset{\pi}{\curvearrowright} E$ is continuous, even the écart $\partial(g, f) = \big| \|\pi(g)\| - \|\pi(f)\| \big|$ need not be continuous on G. Indeed, consider two copies $\mathcal{H}_1$ and $\mathcal{H}_2$ of separable Hilbert space with orthonormal bases $(e_n)_{n=1}^{\infty}$ and $(k_n)_{n=1}^{\infty}$, respectively. We let the Cantor group $(\mathbb{Z}/2\mathbb{Z})^{\mathbb{N}}$ act via bounded operators on $\mathcal{H}_1 \oplus \mathcal{H}_2$ by setting

$$\pi(g)e_n = \begin{cases} 2k_n & \text{if } g_n = 1 \\ e_n & \text{if } g_n = 0 \end{cases} \quad \text{and} \quad \pi(g)k_n = \begin{cases} \frac{e_n}{2} & \text{if } g_n = 1 \\ k_n & \text{if } g_n = 0 \end{cases}$$

and extending linearly to $\mathcal{H}_1 \oplus \mathcal{H}_2$. Then $\|\pi(g)\| = 2$ for all non-identity elements $g \in (\mathbb{Z}/2\mathbb{Z})^{\mathbb{N}}$, whereas nevertheless the action map is jointly continuous.

As the preceding example shows, we are obliged to deal with continuous linear $G \overset{\pi}{\curvearrowright} E$ actions for which the associated homomorphism $G \to \mathrm{GL}(E)$ may not be continuous. In particular, there is a priori no reason to believe that the homomorphism $G \to \mathrm{GL}(E)$ is even bornologous. Despite this, we may still pull back the écart from Example 3.72.

Lemma 3.75 *Suppose $G \overset{\pi}{\curvearrowright} E$ is a continuous linear action of a topological group G on a Banach space E. Then the left-invariant écart*

$$\eta(g, f) = \big| \log\|\pi(g^{-1}f)\| \big| + \big| \log\|\pi(f^{-1}g)\| \big|$$

defines a coarse structure $\mathcal{E}_\eta$ containing the left-coarse structure $\mathcal{E}_L$. In particular, if G has property (PL), then η is either bounded or is coarsely proper.

Proof Because both $\mathcal{E}_L$ and $\mathcal{E}_\eta$ are left-invariant, it suffices to show that every coarsely bounded set is also η-bounded. So let A be a coarsely bounded set in G and fix a symmetric open identity neighbourhood U with $K = \sup_{u \in U} \|\pi(u)\| < \infty$. Set

$$V_n = \{g \in G \mid \ell(\pi(g)) \leqslant 3^n\} \cdot U$$

and note that V_n is open and $V_n^2 \subseteq V_{n+1}$ provided $3^n \geqslant 2 \log K$. As $G = \bigcup_n V_n$, it follows that $A \subseteq V_n$ for some large enough n and hence that A is also η-bounded. Thus, every coarsely bounded set is η-bounded and therefore $\mathcal{E}_L \subseteq \mathcal{E}_\eta$.   □

Despite the face that continuous linear actions on Banach spaces do not correspond to continuous group homomorphisms, as Lemma 3.76 shows, the former concept suffices to faithfully represent the coarse structure of a topological group.

Lemma 3.76 *Suppose d is a continuous left-invariant écart on a topological group G. Then*

(1) *there is a continuous isometric linear action $G \overset{\rho}{\curvearrowright} E$ on a Banach space with an associated continuous cocycle $G \overset{b}{\longrightarrow} E$ satisfying*

$$\|b(g) - b(f)\| = d(g, f)$$

for all $g, f \in G$,

(2) *there is a continuous linear action $G \overset{\pi}{\curvearrowright} E$ on a Banach space so that*

$$\|\pi(f^{-1}g)\| = 1 + d(g, f)$$

for all $g, f \in G$.

Proof The simple proof of part (2) appears in the doctoral dissertation of M. Herbst [37].

Suppose that d is a continuous left-invariant écart on G and let $\mathfrak{L}$ denote the family of real-valued 1-Lipschitz functions on (G, d), i.e., maps $G \overset{F}{\longrightarrow} \mathbb{R}$ so that

$$|F(g) - F(h)| \leqslant d(g, h).$$

Consider the vector space $\mathbb{M}G$ of all finitely supported maps $G \overset{\xi}{\longrightarrow} \mathbb{R}$ of mean 0, i.e., so that $\sum_{g \in G} \xi(g) = 0$, and define a semi-norm on $\mathbb{M}G$ by

$$\|\xi\| = \sup_{F \in \mathfrak{L}} \sum_{g \in G} \xi(g) F(g).$$

Because $\xi \in \mathbb{M}G$ have mean 0, we see that $\sum_{g \in G} \xi(g) F(g) = \sum_{g \in G} \xi(g)$ $F'(g)$ whenever $F - F'$ is constant and hence one can always suppose that $F(1) = 0$. This shows that $\|\xi\|$ is finite. Observe also that, as $d(\cdot, h) \in \mathfrak{L}$, we have

$$\|1_g - 1_h\| = \sup_{F \in \mathfrak{L}} \left(F(g) - F(h) \right) = d(g, h)$$

for all $g, h \in G$. Let now $G \overset{\rho}{\curvearrowright} \mathbb{M}G$ denote the continuous isometric linear action given by

$$\rho(f)\xi = \xi(f^{-1} \cdot)$$

and note that $\rho(f)\left(1_g - 1_h\right) = 1_{fg} - 1_{fh}$. Then

$$b(f) = 1_f - 1_1$$

defines a cocycle $G \overset{b}{\longrightarrow} \mathbb{M}G$ for ρ, because

$$b(fg) = 1_{fg} - 1_1 = \rho(f)\left(1_g - 1_1\right) + \left(1_f - 1_1\right) = \rho(f)b(g) + b(f).$$

In particular,

$$\begin{aligned} \rho(f^{-1})&\left(b(g) - b(f)\right) \\ &= \left(\rho(f^{-1})b(g) + b(f^{-1})\right) - \left(\rho(f^{-1})b(f) + b(f^{-1})\right) \\ &= b(f^{-1}g) - b(f^{-1}f) \\ &= b(f^{-1}g) \end{aligned}$$

and so

$$\|b(f^{-1}g)\| = \|b(g) - b(f)\| = \|1_g - 1_f\| = d(g, f)$$

for all $g, f \in G$. Let now E denote the Banach space completion of the quotient of $\mathbb{M}G$ by the nullspace $N = \{\xi \in \mathbb{M}G \mid \|\xi\| = 0\}$. Then ρ canonically carries over to an isometric linear action of G on E and b can be viewed as a map into E.

For the construction of a linear action of a Banach space, consider the ℓ_∞-sum $E \oplus \mathbb{R}$, that is, the direct sum of E and $\mathbb{R}$ with norm $\|(x, t)\| = \max\left\{\|x\|, |t|\right\}$. Define, for $g \in G$, an element $\pi(g) \in \mathrm{GL}(E \oplus \mathbb{R})$ by

$$\pi(g) = \begin{pmatrix} \rho(g) & b(g) \\ 0 & I \end{pmatrix},$$

that is

$$\pi(g)(x, t) = \left(\rho(g)x + tb(g), t\right).$$

As $\rho(g)$ is an isometry, we observe that

$$\|\pi(g)\| = 1 + \|b(g)\|$$

and so

$$\|\pi(f^{-1}g)\| = 1 + \|b(f^{-1}g)\| = 1 + d(g, f)$$

as claimed.   □

4

Sections, Cocycles and Group Extensions

We will now consider how coarse structure is preserved in short exact sequences

$$1 \to K \xrightarrow{\iota} G \xrightarrow{\pi} H \to 1$$

of topological groups, where K is a closed normal subgroup of G and ι is the inclusion map. This situation is typical and it will be useful to have a general terminology and set of tools to describe and compute the coarse structure of G from those of K and H.

Analogously to the algebraic study of group extensions, a central question is whether the extension splits coarsely. This will require a bornologous section $H \xrightarrow{\phi} G$ for the quotient map $G \xrightarrow{\pi} H$ and, for this reason, we shall explore the tight relationship between sections ϕ and their associated cocycles $H \times H \xrightarrow{\omega_\phi} G$ given by

$$\omega_\phi(x, y) = \phi(xy)^{-1}\phi(x)\phi(y).$$

We have three levels of regularity of ϕ along with useful reformulations in terms of the cocycle ω_ϕ.

(1) ϕ is modest, roughly corresponding to when $\omega_\phi[B \times B]$ is coarsely bounded for all coarsely bounded $B \subseteq H$,
(2) ϕ is bornologous, roughly corresponding to when $\omega_\phi[H \times B]$ is coarsely bounded for all coarsely bounded $B \subseteq H$,
(3) ϕ is a quasi-morphism, corresponding to when $\omega_\phi[H \times H]$ is coarsely bounded.

A second issue, which appears only in the case of Polish or general topological groups and not in the more specific context of locally compact groups, is when a closed subgroup K is coarsely embedded in G. Similarly, the question of when G is locally bounded and whether this can be deduced from the local boundedness of K and H is, of course, not relevant to locally compact groups, but will turn out to be crucial in our study.

4.1 Quasi-morphisms and Bounded Cocycles

A quasi-morphism from a group H into $\mathbb{R}$ is a map $H \xrightarrow{\phi} \mathbb{R}$ so that

$$|\phi(x) + \phi(y) - \phi(xy)| < K$$

for some constant K and all $x, y \in H$. Quasi-morphisms appear naturally in topology and in questions concerning bounded cohomology of groups, but also admit the following generalisation of broader interest.

Definition 4.1 A map $H \xrightarrow{\phi} G$ from a group H to a topological group G is a *quasi-morphism* if the two maps $(x, y) \mapsto \phi(xy)$ and $(x, y) \mapsto \phi(x)\phi(y)$ are close with respect to the coarse structure on G.

Given a map $H \xrightarrow{\phi} G$, we define a map $H \times H \xrightarrow{\omega_\phi} G$ measuring the failure of ϕ to be a homomorphism by

$$\omega_\phi(x, y) = \phi(xy)^{-1}\phi(x)\phi(y).$$

Then we see that ϕ is a quasi-morphism if and only if the *defect* of ϕ,

$$\Delta = \{1\} \cup \omega_\phi[H \times H] \cup \omega_\phi[H \times H]^{-1} = \{1\} \cup \{\phi(xy)^{-1}\phi(x)\phi(y) \mid x, y \in H\}^\pm,$$

is coarsely bounded in G. Note that then $\phi(1) = \phi(1 \cdot 1)^{-1}\phi(1)\phi(1) \in \Delta$, $\phi(xy) \in \phi(x)\phi(y)\Delta$ and

$$\phi(x^{-1}y) = \phi(x)^{-1}\phi(x)\phi(x^{-1}y) \in \phi(x)^{-1}\phi(xx^{-1}y)\Delta = \phi(x)^{-1}\phi(y)\Delta$$

for all $x, y \in H$. Therefore, if $x_i \in H$ and $\epsilon_i = \pm 1$, we have

$$\phi(x_1^{\epsilon_1} x_2^{\epsilon_2} \cdots x_n^{\epsilon_n}) \in \phi(x_1)^{\epsilon_1}\phi(x_2^{\epsilon_2} \cdots x_n^{\epsilon_n})\Delta$$

$$\subseteq \cdots$$

$$\subseteq \phi(x_1)^{\epsilon_1}\phi(x_2)^{\epsilon_2} \cdots \phi(x_{n-1})^{\epsilon_{n-1}}\phi(x_n^{\epsilon_n})\Delta^{n-1}$$

$$\subseteq \phi(x_1)^{\epsilon_1}\phi(x_2)^{\epsilon_2} \cdots \phi(x_n)^{\epsilon_n}\phi(1)\Delta^n$$

$$\subseteq \phi(x_1)^{\epsilon_1}\phi(x_2)^{\epsilon_2} \cdots \phi(x_n)^{\epsilon_n}\Delta^{n+1}.$$

In particular, for all $x, y, g \in H$,

$$\phi(g)^\pm \cdot \phi(xy)^{-1}\phi(x)\phi(y) \cdot \phi(g)^\mp \in \phi(g^\pm(xy)^{-1}xyg^\mp)\Delta^6 = \phi(1)\Delta^6 \subseteq \Delta^7,$$

i.e., $\phi(g)\Delta\phi(g)^{-1} \cup \phi(g)^{-1}\Delta\phi(g) \subseteq \Delta^7$, showing that $\phi[H]$ normalises the subgroup $F = \langle\Delta\rangle$ generated by Δ. Therefore, from ϕ we obtain a canonical homomorphism

$$H \xrightarrow{\tilde{\phi}} N_G(F)/F,$$

where $N_G(F)$ is the normaliser of F in G.

Note also that not every quasi-morphism between Polish groups is bornologous. For example, a linear operator between two Banach spaces is certainly a quasi-morphism, but it is bornologous if and only if it is continuous, i.e., a bounded linear operator.

Proposition 4.2 *Let* $H \xrightarrow{\phi} G$ *be a Baire measurable quasi-morphism between European topological groups. Then ϕ is bornologous.*

Proof Let Δ be the defect of ϕ and observe that, for A a subset of H and $x^{-1}y \in A$, we have $\phi(x)^{-1}\phi(y) \in \phi(x^{-1}y)\Delta \subseteq \phi[A]\Delta$. Therefore,

$$(\phi \times \phi)E_A \subseteq E_{\phi[A]\Delta}.$$

To see that ϕ is bornologous, as Δ is coarsely bounded, it thus suffices to show that $\phi[A]$ is coarsely bounded for any coarsely bounded $A \subseteq H$.

So, fix a coarsely bounded A and a symmetric open identity neighbourhood V in G. As G is European, we may find a chain $1 \in F_1 \subseteq F_2 \subseteq \cdots \subseteq G$ of finite symmetric sets whose union generates G over V. Thus, $F_1 V \subseteq (F_2 V)^2 \subseteq (F_3 V)^3 \subseteq \cdots$ is an exhaustive chain of open subsets of G and $H = \bigcup_n \phi^{-1}((F_n V)^n)$ is a covering of H by countably many sets with the Baire property. Because H is a Baire group, it follows that some $\phi^{-1}((F_n V)^n)$ must be somewhere co-meagre in H and hence, by a lemma of B. J. Pettis [71], that $[\phi^{-1}((F_n V)^n)]^2$ has non-empty interior in H. As A is coarsely bounded, there is a finite set $D \subseteq H$ and an m so that

$$A \subseteq \left(D \cdot [\phi^{-1}((F_n V)^n)]^2\right)^m$$

and thus

$$\phi[A] \subseteq \left(\phi[D] \cdot ((F_n V)^n)^2\right)^m \Delta^{3m-1}.$$

As $\phi[D]$ and F_n are finite and $\Delta \subseteq H$ coarsely bounded, it follows that there is a finite set $F \subseteq G$ and a k so that $\phi[A] \subseteq (FV)^k$. Because V was arbitrary, we conclude that $\phi[A]$ is coarsely bounded and hence that ϕ is bornologous. $\qquad\square$

Suppose G is a topological group so that, whenever $W_1 \subseteq W_2 \subseteq \cdots \subseteq G$ is a countable exhaustive chain of symmetric subsets, then some $W_n^k, n, k \geqslant 1$, has non-empty interior. Then the above proof shows that every quasi-morphism from G into a European topological group is bornologous. Such G include, for example, Polish groups with ample generics or homeomorphism groups of compact manifolds.

As we are primarily interested in weaker conditions on ϕ, it will be useful to establish the following equivalence, similar to Proposition 4.2. We recall that a map $X \xrightarrow{\phi} Y$ between coarse spaces is modest if the image of every coarsely bounded set in X is coarsely bounded in Y.

Proposition 4.3 *Suppose $H \xrightarrow{\phi} G$ is a Baire measurable map from a locally bounded European group H to a European group G. Then ϕ is modest if and only if $\omega_\phi[B \times B]$ is coarsely bounded for every coarsely bounded B.*

Proof Suppose first that ω_ϕ is modest, i.e., that $\omega_\phi[B \times B]$ is coarsely bounded for every coarsely bounded set B, and fix a coarsely bounded set $A \subseteq H$. To see that ϕ is modest, we must show that also $\phi[A]$ is coarsely bounded.

To show this, let V be an arbitrary symmetric open identity neighbourhood in G. Then, as in the proof of Proposition 4.2, there is a finite set F and some n so that $\phi^{-1}\big((VFV)^n\big)$ is co-meagre in some coarsely bounded non-empty open set $W \subseteq H$, whence by Pettis's lemma [71] we have

$$W^2 = \big[\phi^{-1}\big((VFV)^n\big) \cap W\big]^2.$$

As A is coarsely bounded, find a finite set $E \ni 1$ and an m so that $A \subseteq (EW^2)^m$. Set also $B = \bigcup_{l \leqslant 2m}(E \cup W^2)^l$, which is coarsely bounded.

Observe now that, for $e_i \in E$ and $u_i \in W^2$,

$$\phi(e_1 u_1 e_2 u_2 \cdots e_m u_m)$$
$$= \phi(e_1)\phi(u_1 e_2 u_2 \cdots e_m u_m)\omega_\phi(e_1, u_1 e_2 u_2 \cdots e_m u_m)^{-1}$$
$$\in \phi[E]\phi(u_1 e_2 u_2 \cdots e_m u_m)\omega_\phi[B \times B]^{-1}$$
$$\subseteq \phi[E]\phi(u_1)\phi(e_2 u_2 \cdots e_m u_m)\omega_\phi(u_1, e_2 u_2 \cdots e_m u_m)^{-1}\omega_\phi[B \times B]^{-1}$$
$$\subseteq \phi[E]\phi[W^2]\phi(e_2 u_2 \cdots e_m u_m)\omega_\phi[B \times B]^{-1}\omega_\phi[B \times B]^{-1}$$
$$\subseteq \cdots$$
$$\subseteq \big(\phi[E]\phi[W^2]\big)^m \big(\omega_\phi[B \times B]^{-1}\big)^{2m-1}.$$

That is, $\phi[A] \subseteq \big(\phi[E]\phi[W^2]\big)^m \big(\omega_\phi[B \times B]^{-1}\big)^{2m-1}$. But any element $h \in W^2$ can be written as $h = xy$ for some $x, y \in \phi^{-1}\big((VFV)^n\big) \cap W$ and thus

$$\phi(h) = \phi(x)\phi(y)\omega_\phi(x,y)^{-1} \in (VFV)^{2n}\omega_\phi[W \times W]^{-1},$$

that is, $\phi[W^2] \subseteq (VFV)^{2n}\omega_\phi[W \times W]^{-1}$.

Thus, finally,

$$\phi[A] \subseteq \left(\phi[E]\,(VFV)^{2n}\omega_\phi[W \times W]^{-1}\right)^m \left(\omega_\phi[B \times B]^{-1}\right)^{2m-1},$$

where $\omega_\phi[W \times W]$ and $\omega_\phi[B \times B]$ are coarsely bounded and $\phi[E]$ and F are finite, showing that also $\phi[A]$ is coarsely bounded.

The converse implication that ω is modest whenever ϕ is follows directly from the inclusion $\omega_\phi[B \times B] \subseteq \phi[B^2]^{-1}\phi[B]^2$. $\square$

To understand instead when a map ϕ is bornologous, we need only the following simple observation.

Lemma 4.4 *The following are equivalent for a map* $H \xrightarrow{\phi} G$ *between topological groups:*

(1) ϕ *is bornologous,*
(2) ϕ *is modest and* $\omega_\phi[H \times B]$ *is coarsely bounded for every coarsely bounded subset* $B \subseteq H$.

Proof Assume first that ϕ is bornologous and let $B \subseteq H$ be coarsely bounded. As ϕ is bornologous, fix a coarsely bounded set $C \subseteq G$, so that $\phi(x)^{-1}\phi(y) \in C$ whenever $x^{-1}y \in B^{-1}$. Then, for all $x \in H$ and $b \in B$, we have

$$\omega_\phi(x,b) = \phi(xb)^{-1}\phi(x) \cdot \phi(b) \in C \cdot \phi[B].$$

As ϕ is bornologous and thus modest, $C \cdot \phi[B]$ is coarsely bounded and so also $\omega_\phi[H \times B]$ is coarsely bounded.

Conversely, suppose (2) holds and that $B \subseteq H$ is any coarsely bounded set. Note that, for $x \in H$ and $b \in B$, we have

$$\phi(x)^{-1}\phi(xb) = \phi(xb \cdot b^{-1})^{-1}\phi(xb)\phi(b^{-1})\phi(b^{-1})^{-1}$$
$$\in \omega_\phi[H \times B^{-1}] \cdot \phi[B^{-1}]^{-1}.$$

This shows that, whenever $x^{-1}y \in B$, then the product $\phi(x)^{-1}\phi(y)$ belongs to the coarsely bounded set $\omega_\phi[H \times B^{-1}] \cdot \phi[B^{-1}]^{-1}$. So ϕ is bornologous. $\square$

The following is now an immediate consequence of Proposition 4.3 and Lemma 4.4.

Proposition 4.5 *Suppose* $H \xrightarrow{\phi} G$ *is a Baire measurable map from a locally bounded European group* H *to a European group* G. *Then* ϕ *is bornologous if and only if* $\omega_\phi[H \times B]$ *is coarsely bounded for every coarsely bounded* B.

Although Proposition 4.5 does not immediately seem to simplify matters, its utility lies in the situation when ϕ is a section for a quotient map $G \xrightarrow{\pi} H$. For in this case the image of ω_ϕ is contained in the kernel $K = \ker \pi$, whose coarse structure we may know independently of that of G. For example, if, for every coarsely bounded B, $\omega_\phi[H \times B]$ is coarsely bounded in K and thus also in G, then we may conclude that $H \xrightarrow{\phi} G$ is bornologous.

4.2 Local Boundedness of Extensions

Before directly addressing the coarse structure of extensions, we will consider the specific issue of when a Polish group extension is locally bounded. That is, suppose K is a closed normal subgroup of a Polish group G. Under what assumptions on K and G/K is G locally bounded?

Observe first that, as the quotient map $G \xrightarrow{\pi} G/K$ is open, the image of a coarsely bounded identity neighbourhood in G will be a coarsely bounded identity neighbourhood in G/K. So G/K is locally bounded whenever G is. Conversely, one would like to establish local boundedness of G exclusively from knowledge of K and the quotient G/K. Although the following problem is unlikely to have a positive answer, several positive instances will be established.

Problem 4.6 Suppose that K is a closed normal subgroup of a Polish group G and that both K and G/K are locally bounded. Does it follow that also G is locally bounded?

We begin by characterising coarsely bounded identity neighbourhoods in G.

Lemma 4.7 *Suppose K is a closed normal subgroup of a Polish group G and let $G \xrightarrow{\pi} G/K$ denote the quotient map. Then the following are equivalent for an identity neighbourhood V in G.*

(1) *V is coarsely bounded,*
(2) *$\pi[V]$ coarsely bounded in G/K and $(FV)^m \cap K$ is coarsely bounded in G for all finite $F \subseteq G$ and m.*

Proof The implication from (1) to (2) is trivial, so assume instead that (2) holds. To see that V is coarsely bounded in G, assume that $W \subseteq V$ is a symmetric identity neighbourhood in G. Then, as π is an open mapping, $\pi[W]$ is an identity neighbourhood in G/K. Since $\pi[V]$ is coarsely bounded in G/K, there are $F \subseteq G/K$ finite and m so that $\pi[V] \subseteq (F\pi[W])^m$. Choose a finite set $E \subseteq G$ so that $\pi[E] = F$ and note that then $\pi[V] \subseteq \pi[(EW)^m]$, i.e.,

$$V \subseteq (EW)^m \cdot \ker\pi = (EW)^m \cdot K.$$

Thus, every $v \in V$ may be written as $v = xk$, for some $x \in (EW)^m$ and $k \in K$, that is,

$$k = x^{-1}v \in (WE^{-1})^m V \cap K \subseteq (VE^{-1})^m V \cap K.$$

However, because $(VE^{-1})^m V \cap K$ is coarsely bounded in G, there are a finite set $D \subseteq G$ and an n so that $(VE^{-1})^m V \cap K \subseteq (DW)^n$. It follows that

$$V \subseteq (EW)^m \cdot (DW)^n,$$

thus verifying that V is coarsely bounded in G.   □

We now establish an exact equivalence for local boundedness of G.

Proposition 4.8 *Suppose K is a closed normal subgroup of a Polish group G. Then the following conditions are equivalent.*

(1) *G is locally bounded,*
(2) *G/K is locally bounded and there is a compatible left-invariant metric d on G so that $d|_K$ metrises the coarse structure on K induced from G.*

Proof Suppose first that G is locally bounded and fix a coarsely proper metric on G. Then d metrises the coarse structure on G and thus $d|_K$ also metrises the coarse structure on K induced from G. Similarly, as noted above, G/K will be locally bounded.

Conversely, suppose that (2) holds and let d be the given metric. As G/K is locally bounded, fix a coarsely proper metric ∂ on G/K. We claim that the metric

$$D(g, f) = d(g, f) + \partial(\pi(g), \pi(f)),$$

is coarsely proper on G, where $G \xrightarrow{\pi} G/K$ is the quotient map $\pi(g) = gK$. This, in turn, will imply that G is locally bounded.

So, suppose (g_n) is a D-bounded sequence in G and let $\rho \geqslant d$ be any continuous left-invariant écart on G. We must show that (g_n) is also ρ-bounded. Note first that the formula

$$\hat\rho(gK, fK) = \inf_{k \in K} \rho(g, fk)$$

defines a continuous left-invariant metric on the quotient group G/K. As (g_n) is D-bounded, $(\pi(g_n))$ is ∂-bounded and thus coarsely bounded in G/K. In particular, $(\pi(g_n))$ is $\hat\rho$-bounded, whereby

$$\sup_n d(g_n, k_n) \leqslant \sup_n \rho(g_n, k_n) < \infty$$

for some sequence (k_n) in K. As (g_n) is d-bounded, also (k_n) is d-bounded. Because $d|_K$ metrises the coarse structure on K induced from G, we find that (k_n) is coarsely bounded in G and thus also ρ-bounded. It therefore follows that (g_n) is ρ-bounded too. $\qquad\square$

Recall that a closed subgroup K of a Polish group G is coarsely embedded if the coarse structure on K coincides with that induced from G, and that K is well embedded if there is a compatible left-invariant metric d on G so that $d|_K$ is coarsely proper on K.

Corollary 4.9 *Suppose K is a closed normal subgroup of a Polish group G. Then the following two conditions are equivalent.*

(1) *G is locally bounded and K is coarsely embedded in G.*
(2) *G/K is locally bounded and K is well embedded in G.*

Proof If G is locally bounded, then the homomorphic image G/K is also locally bounded by the open mapping theorem. If, moreover, K is coarsely embedded and d is a coarsely proper metric on G, then $d|_K$ is coarsely proper on K and thus K is well embedded.

Conversely, if K is well embedded, there is a compatible left-invariant metric on G so that $d|_K$ is coarsely proper on K. Because also K is coarsely embedded in G, it follows that $d|_K$ metrises the coarse structure on K induced by G. Thus, if in addition G/K is locally bounded, it follows from Proposition 4.8 that G is locally bounded. $\qquad\square$

Thus, if one can construct a coarsely embedded, but not well-embedded, closed normal subgroup K of a Polish group G so that the quotient G/K is locally bounded, then this will provide a counter-example to Problem 4.6.

When G is generated by locally bounded closed subgroups, local boundedness of G easily follows.

Lemma 4.10 *Suppose G is a Polish group generated by closed locally bounded subgroups K and F. Then G is locally bounded.*

Proof Since K and F are locally bounded, they admit coverings $U_1 \subseteq U_2 \subseteq \cdots \subseteq K$ and $V_1 \subseteq V_2 \subseteq \cdots \subseteq F$ by coarsely bounded open identity neighbourhoods. It follows that the $F_n = \overline{(U_n V_n)^n}$ are coarsely bounded closed subsets covering G and hence, by the Baire category theorem, some F_n must have non-empty interior. Thus $F_n^{-1} F_n$ witnesses local boundedness of G. $\qquad\square$

Let Γ be a denumerable discrete group and consider the semi-direct product $G = \mathbb{Z} \ltimes \Gamma^{\mathbb{Z}}$, which is easily seen to be locally bounded, but nevertheless has an open subgroup, namely $\Gamma^{\mathbb{Z}}$, which is not locally bounded. Thus, the local boundedness is really caused by the shift action of $\mathbb{Z}$ on $\Gamma^{\mathbb{Z}}$. On the other hand, there are groups whose local boundedness is truly locally caused. For this, suppose that U is a subset of a group G. We define a product $x_1 x_2 \cdots x_n$ of elements $x_i \in U$ to be U-admissible if there is a way to distribute parentheses so that the product may be evaluated in the local group $(U, \cdot)$. Formally, the set of U-admissible products is the smallest set of products so that

(1) if $x \in U$, the single factor product x is U-admissible,
(2) if $x_1 \cdots x_n$ and $y_1 \cdots y_m$ are U-admissible and $x_1 \cdots x_n \cdot y_1 \cdots y_m \in U$, then also $x_1 \cdots x_n \cdot y_1 \cdots y_m$ is U-admissible.

Strictly speaking, we should be talking about the U-admissibility of the sequence $(x_1, \ldots, x_n)$ rather than of the explicit product $x_1 \cdots x_n$, but this suggestive misuse of language should not cause any confusion.

For example, although all terms and the value of the product $3 \cdot 3^{-1} \cdot 3$ belong to $\{3, 3^{-1}\}$, the product $3 \cdot 3^{-1} \cdot 3$ is not $\{3, 3^{-1}\}$-admissible because for both ways of placing the parentheses, $(3 \cdot 3^{-1}) \cdot 3$ and $3 \cdot (3^{-1} \cdot 3)$, one will get a product $(3 \cdot 3^{-1})$ or $(3^{-1} \cdot 3)$ whose value is not in $\{3, 3^{-1}\}$.

Definition 4.11 A topological group G is *ultralocally bounded* if every identity neighbourhood U contains a further identity neighbourhood V so that, whenever W is an identity neighbourhood, there is a finite set F and a $k \geqslant 1$ for which every element $v \in V$ can be written as a U-admissible product $v = x_1 \cdots x_k$ with terms $x_i \in F \cup W$.

In other words, a group is ultralocally bounded if it is locally bounded and this can be witnessed entirely within arbitrary small identity neighbourhoods. Let us also point out the following series of implications:

$$\text{locally compact} \Rightarrow \text{locally Roelcke pre-compact}$$
$$\Rightarrow \text{ultralocally bounded} \Rightarrow \text{locally bounded.}$$

The only new fact here is that locally Roelcke pre-compact groups are ultralocally bounded. To see this, suppose U is an identity neighbourhood in a locally Roelcke pre-compact group G. Pick a Roelcke pre-compact symmetric identity neighbourhood V so that $V^5 \subseteq U$ and assume a further identity neighbourhood $W \subseteq V$ is given. By Roelcke pre-compactness, there is a finite set $F \subseteq G$ so that $V \subseteq WFW$. Thus, any element $v \in V$ can be written as a product $v = w_1 f w_2$ of elements $f \in F$ and $w_i \in W \subseteq V$, whence also $f = w_1^{-1} v w_2^{-1} \in V^3$. In other words, setting $F' = F \cap V^3$, we still have $V \subseteq WF'W$. Now, since $V^5 \subseteq U$, $W \subseteq V$ and $F' \subseteq V^3$, any product of the type $w_1 f w_2$ for $w_i \in W$ and $f \in F'$ is U-admissible, so this verifies ultralocal boundedness of G.

Similarly, suppose that d is a compatible left-invariant geodesic metric on a topological group G. Then G is ultralocally bounded. Indeed, if U is any identity neighbourhood, pick $\alpha > 0$ so that the open ball $V = B_d(\alpha)$ is contained in U. Suppose now that $W = B_d(\beta)$ is a further identity neighbourhood and that $v \in V$. Then we can write $v = w_1 \cdots w_n$ for some $w_i \in W$ and $n \leqslant \lceil \frac{\alpha}{\beta} \rceil$ minimal so that $n\beta > d(v, 1)$. In particular,

$$d(w_1 \cdots w_i, 1) \leqslant d(w_1, 1) + \cdots + d(w_i, 1) < i \cdot \beta \leqslant d(v, 1) \leqslant \alpha$$

for all $i < n$ and so distributing parentheses to the left shows that $w_1 \cdots w_n$ is a U-admissible product. Evidently, the assumption that d is geodesic may be weakened considerably.

As an easy application of the following proposition, we have that any extension of an ultralocally bounded topological group by a discrete group is locally bounded.

Proposition 4.12 *Suppose Γ is a discrete normal subgroup of a topological group G. Then G is ultralocally bounded if and only if G / Γ is. Similarly, G is locally Roelcke pre-compact if and only if G / Γ is.*

Proof Let $G \xrightarrow{\pi} G / \Gamma$ be the quotient map. Because Γ is discrete, pick a symmetric open identity neighbourhood $U \subseteq G$ so that $\Gamma \cap U^3 = \{1\}$, whence the restriction $U \xrightarrow{\pi} G / \Gamma$ is injective. As π is also an open map, it follows that $\pi[U]$ is an open identity neighbourhood in G / Γ and that the inverse $\pi[U] \xrightarrow{\omega} U$ to $U \xrightarrow{\pi} \pi[U]$ is continuous.

We claim that $U \xrightarrow{\pi} \pi[U]$ is a homeomorphic isomorphism of the local groups $(U, \cdot)$ and $(\pi[U], \cdot)$ with inverse $\pi[U] \xrightarrow{\omega} U$. Specifically, the following two properties hold.

(1) If $u_1, u_2 \in U$ satisfy $u_1 u_2 \in U$, then also $\pi(u_1)\pi(u_2) \in \pi[U]$ and

$$\pi(u_1)\pi(u_2) = \pi(u_1 u_2).$$

(2) If $w_1, w_2 \in \pi[U]$ satisfy $w_1 w_2 \in \pi[U]$, then also $\omega(w_1)\omega(w_2) \in U$ and

$$\omega(w_1)\omega(w_2) = \omega(w_1 w_2).$$

Property (1) is immediate from the fact that π is a group homomorphism. On the other hand, to verify (2), suppose that $w_1, w_2 \in \pi[U]$ satisfy $w_1 w_2 \in \pi[U]$. Then, as

$$\pi\big(\omega(w_1)\omega(w_2)\big) = \pi\big(\omega(w_1)\big)\pi\big(\omega(w_2)\big) = w_1 w_2 = \pi\big(\omega(w_1 w_2)\big),$$

there is $\gamma \in \Gamma$ so that $\omega(w_1)\omega(w_2) = \omega(w_1 w_2)\gamma$ and so $\gamma \in \Gamma \cap U^3 = \{1\}$, i.e., $\omega(w_1)\omega(w_2) = \omega(w_1 w_2) \in U$.

Now, ultralocal boundedness of G and G/Γ is exclusively dependent on the isomorphic topological local groups $(U, \cdot)$ and $(\pi[U], \cdot)$ and therefore G is ultralocally bounded if and only if G/Γ is. Indeed, it suffices to notice that π maps U-admissible products to $\pi[U]$-admissible products and, conversely, ω maps $\pi[U]$-admissible products to U-admissible products.

Similarly, as shown above, a topological group is locally Roelcke precompact exactly when every identity neighbourhood U contains a further identity neighbourhood V with the following property. For any identity neighbourhood W, there is a finite set F so that any element $v \in V$ can be written as a U-admissible product $v = w_1 f w_2$ with $w_i \in W$, $f \in F$. This shows that local Roelcke pre-compactness is preserved under isomorphism of local groups given by identity neighbourhoods. $\qquad\square$

Now, as shown by M. Culler and by the present author in [**78**], if M is a compact manifold of dimension $\geqslant 2$, the group $\mathrm{Homeo}_0(M)$ of isotopically trivial homeomorhisms is not locally Roelcke pre-compact. Nevertheless, by the results of R. D. Edwards and R. C. Kirby [**26**], $\mathrm{Homeo}_0(M)$ is locally contractible and, by [**57**], also locally bounded. We now improve the latter result.

Theorem 4.13 *Let M be a closed manifold. Then the identity component* $\mathrm{Homeo}_0(M)$ *of the homeomorphism group is ultralocally bounded.*

The proof of this relies the following fact.

Lemma 4.14 (Lemma 3.10 in [**56**]) *Fix a compact manifold M with a compatible metric d. Then there is $m \geqslant 1$ (depending only on M) so that, for all $\epsilon > 0$, there is a cover $\{U_1, U_2, \ldots, U_m\}$ of M, where each set U_i is a finite union of interiors $\mathsf{int}(B)$ of disjoint embedded closed balls $B \subseteq M$ of diameter $< \epsilon$.*

We now proceed with the proof of Theorem 4.13.

Proof Fix a compatible metric d on M and let $U \subseteq \mathrm{Homeo}_0(M)$ be an identity neighbourhood. Let also $m \geqslant 1$ be as in Lemma 4.14. Find $\epsilon > 0$ small enough so that $g \in U$ whenever $g \in \mathrm{Homeo}_0(M)$ satisfies

$$d_\infty(g, \mathsf{id}) = \sup_{x \in M} d(g(x), x) < \epsilon.$$

By Lemma 4.14, we may find a sequence

$$B_1, \ldots, B_k$$

of embedded closed balls with $\mathsf{diam}_d(B_i) < \frac{\epsilon}{m}$, whose interiors cover M, and a partition $A_1 \cup \cdots \cup A_m = \{1, \ldots, k\}$ so that $B_i \cap B_j = \emptyset$ for any two distinct i, j belonging to the same A_l. Let also

$$C_l = \bigcup_{i \in A_l} B_i.$$

Observe that, if $g \in \mathrm{Homeo}_0(M)$ satisfies $\mathsf{supp}(g) \subseteq C_l$, then because the B_i with $i \in A_l$ are closed, connected and pairwise disjoint we have $g[B_i] = B_i$ for all $i \in A_l$. For such g, it follows that

$$d_\infty(g, \mathsf{id}) = \sup_{x \in M} d(g(x), x) \leqslant \sup_{x \in C_l} d(g(x), x) \leqslant \max_{i \in A_l} \mathsf{diam}_d(B_i) < \frac{\epsilon}{m}$$

and so, in particular, $g \in U$. Similarly, consider the closed subgroups F_i of $\mathrm{Homeo}_0(M)$ defined by

$$F_i = \{f \in \mathrm{Homeo}_0(M) \mid \mathsf{supp}(f) \subseteq B_i\}.$$

Then, because B_i is an embedded closed ball and $\partial M = \emptyset$, every $f \in F_i$ fixes the boundary ∂B_i pointwise, whereby F_i is isomorphic to the group

$$\mathrm{Homeo}_\partial(\mathbb{B}^r)$$

of homeomorphisms of the standard closed ball $\mathbb{B}^r$ of dimension $r = \dim(M)$ fixing the boundary pointwise. Hence, by Lemma 8 of [57], F_i is coarsely bounded. Moreover, we see that, for $A_l = \{i_1, \ldots, i_q\}$, we have

$$H_l = \{g \in \mathrm{Homeo}_0(M) \mid \mathsf{supp}(g) \subseteq C_l\} = F_{i_1} F_{i_2} \cdots F_{i_q}$$

and so also H_l is coarsely bounded for each $l \leqslant m$. Furthermore, $H_l \subseteq U$.
    We set

$$V = \{g_1 g_2 \cdots g_m \mid \mathsf{supp}(g_i) \subseteq C_i\},$$

which, by the Fragmentation Lemma (see Example 3.19), is an identity neighbourhood in $\mathrm{Homeo}_0(M)$.
    Now, suppose $W \subseteq V$ is any identity neighbourhood in $\mathrm{Homeo}_0(M)$. Because the subgroups $H_l \leqslant \mathrm{Homeo}_0(M)$ are coarsely bounded in themselves and also connected, it follows that there is some n large enough so that

$$H_l = (H_l \cap W)^n$$

for all $l = 1, \ldots, m$.

Now, assume $h \in V$. Then h factors into a product $h = g_1 g_2 \cdots g_m$ with $g_l \in H_l$. It follows that each g_l may be written as a product

$$g_l = g_{l,1} \cdots g_{l,n},$$

where $g_{l,i} \in H_l \cap W$ for all $i \leqslant n$. Because H_l is a group contained in U, we have $g_{l,1} \cdots g_{l,i} \in H_l \subseteq U$ for all $i \leqslant n$ and thus note that the product $g_{l,1} \cdots g_{l,n}$ is U-admissible. Observe now that

$$d_\infty(g_1 g_2 \cdots g_j, \mathrm{id}) \leqslant d_\infty(g_1, \mathrm{id}) + \cdots + d_\infty(g_j, \mathrm{id}) < m \cdot \frac{\epsilon}{m} = \epsilon$$

and hence $g_1 g_2 \cdots g_j \in U$ for all $j \leqslant m$. Therefore, the product $g_1 g_2 \cdots g_m$ and hence also

$$h = g_{1,1} \cdots g_{1,n} \cdot g_{2,1} \cdots g_{2,n} \cdots g_{m,1} \cdots g_{m,n}$$

are U-admissible. So every $h \in V$ can be written as a U-admissible product of length mn with factors in W, thus proving that $\mathrm{Homeo}_0(M)$ is ultralocally bounded. $\square$

Apart from the connection between local boundedness and metrisability of the coarse structure, one reason for our interest in local boundedness of group extensions is that it allows us to construct well-behaved sections for the quotient map. Although, by a result of J. Dixmier [25], every continuous epimorphism between Polish groups admits a Borel section, modesty seems to require additional assumptions.

Proposition 4.15 *Suppose $G \xrightarrow{\pi} H$ is a continuous epimorphism between Polish groups and assume that G is locally bounded. Then there is Borel measurable modest section for π.*

Proof Set $K = \ker \pi$ and fix an increasing exhaustive sequence of coarsely bounded open sets $V_1 \subseteq V_2 \subseteq \cdots \subseteq G$ so that $V_n^2 \subseteq V_{n+1}$. For each n, consider the map $x \in V_n \mapsto V_n \cap xK \in F(V_n)$ into the Effros–Borel space of closed subsets of the Polish space V_n. This will be Borel measurable, because for every open subset $U \subseteq V_n$ we have

$$(V_n \cap xK) \cap U \neq \emptyset \ \Leftrightarrow \ x \in UK.$$

By the selection theorem of K. Kuratowski and C. Ryll-Nardzewski [54], there is a Borel selector $F(V_n) \setminus \{\emptyset\} \xrightarrow{s} V_n$, and so $\sigma(x) = s(V_n \cap xK)$ defines a Borel map $V_n \xrightarrow{\sigma} V_n$.

Observe now that, by the open mapping theorem, the images $U_n = \pi[V_n]$ are open in H and define $U_n \xrightarrow{\phi_n} V_n$ by

$$\phi_n(y) = x \quad \Leftrightarrow \quad \sigma(x) = x \ \& \ \pi(x) = y.$$

As ϕ_n has Borel graph, it is Borel measurable and is clearly a section for $V_n \xrightarrow{\pi} U_n$.

In order to obtain a Borel measurable global section for $G \xrightarrow{\pi} H$, it now suffices to set

$$\phi(y) = \phi_n(y), \text{ where } y \in U_n \setminus U_{n-1}.$$

To see that ϕ is modest, note that the U_n are an increasing exhaustive sequence of open subsets of H satisfying $U_n^2 \subseteq U_{n+1}$. Thus, if A is coarsely bounded in H, it will be contained in some U_n, whereby $\phi[A] \subseteq \phi[U_n] \subseteq V_n$ and thus $\phi[A]$ is coarsely bounded in G. $\qquad\square$

Let us also note that Proposition 4.15 admits a partial converse.

Corollary 4.16 *Suppose $G \xrightarrow{\pi} H$ is a continuous epimorphism between Polish groups and assume that both H and $K = \ker \pi$ are locally bounded. Then the following are equivalent.*

(1) *G is locally bounded,*
(2) *there is modest section for π,*
(3) *there is Borel measurable modest section for π.*

Proof By Proposition 4.15, if G is locally bounded, then there is a modest Borel measurable section for π. This shows (1)$\Rightarrow$(3). Also, (3)$\Rightarrow$(2) is trivial. Finally, for (2)$\Rightarrow$(1), suppose ϕ is a modest section for π. Note that, as K and H are locally bounded, there are increasing exhaustive sequences $A_1 \subseteq A_2 \subseteq \cdots \subseteq K$ and $B_1 \subseteq B_2 \subseteq \cdots \subseteq H$ of coarsely bounded sets. Because also A_n and $\phi[B_n]$ are coarsely bounded in G, it follows that $A_n \cdot \phi[B_n]$ is an increasing sequence of coarsely bounded sets covering G, whence G is locally bounded. $\qquad\square$

The utility of this corollary is nevertheless somewhat restricted by the fact that to verify that a certain section ϕ is modest, one must already have some understanding of the coarse structure of G.

4.3 Refinements of Topologies

We shall now encounter a somewhat surprising phenomenon concerning sections of quotient maps. Indeed, suppose $G \xrightarrow{\pi} H$ is a continuous epimorphism between topological groups and suppose $H' \leqslant H$ is a subgroup equipped with some finer group topology. Then there is a canonical group topology on the group of lifts $G' = \pi^{-1}(H')$, namely, the one in which open sets have the form

$$V \cap \pi^{-1}(U),$$

for V open in G and U open in H'.

We observe that, if G is a Polish group and H' is Polish in its finer group topology, then also G' is Polish in this canonical group topology. To see this, observe first that G' is second countable. Secondly, let d_G and $d_{H'}$ be compatible, complete, but not necessarily left-invariant metrics on G and H' respectively. Then we obtain a compatible metric on G' by setting

$$d_{G'}(g, f) = d_G(g, f) + d_{H'}\big(\pi(g), \pi(f)\big)$$

for $g, f \in G'$. It thus suffices to note that $d_{G'}$ is complete. So suppose (g_n) is $d_{G'}$-Cauchy. Then (g_n) is d_G-Cauchy and $(\pi(g_n))$ is $d_{H'}$-Cauchy, whence $g = \lim_{d_G} g_n$ and $h = \lim_{d_{H'}} \pi(g_n)$ exist. As the topology on H' refines that of H and as $G \xrightarrow{\pi} H$ is continuous, we find that also $h = \lim \pi(g_n) = \pi(g)$ in H. It follows that $g = \lim_{d_{G'}} g_n \in \pi^{-1}(H') = G'$, showing that $d_{G'}$ is complete on G'.

Example 4.17 Let M be a compact differentiable manifold with universal cover $\tilde{M} \xrightarrow{p} M$. Assume that $H = \mathrm{Homeo}(M)$ and that $G \leqslant \mathrm{Homeo}(\tilde{M})$ is the set of lifts of homeomorphisms of M to homeomorphisms of $\tilde{M}$, i.e., homeomorphisms g of $\tilde{M}$ so that $p\big(g(x)\big) = p\big(g(y)\big)$ whenever $p(x) = p(y)$. As every homeomorphism of M admits a lift to $\tilde{M}$, this defines a continuous epimorphism $G \xrightarrow{\pi} H$ by

$$\pi(g)\big(p(x)\big) = p\big(g(x)\big),$$

where G and H are equipped with the Polish topologies of uniform convergence on $\tilde{M}$ and M respectively. If $H' = \mathrm{Diff}^k(M)$, then H' becomes a Polish group in a finer group topology and thus $G' = \pi^{-1}(H')$ is Polish in the lifted group topology described above.

Observe that, if $H \xrightarrow{\phi} G$ is a section for the quotient map $G \xrightarrow{\pi} H$, then ϕ maps H' into G' and thus remains a section for the restricted quotient map $G' \xrightarrow{\pi} H'$. The next result shows that, in common situations, if $H \xrightarrow{\phi} G$ is bornologous, then so is $H' \xrightarrow{\phi} G'$ despite the changes of topology and thus also of coarse structure.

Proposition 4.18 *Suppose $G \xrightarrow{\pi} H$ is a continuous epimorphism between locally bounded Polish groups so that $K = \ker \pi$ is coarsely embedded in G. Assume also that $H' \leqslant H$ is a subgroup equipped with a finer Polish group topology and let $G' = \pi^{-1}(H')$ be the group of lifts equipped with its Polish topology induced from H'. Then, if $H \xrightarrow{\phi} G$ is a bornologous section for π, also $H' \xrightarrow{\phi} G'$ is a bornologous section for the restriction $G' \xrightarrow{\pi} H'$.*

Proof Because the G' topology refines that of G, we see that K is closed in G'. Therefore, as there can be no strictly finer Polish group topology on K, the G and G' topologies must coincide on K and hence the inclusion map $K \to G'$ is a continuous homomorphism. In addition, if a subset of K is coarsely bounded in G', it is also coarsely bounded in G and thus also in K, as K is coarsely embedded in G. This shows that K is a coarsely embedded closed subgroup of G'.

We first show that a section ϕ is modest as a map from H' to G' provided it is modest from H to G. So assume that $B \subseteq H'$ is coarsely bounded in H'. We must show that $\phi[B]$ is coarsely bounded in G'.

So pick an identity neighbourhood $V \subseteq G'$, which is coarsely bounded as a subset of G, and note, as $G' \xrightarrow{\pi} H'$ is an open map, that $\pi[V]$ is also an identity neighbourhood in H'. So find a finite set $E \subseteq G'$ and an n such that $B \subseteq (\pi[E]\pi[V])^n = \pi[(EV)^n]$. Then, for each $x \in B$, we have $\phi(x)k \in (EV)^n$ for some $k \in K$, in fact, $k \in K \cap \phi[B]^{-1}(EV)^n$. As the inclusion map $H' \to H$ is a continuous homomorphism, B is also coarsely bounded in H, and, because $H \xrightarrow{\phi} G$ is modest, we see that $\phi[B]^{-1}(EV)^n$ is coarsely bounded G. Therefore, because K is coarsely embedded in G, we find that the intersection $A = K \cap \phi[B]^{-1}(EV)^n$ is coarsely bounded in K and hence also in G'. It thus follows that $A^{-1} \subseteq (FV)^m$ for some finite set $E \subseteq F \subseteq G'$ and $m \geqslant 1$, whence

$$\phi[B] \subseteq (EV)^n A^{-1} \subseteq (FV)^{n+m}.$$

Given that V is an arbitrarily small identity neighbourhood in G', this implies that $\phi[B]$ is coarsely bounded in G' and hence that $H' \xrightarrow{\phi} G'$ is a modest mapping.

Now, suppose instead that $H \xrightarrow{\phi} G$ is bornologous. Fix also a coarsely bounded subset B of H'. Then B is coarsely bounded in H, as the inclusion $H' \to H$ is continuous. Furthermore, by Lemma 4.4, the subset $\omega_\phi[H \times B]$ of K is coarsely bounded in G and thus also in K, because K is coarsely embedded in G. Therefore, also $\omega_\phi[H' \times B]$ is coarsely bounded in K and hence also in G'. Because $H' \xrightarrow{\phi} G'$ is modest, by Lemma 4.4, we conclude that $H' \xrightarrow{\phi} G'$ is bornologous. $\qquad\square$

4.4 Group Extensions

We return to our stated problem of how coarse structure is preserved in short exact sequences

$$1 \to K \xrightarrow{\iota} G \xrightarrow{\pi} H \to 1,$$

where K is a closed normal subgroup of G and ι is the inclusion map. As our goal is not to enter into a detailed study of group cohomology, we shall restrict ourselves to some basic cases that appear in practice. Oftentimes these are central extensions, but there are several examples of a more general case that still admit a good theory, and we shall therefore develop a slightly wider framework. Indeed, the main setting will be when G is generated by K and its centraliser in G,

$$C_G(K) = \{g \in G \mid \forall k \in K \ gk = kg\},$$

that is, when $G = K \cdot C_G(K)$. Recall that, for future reference,

$$Z(K) = \{g \in K \mid \forall k \in K \ gk = kg\} = K \cap C_G(K)$$

denotes the centre of the group K.

Definition 4.19 Let H and K be groups. Then, a *cocycle* from H to K is a map $H \times H \xrightarrow{\omega} Z(K)$ satisfying the cocycle equation

$$\omega(h_1,h_2)\omega(h_1h_2,h_3) = \omega(h_2,h_3)\omega(h_1,h_2h_3)$$

for all $h_1,h_2,h_3 \in H$.

Observe that, if we apply the cocycle equation to the triples $(h_1,h_2,h_3) = (1,1,x)$ and $(x,1,1)$, we immediately get that

$$\omega(x,1) = \omega(1,x) = \omega(1,1)$$

for all $x \in H$. If, in addition to the cocycle equation, we also have that $\omega(1,1) = 1$, then we say the cocycle is *normalised*. Note that, if ω is any cocycle, then, because ω takes values in the abelian group $Z(K)$,

$$\sigma(x, y) = \omega(1, 1)^{-1}\omega(x, y)$$

is a normalised cocycle. In any case, applying again the cocycle equation to the triple (x, x^{-1}, x), one obtains the equation

$$\omega(x, x^{-1}) = \omega(x^{-1}, x),$$

which thus holds for all cocycles ω.

Example 4.20 (The cocycle defined from a section) Suppose $G \xrightarrow{\pi} H$ is an epimorphism between groups with kernel $K = \ker \pi$ and that $H \xrightarrow{\phi} C_G(K)$ is a section for π. We can then define a cocycle $H \times H \xrightarrow{\omega} Z(K)$ by

$$\omega(h_1, h_2) = \phi(h_1 h_2)^{-1}\phi(h_1)\phi(h_2).$$

To see this, note first that $\omega(h_1, h_2) \in \ker \pi \cap C_G(K) = Z(K)$. Also, for all $h_1, h_2, h_3 \in H$, we have, because ϕ maps into $C_G(K)$, that

$$
\begin{aligned}
\omega(h_1, h_2)\omega(h_1 h_2, h_3) &= \omega(h_1, h_2) \cdot \phi(h_1 h_2 h_3)^{-1}\phi(h_1 h_2)\phi(h_3) \\
&= \phi(h_1 h_2 h_3)^{-1}\phi(h_1 h_2) \cdot \omega(h_1, h_2) \cdot \phi(h_3) \\
&= \phi(h_1 h_2 h_3)^{-1}\phi(h_1 h_2) \cdot \phi(h_1 h_2)^{-1}\phi(h_1)\phi(h_2)\phi(h_3) \\
&= \phi(h_1 h_2 h_3)^{-1}\phi(h_1)\phi(h_2)\phi(h_3) \\
&= \phi(h_1 h_2 h_3)^{-1}\phi(h_1)\phi(h_2 h_3) \cdot \phi(h_2 h_3)^{-1}\phi(h_2)\phi(h_3) \\
&= \omega(h_1, h_2 h_3)\omega(h_2, h_3) \\
&= \omega(h_2, h_3)\omega(h_1, h_2 h_3).
\end{aligned}
$$

So ω satisfies the cocycle equation. Note also that $\omega(1, 1) = \phi(1)$, so ω is normalised if and only if $\phi(1) = 1$.

We remark that, by letting $\psi(x) = \phi(1)^{-1}\phi(x)$, we obtain a new section for π, $H \xrightarrow{\psi} C_G(K)$, satisfying $\psi(1) = 1$. Moreover, if ω_ϕ and ω_ψ are the two associated cocycles, we have, because $\phi(1) = \omega_\phi(1, 1) \in Z(K)$, that

$$\omega_\psi(x, y) = \phi(xy)^{-1}\phi(1)\phi(1)^{-1}\phi(x)\phi(1)^{-1}\phi(y) = \omega_\phi(1, 1)^{-1}\omega_\phi(x, y).$$

In the case that G and H are Polish and π continuous, we see that the passage from ϕ to ψ is entirely innocuous. Namely, if ϕ is modest, bornologous, a quasi-morphism, (uniformly) continuous or Borel measurable, then ψ will be so too.

As we shall see, any cocycle $H \times H \xrightarrow{\omega} Z(K)$ between groups can be obtained in this manner from a section ϕ for some quotient map π.

Although strictly speaking not needed for the following, working with normalised cocycles simplifies computations. Because a normalised cocycle can easily be obtained by either replacing a section ϕ by $\phi(1)^{-1}\phi(\cdot)$ or the cocycle ω by $\omega(1,1)^{-1}\omega(\cdot,\cdot)$, little is lost by restricting the attention to these.

Example 4.21 (Group extensions from cocycles) Suppose $H \times H \xrightarrow{\omega} Z(K)$ is a given normalised cocycle. Using the cocycle equation and the fact that ω takes values in the centre of K, we may define a group multiplication on the cartesian product $K \times H$ by

$$(k_1, h_1) \cdot (k_2, h_2) = \big(k_1 k_2 \omega(h_1, h_2), h_1 h_2\big)$$

with identity element $(1, 1)$ and inverse operation

$$(k, h)^{-1} = \big(k^{-1}\omega(h, h^{-1})^{-1}, h^{-1}\big).$$

We let $K \times_\omega H$ denote the group thus obtained.

Observe that K is naturally homomorphically embedded into $K \times_\omega H$ via the map $\iota(k) = (k, 1)$, whereas $K \times_\omega H \xrightarrow{\pi} H$ given by $\pi(k, h) = h$ is an epimorphism with kernel $\iota[K] = K \times \{1\}$. It follows that $K \times_\omega H$ is an extension of H by K with corresponding exact sequence

$$1 \to K \xrightarrow{\iota} K \times_\omega H \xrightarrow{\pi} H \to 1.$$

Unless ω is trivial, the section $H \xrightarrow{\phi} K \times_\omega H$ of the quotient map π given by $\phi(h) = (1, h)$ is only an injection and not a homomorphism. Observe also that, although K may not be central in $K \times_\omega H$, we have

$$\iota(k)\phi(h) = (k, h) = \phi(h)\iota(k)$$

and hence K is centralised by $\phi[H]$. Identifying K with its image in $K \times_\omega H$ via ι, we see that $K \times_\omega H = K \cdot \phi[H]$ and thus $K \times_\omega H = K \cdot C_{K \times_\omega H}(K)$ and $Z(K) \subseteq Z(K \times_\omega H)$. It follows also that K is central in $K \times_\omega H$ if and only if K is abelian. Moreover, observe that ω and ϕ are related via

$$\omega(h_1, h_2) = \phi(h_1 h_2)^{-1}\phi(h_1)\phi(h_2)$$

or, more formally, via the equation $\iota\big(\omega(h_1, h_2)\big) = \phi(h_1 h_2)^{-1}\phi(h_1)\phi(h_2)$.

Normalised cocycles $H \times H \xrightarrow{\omega} Z(K)$ and sections $H \xrightarrow{\phi} C_G(K)$ with $\phi(1) = 1$ for group extensions $K \to G \to H$ are related to each other in the following manner. Namely, assume that

$$1 \to K \xrightarrow{i} G \xrightarrow{p} H \to 1$$

is an extension of a group H by a group K and that, identifying K with its image by i, we have $G = K \cdot C_G(K)$. Suppose also that $H \xrightarrow{\phi} C_G(K)$ is a section of the quotient map p with $\phi(1) = 1$ and define the normalised cocycle $H \times H \xrightarrow{\omega} Z(K)$ as before by

$$\omega(h_1, h_2) = \phi(h_1 h_2)^{-1} \phi(h_1) \phi(h_2).$$

Then the map $K \times_\omega H \xrightarrow{\alpha} G$ defined by $\alpha(k, h) = k\phi(h)$ is an isomorphism so that the following diagram commutes.

$$
\begin{array}{ccccc}
K & \xrightarrow{\iota} & K \times_\omega H & \xrightarrow{\pi} & H \\
\downarrow{\text{id}} & & \downarrow{\alpha} & & \downarrow{\text{id}} \\
K & \xrightarrow{i} & G & \xrightarrow{p} & H
\end{array}
$$

Now suppose, in addition, that H and K are topological groups. Then, if $H \times H \xrightarrow{\omega} Z(K)$ is a continuous normalised cocycle, the induced group multiplication on the cartesian product $K \times H$ is continuous and thus $K \times_\omega H$ is a topological group.

However, even if G is a central extension of H by K with continuous bonding maps i and p, then G need not be given as $K \times_\omega H$ for some continuous cocycle ω. Indeed, suppose ω is continuous and $K \times_\omega H \xrightarrow{\beta} G$ is a topological isomorphism so that the diagram

$$
\begin{array}{ccccc}
K & \xrightarrow{\iota} & K \times_\omega H & \xrightarrow{\pi} & H \\
\downarrow{\text{id}} & & \downarrow{\beta} & & \downarrow{\text{id}} \\
K & \xrightarrow{i} & G & \xrightarrow{p} & H
\end{array}
$$

commutes. Then $\psi(h) = \beta(1, h)$ is a continuous section $H \xrightarrow{\psi} G$ for the quotient map $G \xrightarrow{p} H$ and this need not exist in general. Consider, for example, the extension $\mathbb{Z} \to \mathbb{R} \to \mathbb{T}$.

Example 4.22 (Bartle–Graves selectors) Let us remark that, by the existence of Bartle–Graves selectors (see Corollary 7.56 in [**30**]), every surjective bounded linear operator $X \xrightarrow{T} Z$ between Banach spaces admits a continuous modest section. Thus, if Y is a closed linear subspace of a Banach space X, then X is isomorphic to a *twisted sum* $Y \times_\omega X/Y$ as above.

Remark 4.23 Although the cocycle ω and thus the description of an extension G of H by K as a product $K \times_\omega H$ depend on the specific choice of section $H \xrightarrow{\phi} C_G(K)$ for the quotient map $G \xrightarrow{\pi} H$, the function $H \times H \xrightarrow{\sigma} G$ defined by $\sigma(h_1, h_2) = [\phi(h_1), \phi(h_2)]$ is independent of ϕ. Also, if $H \cong G/K$ is given the quotient topology, in which case $G \xrightarrow{\pi} H$ is continuous and open, the map $H \times H \xrightarrow{\sigma} G$ becomes continuous.

4.5 External Extensions

The problem of understanding the coarse structure of extensions splits into two tasks. Namely, on the one hand, we must analyse *external* extensions of topological groups K and H given as the skewed product $K \times_\omega H$ for some continuous cocycle $H \times H \xrightarrow{\omega} Z(K)$. On the other hand, there is the more involved task of *internal* extensions given by short exact sequences

$$1 \to K \xrightarrow{\iota} G \xrightarrow{\pi} H \to 1$$

of topological groups, where K is a closed normal subgroup of G and ι is the inclusion map. Observe that, in this case, the section ϕ of the quotient map and the corresponding cocycle ω are not explicitly given and may not, in general, be chosen continuous. Again, in the case of internal extensions, we will mostly consider the case when we have $G = K \cdot C_G(K)$.

Let us stress that, in the case of both external and internal extensions, it is vital to keep track of the range of the maps ϕ and ω when discussing their coarse qualities. For example, because $C_G(K)$ and K may not be coarsely embedded in G, ϕ could be bornologous as a map into G, but not as a map into $C_G(K)$ and, similarly, ω could be bornologous as a map into G, but not as a map into $Z(K)$ or even into K. Of course, when dealing with locally compact groups, this issue does not come up, because closed subgroups are automatically coarsely embedded. For general Polish groups, however, this makes computations substantially more delicate. Nevertheless, in all computations, the only coarse structures that occur are those of K, G and H. On the other hand, the coarse structures of the topological groups $C_G(K)$ and $Z(K)$ are irrelevant.

Let us begin with the easier of the two tasks, namely, external extensions, where we will require the additional assumption of H being locally bounded.

Theorem 4.24 *Suppose $H \times H \xrightarrow{\omega} Z(K)$ is a continuous normalised cocycle from a locally bounded topological group H to the centre of a topological group K and define a section $H \xrightarrow{\phi} K \times_\omega H$ of the quotient map by $\phi(h) = (1, h)$.*

(1) *Assume that $H \times H \xrightarrow{\omega} K$ is modest. Then $H \xrightarrow{\phi} K \times_\omega H$ is modest and K is coarsely embedded in $K \times_\omega H$. Moreover, a subset A is coarsely bounded in $K \times_\omega H$ if and only if the two projections $A_K = \mathsf{proj}_K(A)$ and $A_H = \mathsf{proj}_H(A)$ are coarsely bounded in K and H respectively.*

(2) *Assume that $\omega[H \times B]$ is coarsely bounded in K for all coarsely bounded $B \subseteq H$. Then $H \xrightarrow{\phi} K \times_\omega H$ is bornologous and $K \times_\omega H$ is*

coarsely equivalent to the direct product $K \times H$ via the formal identity $(k,h) \mapsto (k,h)$.

(3) *Finally, if $\omega[H \times H]$ is coarsely bounded in K, then $H \xrightarrow{\phi} K \times_\omega H$ is a quasi-morphism.*

Proof Consider first the case that $H \times H \xrightarrow{\omega} K$ is modest and assume that $A \subseteq K \times_\omega H$ and $C \subseteq H$ are coarsely bounded. Let also $V \subseteq K$ be an arbitrary identity neighbourhood and $U \subseteq H$ a coarsely bounded identity neighbourhood. Pick finite sets $1 \in E \subseteq K$, $F \subseteq H$ and an $m \geq 1$ so that $A \subseteq \left((E \times F) \cdot (V \times U)\right)^m$ and $C \subseteq (FU)^m$.

Then, since elements of $K \times \{1\}$ commute with elements of $\{1\} \times H$, we have

$$\left((E \times F) \cdot (V \times U)\right)^m = \left((E \times \{1\}) \cdot (\{1\} \times F) \cdot (V \times \{1\}) \cdot (\{1\} \times U)\right)^m$$
$$= \left((EV)^m \times \{1\}\right) \cdot \left((\{1\} \times F) \cdot (\{1\} \times U)\right)^m.$$

Also,

$$C \subseteq (FU)^m = \mathsf{proj}_H\left[\left((\{1\} \times F) \cdot (\{1\} \times U)\right)^m\right]$$

and, by the assumption on ω,

$$\left((\{1\} \times F) \cdot (\{1\} \times U)\right)^m \subseteq D \times (FU)^m$$

for some coarsely bounded subset D of K.

In particular,

$$A \subseteq \left((E \times F) \cdot (V \times U)\right)^m \subseteq (EV)^m D \times (FU)^m,$$

showing that the two projections $A_K = \mathsf{proj}_K(A)$ and $A_H = \mathsf{proj}_H(A)$ are coarsely bounded in K and H respectively.

Similarly,

$$\{1\} \times C \subseteq (D^{-1} \times \{1\}) \cdot \left((\{1\} \times F) \cdot (\{1\} \times U)\right)^m$$
$$\subseteq (D^{-1} \times \{1\}) \cdot \left((E \times F) \cdot (V \times U)\right)^m.$$

As $D^{-1} \times \{1\}$ is the homomorphic image of a coarsely bounded set in K, it is coarsely bounded in $K \times_\omega H$ and hence $\{1\} \times C$ is coarsely bounded too. Thus, if B is coarsely bounded in K, then $B \times C = (B \times \{1\}) \cdot (\{1\} \times C)$ is coarsely bounded in $K \times_\omega H$.

We have thus shown that a subset $A \subseteq K \times_\omega H$ is coarsely bounded if and only if the two projections $A_K = \mathsf{proj}_K(A)$ and $A_H = \mathsf{proj}_H(A)$ are coarsely bounded in K and H respectively. In particular, a subset $B \subseteq K$ is coarsely bounded if and only if $B \times \{1\}$ is coarsely bounded in $K \times_\omega H$ and so K is coarsely embedded in $K \times_\omega H$. Similarly, if $C \subseteq H$ is coarsely bounded, then $\phi[C] = \{1\} \times C$ is coarsely bounded in $K \times_\omega H$, showing that $H \xrightarrow{\phi} K \times_\omega H$ is modest.

Consider now the case that $\omega[H \times B]$ is coarsely bounded for every coarsely bounded set B in H. Then $H \times H \xrightarrow{\omega} K$ is modest and hence the coarsely bounded subsets of $K \times_\omega H$ are those contained in products $A \times B$ with A and B are coarsely bounded in K and H respectively.

To see that the formal identity $K \times H \to K \times_\omega H$ is bornologous, suppose that A and B coarsely bounded in K and H respectively. Then, if $k_1^{-1}k_2 \in A$ and $h_1^{-1}h_2 \in B$, also

$$(k_1, h_1)^{-1}(k_2, h_2) = (k_1^{-1}k_2\omega(h_1, h_1^{-1}h_2)^{-1}, h_1^{-1}h_2) \in A\omega[H \times B]^{-1} \times B.$$

As $A\omega[H \times B]^{-1}$ and B are both coarsely bounded, so is $A\omega[H \times B]^{-1} \times B$, as required.

Conversely, to see the inverse formal identity $K \times_\omega H \to K \times H$ is bornologous, suppose that A and B are coarsely bounded in K and H respectively. Then, if

$$(k_1, h_1)^{-1}(k_2, h_2) = (k_1^{-1}k_2\omega(h_1, h_1^{-1}h_2)^{-1}, h_1^{-1}h_2) \in A \times B,$$

also $h_1^{-1}h_2 \in B$ and $k_1^{-1}k_2 \in A\omega(h_1, h_1^{-1}h_2) \subseteq A\omega[H \times B]$.

Thus, the formal identity is a coarse equivalence and, because $H \xrightarrow{\phi} K \times_\omega H$ is the composition of the formal identity with the embedding $H \to K \times H$, also ϕ is bornologous.

Finally, observe that if $\omega[H \times H]$ is coarsely bounded in K, then the defect of ϕ is coarsely bounded in $K \times_\omega H$ and so ϕ is a quasi-morphism. $\square$

For good measure, let us point out that Theorem 4.24, item (1), does not imply that $K \times_\omega H$ is coarsely equivalent to the direct product of K and H; only that they have the same coarsely bounded sets under the natural identification.

Among the intended applications of Theorem 4.24 let us consider a cocycle $\Gamma \times \Gamma \xrightarrow{\omega} Z(K)$ defined on a countable discrete group Γ with values in a topological group K. Then ω is both continuous and modest, because it trivially maps finite sets to finite sets.

Corollary 4.25 *Suppose $\Gamma \times \Gamma \xrightarrow{\omega} Z(K)$ is a normalised cocycle defined on a countable discrete group Γ with values in the centre of a topological group K. Then K is coarsely embedded in $K \times_\omega \Gamma$. Moreover, a subset $A \subseteq K \times_\omega \Gamma$ is coarsely bounded exactly when contained in the product $B \times F$ of a coarsely bounded set B and a finite set F.*

Similarly, with stronger assumptions on Γ, we have the following.

Corollary 4.26 *Suppose* $\Gamma \times \Gamma \xrightarrow{\omega} Z(K)$ *is a normalised cocycle defined on a countable discrete group* Γ *with values in the centre of a topological group* K *and assume that* $\omega[\Gamma \times F]$ *is coarsely bounded for every finite set* F. *Then* $K \times_\omega \Gamma$ *is coarsely equivalent to* $K \times \Gamma$.

4.6 Internal Extensions of Polish Groups

We now come to the more delicate task of understanding internal extensions of Polish groups. Thus, in the present section, we consider a short exact sequence of Polish groups with continuous bonding maps

$$1 \to K \xrightarrow{\iota} G \xrightarrow{\pi} H \to 1.$$

Observe that, because $\iota[K] = \ker \pi$ is closed, by the open mapping theorem for Polish groups, the continuous epimorphism $K \xrightarrow{\iota} \iota[K]$ is open and is thus a topological isomorphism between K and $\iota[K]$. Therefore, to simplify notation, we may identify K with its image under ι and thus assume ι to be the inclusion map. Throughout the rest of the chapter, we shall also assume that G is generated by K and its centraliser, i.e., that $G = K \cdot C_G(K)$.

Example 4.27 An example of this setup is when a Polish group G is generated by a discrete normal subgroup $K = \Gamma$ and a connected subgroup F. Then the conjugation action of G on Γ defines a continuous homomorphism $G \xrightarrow{\text{ad}}$ Aut(Γ). However, as Aut(Γ) is totally disconnected and F is connected, it follows that F is contained in $\ker(\text{ad})$, i.e., that $F \leqslant C_G(\Gamma)$. In other words, $G = \Gamma \cdot C_G(\Gamma)$ and we can let $H = G/\Gamma$.

Working instead exclusively from assumptions on G/Γ, we have the following familiar result.

Lemma 4.28 *Suppose* Γ *is a discrete normal subgroup of a Polish group* G *so that* G/Γ *has no proper open subgroups. Assume also that either* Γ *is finitely generated or that* G/Γ *is locally connected. Then* $G = \Gamma \cdot C_G(\Gamma)$.

Proof Consider the group Aut(Γ) of automorphisms of Γ equipped with the permutation group topology, that is, so that pointwise stabilisers are open. Because Γ is discrete in G, the homomorphism $G \xrightarrow{\text{ad}}$ Aut(Γ) defined by the conjugation action of G on Γ will be continuous.

Assume first that Γ is generated by a finite subset E and note that

$$C_G(\Gamma) = \bigcap_{x \in E} \{g \in G \mid \mathsf{ad}(g)(x) = x\}$$

is the intersection of finitely many open subgroups and thus is itself an open subgroup of G. Because the quotient map $G \xrightarrow{\pi} G/\Gamma$ is open, it follows that $\pi[C_G(\Gamma)]$ is an open subgroup of G/Γ. By assumption on G/Γ, we see that $\pi[C_G(\Gamma)] = G/\Gamma$ and so $G = \Gamma \cdot C_G(\Gamma)$.

Assume now instead that G/Γ is locally connected. Then, as in the proof of Proposition 4.12, we may find an open identity neighbourhood U in G so that the restriction $U \xrightarrow{\pi} \pi[U]$ is a homeomorphism with an open identity neighbourhood in G/Γ. Shrinking $\pi[U]$ if necessary, we may suppose that $\pi[U]$ and thus also U are connected. It follows that $\mathsf{ad}[U] = \mathsf{id}$, i.e., that $U \subseteq C_G(\Gamma)$. So $C_G(\Gamma)$ is an open subgroup of G and, as before, we get $G = \Gamma \cdot C_G(\Gamma)$.   □

Observe that, if G is a Polish group generated by commuting closed subgroups K and F, then $G = K \cdot F$ and the map

$$(k, f) \in K \times F \mapsto kf \in G$$

is a continuous epimorphism from the Polish group $K \times F$ whose kernel is the central subgroup $N = \{(x, x^{-1}) \in K \times F \mid x \in K \cap F\}$. Thus, by the open mapping theorem, we see that G is isomorphic to the quotient group $\frac{K \times F}{N}$.

We now come to the main result of this section, which establishes coarse embeddability of K from assumptions on $H \times H \xrightarrow{\omega_\phi} K$, where ϕ is a section for the quotient map. As always, K will be a closed subgroup of a Polish group G so that $G = K \cdot C_G(K)$ and $G \xrightarrow{\pi} H$ is the quotient map to $H = G/K$.

Theorem 4.29 *Assume that H is locally bounded and that $H \xrightarrow{\phi} C_G(K)$ is a C-measurable section for π. Then the following conditions are equivalent.*

(1) $H \xrightarrow{\phi} G$ *is modest and K is coarsely embedded in G,*
(2) $\omega_\phi[B \times B]$ *is coarsely bounded in K for all coarsely bounded $B \subseteq H$.*

Moreover, if these hold, then a subset $A \subseteq G$ is coarsely bounded in G if and only if there are $B \subseteq K$ and $C \subseteq C_G(K)$ both coarsely bounded in G so that $A \subseteq BC$.

Before commencing the proof, let us point out a crucial feature of Theorem 4.29. Namely, in (2), $\omega_\phi[B \times B]$ is assumed to be coarsely bounded in K and not just in G. On the one hand, unless we already know something about G, this would probably be our only way to show that $\omega_\phi[B \times B]$ is coarsely bounded in G, but, on the other hand, it is formally stronger and accounts for K being coarsely embedded in G. Notwithstanding this, the non-trivial implication from (2) to (1) allows us to gather important structural information about G and the position of K within G just from knowledge of K, H and an appropriate map between them.

Proof Suppose first that $H \xrightarrow{\phi} G$ is modest and K is coarsely embedded in G. Then, for every coarsely bounded $B \subseteq H$, we have

$$\omega_\phi[B \times B] \subseteq \left(\phi[B^2]^{-1}\phi[B]^2\right) \cap K,$$

whence $\omega_\phi[B \times B]$ is coarsely bounded in G and thus also in K, as the latter is coarsely embedded in G. This verifies (1)$\Rightarrow$(2).

Assume instead that (2) holds. Replacing, if necessary, ϕ by the section $\phi' = \phi(1)^{-1}\phi(\cdot)$ and hence the associated cocycle ω_ϕ by $\omega_{\phi'} = \omega_\phi(1,1)^{-1}\omega_\phi(\cdot,\cdot)$, we may, without loss of generality, assume that $\phi(1) = 1$ and that ω_ϕ is normalised.

So set $\beta = \phi \circ \pi$ and let $\alpha(g) = g\beta(g)^{-1}$, whence $g = \alpha(g)\beta(g)$ is the canonical factorisation of any $g \in G$ into $\alpha(g) \in K$ and $\beta(g) \in C_G(K)$. In particular, $\alpha(k) = k\beta(k)^{-1} = k\phi(1)^{-1} = k$ for all $k \in K$. Note then that, for all $x, y \in G$, we have

$$\alpha(xy)\beta(xy) = xy = \alpha(x)\beta(x)\alpha(y)\beta(y) = \alpha(x)\alpha(y)\beta(x)\beta(y)$$

and hence, as $\omega_\beta(x, y) \in K$ and $\beta(xy) \in C_G(K)$,

$$\begin{aligned}
\omega_\alpha(x, y)^{-1} &= \alpha(y)^{-1}\alpha(x)^{-1}\alpha(xy) \\
&= \beta(x)\beta(y)\beta(xy)^{-1} \\
&= \beta(xy) \cdot \omega_\beta(x, y) \cdot \beta(xy)^{-1} \\
&= \omega_\beta(x, y).
\end{aligned}$$

Observe also that

$$\begin{aligned}
\omega_\beta(x, y) &= \beta(xy)^{-1}\beta(x)\beta(y) \\
&= \phi(\pi(xy))^{-1}\phi(\pi(x))\phi(\pi(y)) \\
&= \phi(\pi(x)\pi(y))^{-1}\phi(\pi(x))\phi(\pi(y)) \\
&= \omega_\phi(\pi(x), \pi(y)).
\end{aligned}$$

It thus follows that, if $A \subseteq G$ is coarsely bounded in G, then $\pi[A]$ is coarsely bounded in H and therefore

$$\omega_\alpha[A \times A] = \omega_\beta[A \times A]^{-1} = \omega_\phi\left[\pi[A] \times \pi[A]\right]^{-1}$$

are all coarsely bounded in K. Applying Proposition 4.3, we find that the C-measurable and thus a fortiori Baire measurable maps

$$G \xrightarrow{\alpha} K, \qquad G \xrightarrow{\beta} G, \qquad H \xrightarrow{\phi} G$$

are all modest. As α is a modest retraction of G onto K, we see that K is coarsely embedded in G.

For the last point, note that, if $A \subseteq G$ is coarsely bounded, then so are $B = \alpha[A]$ and $C = \beta[A]$, whereas $A \subseteq BC$. □

Combining Theorem 4.29 with Corollary 4.16, we have the following.

Corollary 4.30 *Assume that K and H are locally bounded and suppose that $H \xrightarrow{\phi} C_G(K)$ is a C-measurable section for π so that $\omega_\phi[B \times B]$ is coarsely bounded in K for all coarsely bounded $B \subseteq H$. Then G is locally bounded, K is coarsely embedded in G and $H \xrightarrow{\phi} G$ is modest.*

Similarly to Theorem 4.29, we also obtain a criterion for bornologous sections.

Corollary 4.31 *Assume that H is locally bounded and that $H \xrightarrow{\phi} C_G(K)$ is a C-measurable section for π. Then the following conditions are equivalent:*

(1) *$H \xrightarrow{\phi} G$ is bornologous and K is coarsely embedded in G,*
(2) *$\omega_\phi[H \times B]$ is coarsely bounded in K for all coarsely bounded $B \subseteq H$.*

Proof Suppose first that $H \xrightarrow{\phi} G$ is bornologous and $B \subseteq H$ is coarsely bounded. Then, by Lemma 4.4, $\omega_\phi[H \times B]$ is coarsely bounded in G. So, if also K is coarsely embedded in G, then $\omega_\phi[H \times B]$ is coarsely bounded in K too.

Conversely, assume $\omega_\phi[H \times B]$ is coarsely bounded in K for all coarsely bounded $B \subseteq H$. Then, by Theorem 4.29, $H \xrightarrow{\phi} G$ is modest and K is coarsely embedded in G. Applying Lemma 4.4 once again, we find that $H \xrightarrow{\phi} G$ is actually bornologous. □

Occasionally, there may be other ways to see that K is coarsely embedded in G rather than by employing the cocycle ω_ϕ. Also, in various cases, it may be more natural to deal with subgroups F of the centraliser $C_G(K)$ instead of all of it.

Lemma 4.32 *Suppose G is a Polish group generated by commuting closed subgroups K and F, i.e., so that $kf = fk$ for all $k \in K$ and $f \in F$. Assume also that one of the following holds.*

(1) $K \cap F$ is well embedded in F,
(2) $K \cap F$ is coarsely embedded in G and K is locally bounded,
(3) K is locally bounded and $K \cap F$ is coarsely embedded in both K and F.

Then K is coarsely embedded in G.

Proof Note that, because K and F commute and generate G, we have that $K \cap F$ is central in G. Observe also that, if λ and ℓ are continuous length functions on K and F respectively, then we obtain a continuous length function L on G by setting

$$L(kf) = \inf_{x \in K \cap F} \lambda(kx) + \ell(fx^{-1})$$

for $k \in K$ and $f \in F$. First, to see that this is well defined, note that, for $k \in K, z \in K \cap F$ and $f \in F$,

$$\inf_{x \in K \cap F} \lambda(kzx) + \ell(fx^{-1}) = \inf_{y \in K \cap F} \lambda(ky) + \ell(fy^{-1}z)$$

$$= \inf_{y \in K \cap F} \lambda(ky) + \ell(zfy^{-1}).$$

So the definition of $L(kzf)$ is independent of the choice of decomposition $kzf = kz \cdot f$ or $kzf = k \cdot zf$ with $kz, k \in K$ and $f, zf \in F$. Secondly, if $k_1, k_2 \in K$, $f_1, f_2 \in F$ and $\epsilon > 0$, find $x_1, x_2 \in K \cap F$ so that $\lambda(k_i x_i) + \ell(f_i x_i^{-1}) < L(k_i f_i) + \epsilon$. Then

$$\begin{aligned}
L(k_1 f_1 \cdot k_2 f_2) &= L(k_1 k_2 f_1 f_2) \\
&\leqslant \lambda(k_1 k_2 x_1 x_2) + \ell(f_1 f_2 x_2^{-1} x_1^{-1}) \\
&= \lambda(k_1 x_1 k_2 x_2) + \ell(f_1 x_1^{-1} f_2 x_2^{-1}) \\
&\leqslant \lambda(k_1 x_1) + \lambda(k_2 x_2) + \ell(f_1 x_1^{-1}) + \ell(f_2 x_2^{-1}) \\
&< L(k_1 f_1) + L(k_2 f_2) + 2\epsilon,
\end{aligned}$$

which shows that $L(k_1 f_1 \cdot k_2 f_2) \leqslant L(k_1 f_1) + L(k_2 f_2)$.

Finally, to see that L is continuous, note that the mapping $(k, f) \in K \times F \mapsto kf \in G$ is a continuous epimorphism and thus an open mapping. It follows that sets of the form VW, with V and W identity neighbourhoods in K and F respectively, form a neighbourhood basis at the identity in G. As λ and ℓ are continuous length functions on K and F, we see that L is continuous at $1 \in G$ and hence is continuous everywhere.

(1) Assume first that $K \cap F$ is well embedded in F and pick a continuous length function ℓ on F so that $\ell|_{K \cap F}$ is coarsely proper. Suppose towards a contradiction that K is not coarsely embedded in G. We may then find a sequence $k_n \in K$, which is coarsely bounded in G, but so that $\lim_n \lambda(k_n) = \infty$

for some continuous length function λ on K. Let L be the continuous length function on G defined as above.

Because $\{k_n\}_n$ is coarsely bounded in G, $L(k_n)$ is bounded. Thus, we may find a sequence $x_n \in K \cap F$ so that $\lambda(k_n x_n) + \ell(x_n^{-1})$ is bounded. As ℓ is coarsely proper on $K \cap F$, this means that $\{x_n^{-1}\}_n$ is coarsely bounded in $K \cap F$ and thus in K too. It follows that $\lambda(x_n^{-1})$ and $\lambda(k_n x_n)$ are bounded, while $\lambda(k_n)$ is unbounded, which is absurd.

(2) Suppose that K is locally bounded and that $K \cap F$ is coarsely embedded in G. Fix a coarsely proper length function λ on K. Let ℓ be any continuous length function on F and let L be defined as above. Suppose towards a contradiction that K is not coarsely embedded in G. We may then find a sequence $k_n \in K$, which is coarsely bounded in G, but so that $\lim_n \lambda(k_n) = \infty$. Then $L(k_n)$ is bounded, which means that there is some sequence $x_n \in K \cap F$ so that $\lambda(k_n x_n) + \ell(x_n^{-1})$ is bounded. In particular, $\lambda(k_n x_n)$ is bounded, whence $\{k_n x_n\}_n$ is coarsely bounded in K and thus in G. Since both $\{k_n\}_n$ and $\{k_n x_n\}_n$ are coarsely bounded in G, we finally see that $\{x_n\}_n$ is coarsely bounded in G and hence also in $K \cap F$. It follows again that $\lambda(x_n)$ and $\lambda(k_n x_n)$ are bounded, while $\lambda(k_n)$ is unbounded, which is absurd.

(3) Finally, suppose that K is locally bounded and $K \cap F$ is coarsely embedded in both K and F. In particular, this means that $K \cap F$ is well embedded in K and so, by case (1), we see that F is coarsely embedded in G and hence also that $K \cap F$ is coarsely embedded in G. The conclusion now follows from (2). $\square$

For good measure, let us also note the following easy fact.

Lemma 4.33 *Suppose G is a Polish group generated by commuting subgroups K and F with F open in G. Then F is coarsely embedded. In particular, every open subgroup of an abelian Polish group is coarsely embedded.*

Proof Suppose a subset $A \subseteq F$ is coarsely bounded in G. We must show that A is also coarsely bounded in F. So assume V is an identity neighbourhood in F. As A is coarsely bounded in G and F is open in G, there is a finite set $E \subseteq G$ and some n so that $A \subseteq (EV)^n$. Now, as $G = K \cdot F$, find finite sets $B \subseteq K$ and $D \subseteq F$ so that $E \subseteq BD$, whereby

$$A \subseteq (EV)^n \cap F \subseteq (BDV)^n \cap F = \left[B^n \cdot (DV)^n\right] \cap F = [B^n \cap F] \cdot (DV)^n.$$

As $B^n \cap F$ is finite, this shows that A is coarsely bounded in F. $\square$

Whereas in Corollary 4.16 modest sections were used to prove local boundedness of G, thus far we have not employed the added strength of having a bornologous section. This is done in the following.

Proposition 4.34 *Let* $G \xrightarrow{\pi} H$ *be a continuous epimorphism between topological groups. Assume also that the kernel* $K = \ker \pi$ *is coarsely embedded in* G *and that* $H \xrightarrow{\phi} C_G(K)$ *is a section for* π, *which is bornologous as a map* $H \xrightarrow{\phi} G$. *Then* G *is coarsely equivalent to* $K \times H$.

Proof Observe that, because ϕ takes values only in $C_G(K)$, we automatically have $G = K \cdot C_G(K)$. Also, because both $G \xrightarrow{\pi} H$ and $H \xrightarrow{\phi} G$ are bornologous, so is the composition $\beta = \phi \circ \pi$. For simplicity of notation, we let $\alpha(g) = g \cdot \beta(g)^{-1}$, whereby $g = \alpha(g) \cdot \beta(g)$ is the associated decomposition of any element $g \in G$ with $\alpha(g) \in K$ and $\beta(g) \in C_G(K)$. As in the proof of Theorem 4.29, we see that $\omega_\alpha(x, y) = \omega_\beta(x, y)^{-1}$.

Suppose first that $A \subseteq G$ is coarsely bounded. Then $\beta[A]$ is coarsely bounded in G and $(A \cdot \beta[A]^{-1}) \cap K$ is coarsely bounded in G and hence also in K, because the latter is coarsely embedded in G. As $\alpha[A] \subseteq (A \cdot \beta[A]^{-1}) \cap K$, this shows that $G \xrightarrow{\alpha} K$ is modest.

Assume again that $A \subseteq G$ is coarsely bounded. Because $G \xrightarrow{\beta} G$ is bornologous, Lemma 4.4 implies that $\omega_\beta[G \times A]$ is coarsely bounded in G and so $\omega_\alpha[G \times A] = \omega_\beta[G \times A]^{-1}$ is coarsely bounded in G and hence also in K. As $G \xrightarrow{\alpha} K$ is also modest, another application of Lemma 4.4 shows that $G \xrightarrow{\alpha} K$ is bornologous.

Note now that the map $K \times H \xrightarrow{\Theta} G$ defined by $\Theta(k, h) = k\phi(h)$ is the inverse of the bornologous map $g \in G \mapsto (\alpha(g), \pi(g)) \in K \times H$. We claim that Θ is bornologous. For note that, for all $k, a \in K$ and $h, b \in H$,

$$\Theta(k, h)^{-1}\Theta(ka, hb) = (k\phi(h))^{-1}ka\phi(hb) = \phi(h)^{-1}k^{-1}ka\phi(hb)$$
$$= a\phi(h)^{-1}\phi(hb).$$

So, if we restrict a to a coarsely bounded set A in K and b to a coarsely bounded set B in H, then, since ϕ is bornologous, we see that $a\phi(h)^{-1}\phi(hb)$ is an element of a coarsely bounded set in G depending only on A and B. It follows that Θ is bornologous with bornologous inverse and thus a coarse equivalence between $K \times H$ and G as claimed. $\qquad\square$

Combing now Corollary 4.31 and Proposition 4.34, we arrive at the second main result of this section.

Theorem 4.35 *Suppose* $G \xrightarrow{\pi} H$ *is a continuous epimorphism between Polish groups with kernel* K. *Assume also that* H *is locally bounded and that* $H \xrightarrow{\phi} C_G(K)$ *is a C-measurable section so that* $\omega_\phi[H \times B]$ *is coarsely* *bounded in K for every coarsely bounded set $B \subseteq H$. Then G is coarsely equivalent to $K \times H$.*

Focusing instead on extensions by discrete groups, we establish the following result.

Theorem 4.36 *Let G be a Polish group generated by a discrete normal subgroup Γ and a connected closed subgroup F. Assume also that $\Gamma \cap F$ is coarsely embedded in F and that $G/\Gamma \xrightarrow{\phi} G$ is a bornologous lift of the quotient map with $\operatorname{im}\phi \subseteq C_G(\Gamma)$. Then G is coarsely equivalent with $\Gamma \times G/\Gamma$.*

Proof Observe first that by Example 4.27, F will automatically be a subgroup of the centraliser $C_G(\Gamma)$. Being discrete, Γ is also locally bounded and thus Lemma 4.32 applies to show that Γ is coarsely embedded in G. By Proposition 4.34, we now see that G is coarsely equivalent to $\Gamma \times G/\Gamma$. $\square$

Note that the choice of dealing with the connected subgroup $F \leqslant C_G(K)$ allows us some additional flexibility, which may facilitate the verification that $K \cap F$ is coarsely embedded in F.

4.7 A Further Computation for General Extensions

We now provide some computations for general extensions of Polish groups, i.e., without assuming centrality.

In a short exact sequence of topological groups

$$1 \to K \xrightarrow{\iota} G \xrightarrow{\pi} H \to 1$$

it is natural to conjecture that, if one of the two groups K and H is trivial, the middle term G should essentially be equal to the other of K and H. In the case where K is the trivial term, this is verified for coarse structure by the following simple fact.

Proposition 4.37 *Suppose K is a closed normal subgroup of a Polish group G. Then the quotient map $G \xrightarrow{\pi} G/K$ is a coarse equivalence if and only if K is coarsely bounded in G.*

Proof Since π is a continuous epimorphism, it is bornologous and evidently also cobounded. Therefore, π is a coarse equivalence if and only if it is expanding, which, because π is a homomorphism, is equivalent to π being coarsely proper. Now, $\pi[K] = \{1\}$ is coarsely bounded, so, if π is coarsely proper, then K must be coarsely bounded in G.

Conversely, suppose K is coarsely bounded in G. Assume $A \subseteq G$ is coarsely unbounded in G and fix a continuous left-invariant écart d on G so that A has infinite d-diameter. Then the Hausdorff distance d_H on the quotient group G/K, defined by

$$d_H(gK, fK) = \max\left\{ \sup_{a \in gK} \inf_{b \in fK} d(a,b), \ \sup_{b \in fK} \inf_{a \in gK} d(a,b) \right\},$$

is a continuous left-invariant écart and, moreover, satisfies

$$d_H(gK, fK) = \inf_{k \in K} d(g, fk) = \inf_{k \in K} d(gk, f).$$

Because K is coarsely bounded in G, it has finite d-diameter, $\mathsf{diam}_d(K) = C$. Also, as A has infinite d-diameter, there are $x_n \in A$ with $d(x_n, x_1) > n$, whence

$$
\begin{aligned}
d_H(x_n K, x_1 K) &= \inf_{k \in K} d(x_n, x_1 k) \\
&\geqslant \inf_{k \in K} \big(d(x_n, x_1) - d(x_1, x_1 k) \big) \\
&= \inf_{k \in K} \big(d(x_n, x_1) - d(1, k) \big) \\
&\geqslant n - C.
\end{aligned}
$$

So $\pi[A]$ has infinite d_H-diameter and therefore fails to be coarsely bounded in G/K. In other words, π is coarsely proper. $\qquad\square$

4.8 Coarse Structure of Covering Maps

In the following we fix a path-connected, locally path-connected and locally compact, metrisable space X. We moreover assume that X is *semi-locally simply connected*, i.e., that every point $x \in X$ has a neighbourhood V so that any loop lying in V is nullhomotopic in X.

Assume also that Γ is a finitely generated group acting freely and cocompactly by homeomorphisms on X. Furthermore, suppose that the action is *proper*, i.e., that the set $\{a \in \Gamma \mid a \cdot K \cap K \neq \emptyset\}$ is finite for every compact set $K \subseteq X$. We let $M = X/\Gamma$ denote the compact metrisable quotient space and define

$$X \xrightarrow{\ p\ } M$$

to be the corresponding covering map. Observe that, as p is locally a homeomorphism, M is also path-connected, locally path-connected and semi-locally simply connected. In particular, as X and M are path-connected, their fundamental groups are independent up to isomorphism of the choice of base point.

By a result of R. Arens [2], because X is locally connected, the homeomorphism group of the locally compact space X is a topological group when equipped with the compact-open topology, i.e., given by the subbasic open sets

$$O_{C,U} = \{g \in \mathrm{Homeo}(X) \mid g[C] \subseteq U\},$$

where $C \subseteq X$ is compact and $U \subseteq X$ open. This is simply the induced topology on $\mathrm{Homeo}(X)$ when viewed as a closed subgroup of $\mathrm{Homeo}(\hat{X})$, where $\hat{X}$ is the Alexandroff one-point compactification and $\mathrm{Homeo}(\hat{X})$ is equipped with the compact-open topology of $\hat{X}$. Similarly, $\mathrm{Homeo}(M)$ will be equipped with the compact-open topology. As Γ acts freely on X, we may identify Γ with its image in $\mathrm{Homeo}(X)$. Observe that, because Γ acts properly on X, it will be a discrete subgroup of $\mathrm{Homeo}(X)$.

Recall that a homeomorphism $\tilde{h} \in \mathrm{Homeo}(X)$ is a *lift* of a homeomorphism $h \in \mathrm{Homeo}(M)$ provided that the diagram

$$
\begin{array}{ccc}
X & \xrightarrow{\;\tilde{h}\;} & X \\
\downarrow{\scriptstyle p} & & \downarrow{\scriptstyle p} \\
M & \xrightarrow{\;h\;} & M
\end{array}
$$

commutes. As X may not be simply connected, some $h \in \mathrm{Homeo}(M)$ may not admit a lift. For example, if $X = \mathbb{R} \times \mathbb{S}^1$ is the open annulus and $M = \mathbb{S}^1 \times \mathbb{S}^1$ the torus with $p(t,x) = (e^{2\pi t i}, x)$, then the homeomorphism $h(x,y) = (y,x)$ of M evidently has no lift to X.

However, if $\tilde{h}$ is the lift of some homeomorphism h, then $\pi(\tilde{h}) = h$ is uniquely defined from $\tilde{h}$ by $hp = p\tilde{h}$. Moreover, as the set of all lifts of homeomorphisms of M forms a subgroup L of $\mathrm{Homeo}(X)$, this gives us a homomorphism $L \xrightarrow{\pi} \mathrm{Homeo}(M)$.

We claim that $\ker \pi = \Gamma$. Indeed, that $\Gamma \leqslant L$ and $\pi(a) = \mathrm{id}_M$ for all $a \in \Gamma$ is immediate. Conversely, suppose that $f \in \ker \pi$. Then, as Γ acts freely on X, the closed sets $X_a = \{x \in X \mid f(x) = a(x)\}$ for $a \in \Gamma$ form a partition of X. However, as Γ also acts properly, each X_a will be open. Thus, as X is connected, we have $X = X_a$ for some $a \in \Gamma$, i.e., $f = a \in \Gamma$.

As $\Gamma = \ker \pi$ is normal in L, this shows that L is contained in the normaliser $N_{\mathrm{Homeo}(X)}(\Gamma)$ of Γ inside $\mathrm{Homeo}(X)$. Conversely, if $f \in \mathrm{Homeo}(X)$

normalises Γ, we may define a homeomorphism $h \in \mathrm{Homeo}(M)$ by $hp(x) = pf(x)$ having f as a lift and thus showing that $f \in L$. In other words, $L = N_{\mathrm{Homeo}(X)}(\Gamma)$. Observe also that, as Γ is discrete in $\mathrm{Homeo}(X)$, its normaliser is closed.

Lemma 4.38 *The set*

$$Q = \{h \in \mathrm{Homeo}(M) \mid h \text{ admits a lift } \tilde{h} \in \mathrm{Homeo}(X)\}$$

is an open subgroup of $\mathrm{Homeo}(M)$.

Proof We first show that there is an identity neighbourhood V in $\mathrm{Homeo}(M)$ so that, for any loop $\sigma : \mathbb{S}^1 \to M$ and homeomorphism $f \in V$, the two maps

$$\sigma : \mathbb{S}^1 \to M \quad \text{and} \quad f\sigma : \mathbb{S}^1 \to M$$

are homotopic. In particular, by the lifting of homotopies, σ admits a lift $\tilde{\sigma} : \mathbb{S}^1 \to X$ if and only if $f\sigma$ admits a lift $\widetilde{f\sigma} : \mathbb{S}^1 \to X$.

Indeed, let d be a compatible metric on M and fix a covering $\mathcal{U}$ of M by path-connected open sets U so that any loop in U is nullhomotopic in M. Let also $\epsilon > 0$ be a Lebesgue number for $\mathcal{U}$, i.e., so that any set of diameter ϵ is contained in some $U \in \mathcal{U}$. Fix a covering $\mathcal{W}$ of M by path-connected open sets of diameter $< \frac{\epsilon}{2}$ and let $0 < \delta < \frac{\epsilon}{2}$ be a Lebesgue number for $\mathcal{W}$. We set

$$V = \left\{ f \in \mathrm{Homeo}(M) \mid \sup_{x \in M} d(f(x), x) < \delta \right\}.$$

Now suppose that $\sigma : \mathbb{S}^1 \to M$ and $f \in V$. Find a finite subset $D \subseteq \mathbb{S}^1$ so that, for neighbours $x, y \in D$ in the induced circular ordering, the σ-image of the shortest circular arc $I_{x,y} \subseteq \mathbb{S}^1$ from x to y has diameter $< \delta$. For each $x \in D$, observe that $d(\sigma(x), f\sigma(x)) < \delta$ and so $\sigma(x)$ and $f\sigma(x)$ belong to some common $W \in \mathcal{W}$, which means that there is a path $\gamma_x \subseteq W$ from $\sigma(x)$ to $f\sigma(x)$. Then, if $x, y \in D$ are neighbours in the circular ordering and $I_{x,y}$ is as above,

$$\gamma_x \cdot f\sigma[I_{x,y}] \cdot \overline{\gamma_y} \cdot \overline{\sigma[I_{x,y}]}$$

is a loop of diameter $< \delta + \frac{\epsilon}{2} < \epsilon$ and thus is contained in some single $U \in \mathcal{U}$, whence nullhomotopic. As σ and $f\sigma$ are the concatenations of the $\sigma[I_{x,y}]$ and $f\sigma[I_{x,y}]$ respectively with $x, y \in D$, this shows that σ and $f\sigma$ are homotopic in M.

Observe now that, because $X \xrightarrow{p} M$ is a regular covering, i.e., Γ acts transitively on each fibre of p, the criterion for lifting of homeomorphisms of M to X can be formulated as follows. A homeomorphism $h \in \text{Homeo}(M)$ admits a lift $\tilde{h} \in \text{Homeo}(X)$ if and only if, for every loop $\sigma : \mathbb{S}^1 \to X$, the loop

$$hp\sigma : \mathbb{S}^1 \to M$$

admits a lift $\widetilde{hp\sigma} : \mathbb{S}^1 \to X$.

Now suppose $h \in \text{Homeo}(M)$ admits a lift and $f \in V$. Then also fh admits a lift. For, if $\sigma : \mathbb{S}^1 \to X$, then $hp\sigma : \mathbb{S}^1 \to M$ admits a lift $\widetilde{hp\sigma} : \mathbb{S}^1 \to X$ and so, by what was established above, also $fhp\sigma : \mathbb{S}^1 \to M$ admits a lift $\widetilde{fhp\sigma} : \mathbb{S}^1 \to X$.

Because V is symmetric, this show that, for $f \in V$, a homeomorphism $h \in \text{Homeo}(M)$ admits a lift if and only if fh does. Thus, as V is open, the set of homeomorphisms with lifts is an open subgroup of $\text{Homeo}(M)$. $\quad\square$

We therefore have a natural epimorphism $N_{\text{Homeo}(X)}(\Gamma) \xrightarrow{\pi} Q$ with kernel Γ and hence a short exact sequence of Polish groups

$$1 \longrightarrow \Gamma \longrightarrow N_{\text{Homeo}(X)}(\Gamma) \xrightarrow{\pi} Q \longrightarrow 1.$$

Now, as Γ is finitely generated, its automorphism group $\text{Aut}(\Gamma)$ is countable and so the kernel of the representation by conjugation

$$N_{\text{Homeo}(X)}(\Gamma) \xrightarrow{\text{ad}} \text{Aut}(\Gamma)$$

is a countable index open normal subgroup, namely the centraliser $\text{ker}(\text{ad}) = C_{\text{Homeo}(X)}(\Gamma)$ of Γ in $\text{Homeo}(X)$. We set

$$Q_0 = \pi\big[C_{\text{Homeo}(X)}(\Gamma)\big],$$

which, as π is open, is an open normal subgroup of Q. Observe that, because $C_{\text{Homeo}(X)}(\Gamma)$ is open in $N_{\text{Homeo}(X)}(\Gamma)$, also $\Gamma \cdot C_{\text{Homeo}(X)}(\Gamma)$ is an open normal subgroup of $N_{\text{Homeo}(X)}(\Gamma)$, which gives us the exact sequence

$$1 \longrightarrow \Gamma \longrightarrow \Gamma \cdot C_{\text{Homeo}(X)}(\Gamma) \xrightarrow{\pi} Q_0 \longrightarrow 1.$$

To keep track of our discussion, let us recapitulate the situation so far.

Proposition 4.39 *Suppose that $\Gamma \curvearrowright X$ is a proper, free and cocompact action of a finitely generated group Γ on a path-connected, locally path-connected and semi-locally simply connected, locally compact metrisable space X. Let $M = X/\Gamma$ and consider $\text{Homeo}(M)$ and $\text{Homeo}(X)$ with the compact-open topologies.*

Then the normaliser

$$N_{\text{Homeo}(X)}(\Gamma) = \{\tilde{h} \in \text{Homeo}(X) \mid \tilde{h} \text{ is a lift of a homeomorphism } \pi(h) \text{ of } M\}$$

is a closed subgroup of Homeo(X), *whereas*

$$Q = \{h \in \text{Homeo}(M) \mid h \text{ admits a lift to } X\}$$

is an open subgroup of Homeo(M). *We thus have a short exact sequence of Polish groups*

$$1 \longrightarrow \Gamma \longrightarrow N_{\text{Homeo}(X)}(\Gamma) \xrightarrow{\;\pi\;} Q \longrightarrow 1.$$

Furthermore, the centraliser $C_{\text{Homeo}(X)}(\Gamma)$ *is open in* $N_{\text{Homeo}(X)}(\Gamma)$ *and gives rise to a short exact sequence*

$$1 \longrightarrow \Gamma \longrightarrow \Gamma \cdot C_{\text{Homeo}(X)}(\Gamma) \xrightarrow{\;\pi\;} Q_0 \longrightarrow 1,$$

where $Q_0 = \pi\big[C_{\text{Homeo}(X)}(\Gamma)\big]$ *is an open subgroup of* Q.

Our next step is now to consider the setup of Proposition 4.39 under the added assumption that the quotient map $\Gamma \to \Gamma/Z(\Gamma)$, where $Z(\Gamma) = \Gamma \cap C_{\text{Homeo}(X)}(\Gamma)$ is the centre of Γ, admits a bornologous section

$$\Gamma/Z(\Gamma) \xrightarrow{\;\psi\;} \Gamma.$$

Example 4.40 (Bornologous sections for quotient maps of discrete groups) Suppose Λ is a normal subgroup of a countable discrete group Δ. If Λ is either finite or has finite index in Δ, then there is a bornologous section $\Delta/\Lambda \xrightarrow{\;\phi\;} \Delta$ for the quotient map $\Delta \xrightarrow{\;\pi\;} \Delta/\Lambda$. Indeed, if Λ has finite index in Δ, then Δ/Λ is finite and clearly every map from a finite group to a topological group will be bornologous. Suppose, on the other hand, Λ is finite and ϕ is a section for π. Then, for any finite set $A \subseteq \Delta/\Lambda$, we have that

$$(\phi \times \phi)E_A \subseteq E_{\pi^{-1}(A)}.$$

As $\pi^{-1}(A)$ is finite, this shows that ϕ is bornologous.

So suppose that ψ is a bornologous section for $\Gamma \to \Gamma/Z(\Gamma)$. By Proposition 4.34, the map

$$(a, s) \in Z(\Gamma) \times \Gamma/Z(\Gamma) \mapsto a\psi(s) \in \Gamma$$

is a coarse equivalence and hence

$$a\psi(s) \mapsto a$$

is a bornologous map from $\Gamma = Z(\Gamma) \cdot \text{im}(\psi)$ onto $Z(\Gamma)$.

Pick a relatively compact *fundamental domain* $D \subseteq X$ for the action $\Gamma \curvearrowright X$, that is, D contains a unique point from every Γ-orbit, and fix a point $x_0 \in D$.

Then, because $\mathsf{im}(\psi)$ is a transversal for $Z(\Gamma)$ in Γ and Γ acts freely on X, we see that $\mathsf{im}(\psi) \cdot D$ is a fundamental domain for the action $Z(\Gamma) \curvearrowright X$.

Suppose also that ρ is a left-invariant proper metric on Γ. We define an écart d on X by simply letting

$$d(aD \times bD) = \rho(a,b)$$

for $a, b \in \Gamma$. That is, for $x, y \in D$ and $a, b \in \Gamma$, we set $d\big(a(x), b(y)\big) = \rho(a, b)$. Clearly, d is invariant under the action by Γ on X. Moreover, for elements $g, f \in C_{\mathsf{Homeo}(X)}(\Gamma)$,

$$\sup_{x \in X} d\big(g(x), f(x)\big) = \sup_{x \in D} \sup_{a \in \Gamma} d(ga(x), fa(x)) = \sup_{x \in D} d(g(x), f(x)) < \infty,$$

because $a \in \Gamma$ commute with g, f and act by isometries. It thus follows that

$$d_\infty(g, f) = \sup_{x \in X} d\big(g^{-1}(x), f^{-1}(x)\big)$$

is a (generally discontinuous) left-invariant écart on $C_{\mathsf{Homeo}(X)}(\Gamma)$.

Claim 4.41 Every $h \in Q_0$ has a unique lift $\phi(h) \in C_{\mathsf{Homeo}(X)}(\Gamma)$ so that

$$\phi(h)^{-1}(x_0) \in \mathsf{im}(\psi) \cdot D.$$

Indeed, if $\tilde{h} \in C_{\mathsf{Homeo}(X)}(\Gamma)$ is any lift of h, note that, because D is a fundamental domain for the action $\Gamma \curvearrowright X$, there are $a \in Z(\Gamma)$ and $s \in \Gamma/Z(\Gamma)$ so that $\tilde{h}^{-1}(x_0) \in a\psi(s) \cdot D$. So, if we let $\phi(h) = \tilde{h}a$, then

$$\phi(h)^{-1}(x_0) = a^{-1}\tilde{h}^{-1}(x_0) \in \psi(s)D \subseteq \mathsf{im}(\psi) \cdot D.$$

Uniqueness follows from the fact that any two lifts of $h \in Q_0$ in $C_{\mathsf{Homeo}(X)}(\Gamma)$ differ by an element of $Z(\Gamma)$ and that $\mathsf{im}(\psi) \cdot D$ is a fundamental domain for the action of $Z(\Gamma)$ on X.

Lemma 4.42 *For every constant c, there is a finite set $A \subseteq Z(\Gamma)$ so that*

$$d_\infty\big(\phi(h), \phi(g)a\big) \leqslant c \quad \Rightarrow \quad a \in A$$

for all $h, g \in Q_0$ and $a \in Z(\Gamma)$.

Proof Fix the constant c. Then, as ρ is a proper left-invariant metric on Γ and the map $a\psi(s) \mapsto a$ from $\Gamma = Z(\Gamma) \cdot \mathsf{im}(\psi)$ to $Z(\Gamma)$ is bornologous, there is a finite set $A \subseteq Z(\Gamma)$ so that

$$\rho\big(a\psi(s), b\psi(t)\big) \leqslant c \quad \Rightarrow \quad b^{-1}a \in A$$

for all $a, b \in Z(\Gamma)$ and $s, t \in \Gamma/Z(\Gamma)$.

Suppose now that $h, g \in Q_0$ and $a \in Z(\Gamma)$ with $d_\infty(\phi(h), \phi(g)a) \leqslant c$. Choose $s, t \in \Gamma/Z(\Gamma)$ so that $\phi(h)^{-1}(x_0) \in \psi(s)D$ and $\phi(g)^{-1}(x_0) \in \psi(t)D$. Then $a\phi(h)^{-1}(x_0)$ and $a\psi(s)(x_0)$ both belong to $a\psi(s) \cdot D$ and thus have d-distance 0. Similarly, $d(\psi(t)(x_0), \phi(g)^{-1}(x_0)) = 0$. It follows that

$$\begin{aligned} \rho(a\psi(s), \psi(t)) &= d(a\psi(s)(x_0), \psi(t)(x_0)) \\ &= d(a\phi(h)^{-1}(x_0), \phi(g)^{-1}(x_0)) \\ &\leqslant d_\infty(\phi(h)a^{-1}, \phi(g)) \\ &= d_\infty(\phi(h), \phi(g)a) \\ &\leqslant c \end{aligned}$$

and so $a \in A$ as required.   □

Suppose now that H is a subgroup of Q_0, which is Polish in some finer group topology τ_H. For example, M could be a manifold and H could be the symmetries of some additional structure of M, e.g., a volume form, a differentiable or symplectic structure, and τ_H could be a canonical topology defined from this additional structure. Note that H will be closed in $\mathrm{Homeo}(M)$ if and only if its Polish topology coincides with that induced from $\mathrm{Homeo}(M)$. Let also $G = \pi^{-1}(H)$ be the group of all lifts of elements in H, whence

$$\Gamma \leqslant G \leqslant \Gamma \cdot C_{\mathrm{Homeo}(X)}(\Gamma).$$

As in Section 4.3, G is given its canonical topology lifted from H.

We now arrive at the main result of this section, with antecedents in an earlier result of K. Mann and the author (Theorem 30 in [57]). The latter dealt exclusively with the fundamental group $\Gamma = \pi_1(M)$ of a compact manifold M acting by deck-transformations on the universal cover $X = \tilde{M}$. However, even in this case, the assumptions on Γ were slightly different from those below, because one required a bornologous section $\Gamma/A \xrightarrow{\phi} \Gamma$, where A is a specific geometrically defined central subgroup of Γ. Moreover, in that result, the subgroup H was simply $\mathrm{Homeo}_0(M)$ itself.

So, to state the theorem, let us briefly summarise the setup. We are given a proper, free and cocompact action $\Gamma \curvearrowright X$ of a finitely generated group Γ on a path-connected, locally path-connected and semi-locally simply connected, locally compact metrisable space X. Then the centraliser $C_{\mathrm{Homeo}(X)}(\Gamma)$ is an open subgroup of the normaliser $N_{\mathrm{Homeo}(X)}(\Gamma)$, which, in turn, is the group of all lifts of homeomorphisms of $M = X/\Gamma$ to X. Let

$$N_{\mathrm{Homeo}(X)}(\Gamma) \xrightarrow{\pi} \mathrm{Homeo}(M)$$

be the corresponding quotient map and let

$$Q_0 = \pi\big[C_{\mathrm{Homeo}(X)}(\Gamma)\big]$$

be the open subgroup of $\mathrm{Homeo}(M)$ consisting of homeomorphisms admitting lifts in $C_{\mathrm{Homeo}(X)}(\Gamma)$.

Theorem 4.43 *Suppose that there is a bornologous section $\Gamma/Z(\Gamma) \xrightarrow{\psi} \Gamma$ for the quotient map. Assume that H is a subgroup of Q_0, which is Polish in some finer group topology, and let $G = \pi^{-1}(H)$ be the group of lifts of elements of H with the topology lifted from H, whence the exact sequence*

$$1 \to \Gamma \to G \xrightarrow{\pi} H \to 1.$$

Then there is a bornologous section $H \xrightarrow{\phi} C_G(\Gamma)$ for the quotient map π and we have the following coarse equivalences

$$G \approx_{\mathrm{coarse}} H \times \Gamma, \qquad C_G(\Gamma) \approx_{\mathrm{coarse}} H \times Z(\Gamma).$$

Proof Let ϕ be the section defined in Claim 4.41. Suppose that $h \in H$. Then, because $\phi(h) \in C_{\mathrm{Homeo}(X)}(\Gamma)$ is a lift of h and $\ker \pi = \Gamma \leqslant G$, we see that $\phi(h)$ belongs to the closed subgroup $C_G(\Gamma) = G \cap C_{\mathrm{Homeo}(X)}(\Gamma)$ of G.

We claim that $H \xrightarrow{\phi} C_G(\Gamma)$ and a fortiori $H \xrightarrow{\phi} G$ is bornologous. To see this, suppose that $B \subseteq H$ is coarsely bounded and $V \subseteq C_G(\Gamma)$ is an identity neighbourhood. We must find a finite set $F \subseteq C_G(\Gamma)$ and an $n \geqslant 1$ so that, for $h, g \in H$,

$$h^{-1}g \in B \quad \Rightarrow \quad \phi(h)^{-1}\phi(g) \in (FV)^n.$$

Let first $U \subseteq X$ be a relatively compact open set containing $\overline{D}$, where D is the fundamental domain for the action $\Gamma \curvearrowright X$. Thus, by the properness of the action, the set $\{a \in \Gamma \mid a \cdot D \cap U \neq \varnothing\}$ is finite and so U has finite d-diameter. Also, as the topology on G and thus also on $C_G(\Gamma)$ refine the compact-open topology from the action on X, the set $W = \{f \in C_G(\Gamma) \mid f^{-1}[\overline{D}] \subseteq U\}$ is an identity neighbourhood in $C_G(\Gamma)$. Observe that, if $f \in W$, then

$$d_\infty(f, \mathrm{id}) = \sup_{x \in D} d(f^{-1}x, x) \leqslant \mathrm{diam}_d(U).$$

That is, W has finite d_∞-diameter.

Now, $V \cap W$ is a identity neighbourhood in $C_G(\Gamma)$, so, as $C_G(\Gamma) \xrightarrow{\pi} H$ is a continuous epimorphism and therefore an open map, $\pi[V \cap W]$ is an identity neighbourhood in H. Thus, as B is coarsely bounded, there is a finite set

$E \subseteq H$ and some n so that $B \subseteq (E \cdot \pi[V \cap W])^n$. Write $E = \pi[F]$ for some finite set $F \subseteq C_G(\Gamma)$, whereby

$$B \subseteq (\pi[F] \cdot \pi[V \cap W])^n = \pi[(E \cdot (V \cap W))^n].$$

As W has finite d_∞-diameter and d_∞ is left-invariant, also

$$c = \mathrm{diam}_{d_\infty}[(E \cdot (V \cap W))^n] < \infty.$$

We let $A \subseteq Z(\Gamma)$ be the finite set associated with c by Lemma 4.42.

Now suppose $h, g \in H$ with $h^{-1}g \in B$. Then, as $\phi(h)^{-1}\phi(g) \in C_G(\Gamma)$ is a lift of $h^{-1}g$, there is some $a \in \Gamma \cap C_G(\Gamma) = Z(\Gamma)$ so that $\phi(h)^{-1}\phi(g)a \in (E \cdot (V \cap W))^n$, whereby

$$d_\infty(\phi(h), \phi(g)a) \leqslant c$$

and so $a \in A$. It follows that, for $h, g \in H$,

$$h^{-1}g \in B \implies \phi(h)^{-1}\phi(g) \in (E \cdot (V \cap W))^n A^{-1},$$

showing that $H \xrightarrow{\phi} C_G(\Gamma)$ is bornologous.

We now show that $Z(\Gamma) = \Gamma \cap C_G(\Gamma)$ is coarsely embedded in $C_G(\Gamma)$. To see this, note that the set W above is an identity neighbourhood in $C_G(\Gamma)$ of finite d_∞-diameter. As d_∞ is a left-invariant écart on $C_G(\Gamma)$, it follows that $(FW)^n$ has finite d_∞-diameter for all finite sets F and $n \geqslant 1$. So every coarsely bounded set in $C_G(\Gamma)$ has finite d_∞-diameter. On the other hand, by Lemma 4.42, every infinite subset of $Z(\Gamma)$ has infinite d_∞-diameter. This shows that $Z(\Gamma)$ is coarsely embedded in $C_G(\Gamma)$.

Now, as Γ is locally bounded and $Z(\Gamma) = \Gamma \cap C_G(\Gamma)$ is coarsely embedded in both Γ and $C_G(\Gamma)$, by Lemma 4.32, we conclude that also Γ is coarsely embedded in G. Therefore, we may now apply Proposition 4.34 to see that $C_G(\Gamma)$ is coarsely equivalent to $Z(\Gamma) \times H$, whereas G is coarsely equivalent to $\Gamma \times H$. $\square$

Suppose a compact manifold M is given with universal cover $X = \tilde{M}$. Then $\Gamma = \pi_1(M)$ acts freely, properly and cocompactly by deck-transformations on X and $M = X/\Gamma$. Moreover, as X is simply connected, every homeomorphism of M lifts to X and so $Q = \mathrm{Homeo}(M)$. Furthermore, because the group $\mathrm{Homeo}_0(M)$ of isotopically trivial homeomorphisms by definition is path-connected, it will be contained in the open subgroup Q_0 of $\mathrm{Homeo}(M)$.

Theorem 4.44 *Suppose M is a compact manifold so that the quotient map*

$$\pi_1(M) \longrightarrow \pi_1(M)/Z(\pi_1(M))$$

admits a bornologous section. Assume also that H is a subgroup of $\text{Homeo}_0(M)$, which is Polish in some finer group topology, and let G be the group of all lifts of elements in H to homeomorphisms of the universal cover $\tilde{M}$. Then G is coarsely equivalent to the direct product group $\pi_1(M) \times H$.

Perhaps equally important is the fact that the coarse equivalence between G and $\pi_1(M) \times H$ is given by a bornologous section $H \xrightarrow{\phi} G$ of the quotient map, where ϕ is the map defined above.

5

Polish Groups of Bounded Geometry

5.1 Gauges and Groups of Bounded Geometry

In [75], J. Roe considers the coarse spaces of bounded geometry, which are a natural generalisation of metric spaces of bounded geometry. For this, suppose $(X, \mathcal{E})$ is a coarse space and $E \in \mathcal{E}$ a symmetric entourage. Then, in analogy with A. N. Kolmogorov's notions of metric entropy and capacity [51, 52], we define the *E-capacity* and *E-entropy* of a subset $A \subseteq X$ by

$$\mathrm{cap}_E(A) = \sup \left(k \mid \exists a_1, \ldots, a_k \in A \colon (a_i, a_j) \notin E \text{ for } i \neq j \right)$$

and

$$\mathrm{ent}_E(A) = \min \left(|B| \mid A \subseteq E[B] \right) = \min \left(|B| \mid \forall a \in A \; \exists b \in B \; (a, b) \in E \right).$$

The following inequalities are then straightforward to verify

$$\mathrm{cap}_{E \circ E} \leqslant \mathrm{ent}_E \leqslant \mathrm{cap}_E.$$

Also, if $E \subseteq E'$, then clearly $\mathrm{ent}_{E'} \leqslant \mathrm{ent}_E$ and $\mathrm{cap}_{E'} \leqslant \mathrm{cap}_E$.

Definition 5.1 A coarse space $(X, \mathcal{E})$ has *bounded geometry*[1] if there is a symmetric entourage $E \in \mathcal{E}$ so that, for every entourage $F \in \mathcal{E}$,

$$\sup_{x \in X} \mathrm{ent}_E(F_x) < \infty,$$

where $F_x = \{y \in X \mid (y, x) \in F\}$.

---

[1] The actual definition given by Roe in [75] is, for various reasons, more complicated and expressed in terms of the capacity, but can easily be checked to be equivalent to ours.

For example, a metric space (X,d) has bounded geometry if and only if there is a finite diameter α with the property that, for every β, there is a k so that every set of diameter $\leqslant \beta$ can be covered by k sets of diameter α.

When dealing with discrete metric spaces, it is often useful to employ a slightly stronger notion, namely, that of uniformly locally finite spaces, which unfortunately is also denoted bounded geometry in the literature. To avoid ambiguity, we keep a separate terminology.

Definition 5.2 A metric space (X,d) is *locally finite* if every set of finite diameter is finite. Also, (X,d) is *uniformly locally finite* if, for every diameter β, there is a K so that subsets of diameter $\leqslant \beta$ have cardinality at most K.

Also, bounded geometry is a *coarse invariant*, i.e., any coarse space $(Y,\mathcal{F})$ coarsely equivalent to a coarse space $(X,\mathcal{E})$ of bounded geometry is itself coarsely bounded. In fact, we have the following.

Lemma 5.3 *Suppose* $Y \xrightarrow{\phi} X$ *is a coarse embedding of a coarse space* $(Y,\mathcal{F})$ *into a coarse space* $(X,\mathcal{E})$ *of bounded geometry. Then* $(Y,\mathcal{F})$ *has bounded geometry.*

Proof Because $(X,\mathcal{E})$ has bounded geometry, there is a symmetric entourage $E \in \mathcal{E}$ so that, for all other $E' \in \mathcal{E}$, there is a constant k for which

$$\forall x \; \exists z_1, \ldots, z_k \; \forall u \; (uE'x \Rightarrow \exists i \; uEz_i).$$

As ϕ is a coarse embedding, the set

$$F = \{(a,b) \in Y \times Y \mid \big(\phi(a),\phi(b)\big) \in E \circ E\}$$

is a symmetric entourage on Y.

Now, suppose $F' \in \mathcal{F}$. Then we have $E' = (\phi \times \phi)F' \in \mathcal{E}$ and can therefore choose k as above. Thus, given $a \in Y$, find $z_1, \ldots, z_k \in X$ so that, for all $u \in X$,

$$uE'\phi(a) \implies \exists i \; uEz_i.$$

For each z_i, find if possible some $b_i \in Y$ so that $\phi(b_i)Ez_i$ and choose b_i arbitrarily otherwise. Then, we see that, for all $c \in Y$,

$$
\begin{aligned}
cF'a &\Rightarrow \phi(c)E'\phi(a) \\
&\Rightarrow \exists i \; \phi(c)Ez_i \\
&\Rightarrow \exists i \; \phi(c)Ez_i \text{ and } \phi(b_i)Ez_i \\
&\Rightarrow \exists i \; \phi(c)E^2\phi(b_i) \\
&\Rightarrow \exists i \; cFb_i.
\end{aligned}
$$

In other words, for every $F' \in \mathcal{F}$, there is a constant k, so that

$$\forall a \; \exists b_1, \ldots, b_k \; \forall c \; (cF'a \Rightarrow \exists i \; cFb_i),$$

showing that $(Y, \mathcal{F})$ has bounded geometry. $\qquad\qquad\qquad\square$

Now, if G is a topological group, the basic entourages of the form $E_A = \{(x, y) \mid x^{-1}y \in A\}$ form a basis for the coarse structure on G and, with this observation, the following lemma is straightforward to verify.

Lemma 5.4 *A topological group G has bounded geometry if and only if there is a coarsely bounded set $A \subseteq G$ that covers every other coarsely bounded set B by finitely many left-translates, i.e., so that $B \subseteq FA$ for some finite set $F \subseteq G$.*

It will be useful to have a name for the sets A appearing in Lemma 5.4, as these will appear throughout this section.

Definition 5.5 A subset A of a topological group G is said to be a *gauge* for G if A is coarsely bounded, symmetric, $1 \in A$ and, for every coarsely bounded set $B \subseteq G$, there is a finite set F so that $B \subseteq FA$.

Of course, the quintessential example of a Polish group with bounded geometry is a locally compact group. In fact, in a locally compact Polish group G, every symmetric relatively compact identity neighbourhood V is a gauge, because every relatively compact set can be covered by finitely many left-translates of V.

Our first observation is that Polish groups with bounded geometry are automatically locally bounded, although not necessarily locally compact.

Lemma 5.6 *Polish groups with bounded geometry are locally bounded.*

Proof Let A be a gauge for a Polish group G. Replacing A by its closure, we may assume that A is closed, whereby all powers A^k are analytic sets and thus have the property of Baire. Obviously, we may also assume that G is uncountable and hence a perfect space.

We claim that A^4 is a coarsely bounded identity neighbourhood. To see this, observe that, if A^2 is non-meagre, then by Pettis's Lemma [71] $A^4 = (A^2)^{-1}A^2$ is an identity neighbourhood. On the other hand, if A^2 is meagre, then, because the mapping $(x, y) \in G \times G \mapsto x^{-1}y \in G$ is surjective, continuous and open, the subset $\{(x, y) \in G \times G \mid x^{-1}y \notin A^2\}$ is co-meagre in $G \times G$. By Mycielski's Independence Theorem [65], we can thus find a homeomorphic copy $C \subseteq G$ of Cantor space so that $x^{-1}y \notin A^2$ and hence $xA \cap yA = \emptyset$ for all distinct $x, y \in C$.

Now, because C is compact and hence coarsely bounded in G, there is a finite subset $F \subseteq G$ so that $C \subseteq FA$. There are thus distinct $x, y \in C$ belonging to some fA with $f \in F$, i.e., $f \in xA \cap yA$, contradicting the assumption on C. □

By Lemma 5.6, every Polish group of bounded geometry admits an open gauge. Indeed, if A is a gauge and V is a coarsely bounded symmetric open identity neighbourhood, which exists by Lemma 5.6, then VAV is also a gauge. Also, as mentioned above, every symmetric relatively compact identity neighbourhood in a locally compact Polish group is a gauge. Coversely, if G is a Polish group with an identity neighbourhood U that can be covered by finitely many left-translates by any smaller identity neighbourhood, then U will in fact be relatively compact and thus G is locally compact. So gauges in general Polish groups of bounded geometry can be neither too big, nor too small.

Suppose G is a monogenic Polish group. Then G will have bounded geometry if and only if G is generated by a coarsely bounded set A with the property that $A^2 \subseteq FA$ for some finite F. Indeed, if G has bounded geometry, then we may find some gauge A generating G. Because also A^2 is coarsely bounded, we have $A^2 \subseteq FA$ for some finite subset $F \subseteq G$. Conversely, if A is a coarsely bounded generating set so that $A^2 \subseteq FA$ for some finite F, then also $(\overline{A})^2 \subseteq \overline{A^2} \subseteq \overline{FA} = F\overline{A}$ and so $\overline{A}^n \subseteq F^{n-1}\overline{A}$. As the $\overline{A}^n$ form a basis for the ideal of coarsely bounded sets in G, we find that $\overline{A}$ covers any coarsely bounded set by a finite number of left-translates.

Now recall that a Polish group G is locally bounded if and only if it admits a coarsely proper metric. Therefore, if it has bounded geometry, it has a coarsely proper metric d and, after rescaling this, we can suppose that the open unit ball is a gauge for G.

Definition 5.7 A compatible left-invariant metric d on a topological group G is a *gauge metric* if it is coarsely proper and the open unit ball is a gauge for G.

A natural and immediate concern is, of course, to determine the permanence properties of the class of Polish groups of bounded geometry, for example, preservation under quotients and group extensions. In other words, suppose

$$1 \to K \to G \to H \to 1$$

is a short exact sequence of Polish groups, does G have bounded geometry if and only if K and H do?

Note first that the universal Polish group $\mathrm{Homeo}([0,1]^{\mathbb{N}})$ has bounded geometry in virtue of being coarsely bounded in itself [77]. Since there are

Polish group groups without bounded geometry, this shows that the class of groups with bounded geometry is not closed under passing to arbitrary closed subgroups. Of course, by Lemma 5.3, it is closed under passing to coarsely embedded closed subgroups. Also, the following implication holds.

Lemma 5.8 *Suppose* $G \xrightarrow{\pi} H$ *is a continuous epimorphism between Polish groups. Then, if G has bounded geometry, so does H.*

Proof By Lemma 5.6, G must be locally bounded. Therefore, by Proposition 4.15, the map π admits a modest section ϕ. Now pick a gauge A for G. Then, if $B \subseteq H$ is coarsely bounded, so is $\phi[B]$, whence $\phi[B] \subseteq FA$ for some finite subset $F \subseteq G$. It therefore follows that

$$B = \pi \circ \phi[B] \subseteq \pi[FA] = \pi[F] \cdot \pi[A],$$

which shows that $\pi[A]$ is a gauge for H. □

5.2 Dynamic and Geometric Characterisations of Bounded Geometry

Postponing for the moment the presentation of specific examples of groups of bounded geometry, we instead give alternate descriptions of these and thus point to how they might arise. The setting we will consider here is continuous actions $G \curvearrowright X$ of a topological group G on a locally compact Hausdorff space X. We begin by introducing some basic terminology.

Definition 5.9 (Coarsely proper, modest and cocompact actions) Assume that $G \curvearrowright X$ is a continuous action of a topological group on a locally compact Hausdorff space.

(1) The action is *coarsely proper* if the set $\{g \in G \mid g \cdot K \cap K \neq \emptyset\}$ is coarsely bounded in G for all compact subsets $K \subseteq X$.
(2) The action is *modest* if $\overline{B \cdot K}$ is compact for all compact $K \subseteq X$ and coarsely bounded sets $B \subseteq G$.
(3) The action is *cocompact* if $X = G \cdot K$ for some compact $K \subseteq X$.

Observe that, if G is locally compact σ-compact and thus the coarsely bounded sets in G are just the relatively compact sets, then a continuous action $G \curvearrowright X$ as above is always modest. Thus, modesty is really only a restriction for groups beyond the locally compact σ-compact ones. Also, an equivalent way of stating that an action is coarsely proper is just that, for every compact set $K \subseteq X$, the set

$$E_K = \{(g, f) \in G \times G \mid gK \cap fK \neq \emptyset\}$$

is a coarse entourage in G.

Recall that a compact Hausdorff space K has a unique compatible uniformity, that is, a unique uniformity $\mathcal{U}$ inducing the topology on K. Thus, one has an unambiguous notion of Cauchy nets in K. On the other hand, in a locally compact Hausdorff space X, one has an unambiguous notion of nets tending to infinity, namely, $x_i \to \infty$ if the x_i eventually leave every compact set in X. Still this latter does not appear to arise from any canonical coarse structure on X. Nevertheless, the following proposition, which also shows that the notions of coarse properness and modesty are appropriate for our setting, indicates how certain actions on locally compact spaces can give rise to a coarse structure on G.

Proposition 5.10 *Suppose $G \curvearrowright X$ is a modest continuous action of a topological group G on a locally compact Hausdorff space X. Then the following are equivalent.*

(1) *The action $G \curvearrowright X$ is coarsely proper.*
(2) *The sets*

$$E_K = \{(g, f) \in G \times G \mid gK \cap fK \neq \emptyset\}$$

form a basis for the left-coarse structure $\mathcal{E}_L$ as K varies over compact subsets of X.

Proof As observed above, the action is coarsely proper if and only if the sets

$$E_K = \{(g, f) \in G \times G \mid gK \cap fK \neq \emptyset\},$$

with K compact, are coarse entourages in G. In particular, (2) implies (1).

For the implication (1)$\Rightarrow$(2), suppose that the action is modest and coarsely proper. To show (2), assume also that $A \subseteq G$ is a coarsely bounded set and consider the coarse entourage

$$E_A = \{(g, f) \in G \times G \mid g^{-1}f \in A\}.$$

We must show that $E_A \subseteq E_K$ for some compact $K \subseteq X$. So, pick any $x \in X$ and let $K = \{x\} \cup \overline{A \cdot x}$. As the action is modest, K is compact. Also,

$$g^{-1}f \in A \implies g^{-1}f \cdot x \in K \implies gK \cap fK \neq \emptyset,$$

i.e., $E_A \subseteq E_K$ as required. $\square$

Note that, when G is locally compact σ-compact, then, as modesty of the action is automatic, we see that a continuous action $G \curvearrowright X$ is coarsely proper exactly when the sets E_K form a basis for $\mathcal{E}_L$.

We observe also that without either modesty or coarse properness of the action, it is not clear that the composition

$$E_K \circ E_K$$

is contained in any E_C with $C \subseteq X$ compact and so the sets E_K may not form a basis for any coarse structure on G.

We shall now characterise cocompactness of the action in the context of modesty and coarse properness.

Lemma 5.11 *Suppose $G \curvearrowright X$ is a continuous cocompact action of a topological group G on a locally compact Hausdorff space X and assume that, for some $x \in X$ and every subset $A \subseteq G$,*

$$A \cdot x \text{ is relatively compact} \iff A \text{ is coarsely bounded.}$$

Then G has bounded geometry.

Proof Let $K \subseteq X$ be a compact set with $G \cdot K = X$ and $x \in K$. As K is compact and X locally compact, we may find a relatively compact open set $U \supseteq K$. Let also $A = \{g \in G \mid gx \in U\}$, which is coarsely bounded.

Assume that $B \subseteq G$ is coarsely bounded. Then $\overline{Bx}$ is compact, so, as $X = G \cdot K = G \cdot U$, there is a finite set $F \subseteq G$ with $\overline{Bx} \subseteq F \cdot U$. But then, for $g \in B$, there is $f \in F$ with $gx \in fU$, i.e., $f^{-1}g \in A$, showing that $B \subseteq FA$. So A covers every coarsely bounded set by finitely many left-translates and G has bounded geometry. □

Suppose that $G \curvearrowright X$ is a continuous action of a topological group G on a locally compact Hausdorff space X and $Y \subseteq X$ is a closed G-invariant closed subspace. Then we see that each of the properties of modesty, coarse properness and cocompactness pass from the action $G \curvearrowright X$ to the action $G \curvearrowright Y$. Thus, given an action $G \curvearrowright X$ with all three properties, if we let $Y = \overline{G \cdot x}$ for some $x \in X$, we obtain a new action $G \curvearrowright Y$ with all three properties and, furthermore, having a dense orbit.

Proposition 5.12 *Let $G \curvearrowright X$ be a modest and coarsely proper continuous action of a topological group on a locally compact Hausdorff space and assume that there is a dense orbit. Then G has bounded geometry if and only if the action is cocompact.*

Proof Pick some $x \in X$ with a dense orbit. Assume first that G has bounded geometry and fix a gauge A for G. Set $K = \overline{A \cdot x}$, which is compact by modesty of the action. Now, suppose that $y \in X$ is any point and fix some compact set N containing x and that is a neighbourhood of y. Let

$$B = \{g \in G \mid gx \in N\},$$

which is coarsely bounded, as it is contained in the coarsely bounded set

$$\{g \in G \mid g \cdot N \cap N \neq \emptyset\}.$$

There are therefore $f_1, \ldots, f_n \in G$ so that $B \subseteq \bigcup_{i=1}^n f_i A$. Also, as the orbit $G \cdot x$ is dense in X, we have

$$y \in \overline{B \cdot x} \subseteq \overline{\bigcup_{i=1}^n f_i A \cdot x} = \bigcup_{i=1}^n \overline{f_i A \cdot x} = \bigcup_{i=1}^n f_i \overline{A \cdot x} = \bigcup_{i=1}^n f_i K$$

and so $y \in G \cdot K$. This shows that the action is cocompact.

Conversely, assume that the action is cocompact. To see that G has bounded geometry, we verify the hypotheses of Lemma 5.11 for the chosen point x. Now, if $B \subseteq G$ is any coarsely bounded set, then $B \cdot x$ is relatively compact by modesty of the action. Conversely, if $B \subseteq G$ is a subset so that $B \cdot x$ is relatively compact, then $C = \overline{B \cdot x} \cup \{x\}$ is compact and

$$g \cdot C \cap C \neq \emptyset$$

for all $g \in B$. Because the action is coarsely proper, it follows that B is coarsely bounded in G. Thus, by Lemma 5.11, G has bounded geometry. $\square$

Recall that, if G is a Polish group with a compatible left-invariant metric d, the Roelcke uniformity on G is that induced by the metric

$$d_\wedge(g, f) = \inf_{h \in G} d(g, h) + d(h^{-1}, f^{-1}).$$

Also, a subset $A \subseteq G$ is Roelcke pre-compact if it is $d_\wedge$-totally bounded or, alternatively, if, for every identity neighbourhood $V \subseteq G$, there is a finite set $F \subseteq G$ so that $A \subseteq VFV$.

Now, a Polish group G is locally Roelcke precompact if it has a Roelcke pre-compact identity neighbourhood. For these, the fundamental result by J. Zielinski, Theorem 3.11, states that a subset is coarsely bounded if and only if it is Roelcke pre-compact and so, in particular, the Roelcke pre-compact sets are closed under multiplication. Moreover, the completion $X = \overline{(G, d_\wedge)}$ of a locally Roelcke pre-compact Polish group with respect to its Roelcke uniformity is in fact locally compact.

Before stating the next result, we must recall that a metric space (X, d) is *proper* if all closed sets of finite diameter are compact.

Lemma 5.13 *Suppose G is a locally Roelcke pre-compact Polish group and let X be the Roelcke completion of G. Then the left- and right-shift actions*

$\lambda\colon G \curvearrowright G$ and $\rho\colon G \curvearrowright G$ have unique extensions to commuting, coarsely proper, modest, continuous actions $\lambda\colon G \curvearrowright X$ and $\rho\colon G \curvearrowright X$.

Proof It is well known and easy to check that the two shift actions extend uniquely to commuting continuous actions on the completion X, so we must verify that they are modest and coarsely proper. We do only the argument for the left-shift action as the argument for ρ is symmetric.

To see that the action is modest, suppose $K \subseteq X$ is compact and $B \subseteq G$ is coarsely bounded. Then there is a relatively compact open set $U \subseteq X$ containing K, whence, as G is dense in X, we have $K \subseteq \overline{U \cap G}$. Moreover, because $U \cap G$ is Roelcke pre-compact in G, it must be coarsely bounded, whereby $B \cdot (U \cap G)$ is coarsely bounded and thus Roelcke pre-compact. As

$$B \cdot K \subseteq B \cdot \overline{U \cap G} \subseteq \overline{B \cdot (U \cap G)},$$

also $B \cdot K$ is relatively compact in X.

Similarly, to see that the action is coarsely proper, let $K \subseteq X$ be compact and pick a relatively compact open set $U \subseteq X$ containing K. Assume $f \in G$ and $fK \cap K \neq \emptyset$. Then $fU \cap U \neq \emptyset$, so, as $U \cap G$ is dense in U, we have $f \cdot (U \cap G) \cap U \neq \emptyset$ and thus also $f \cdot (U \cap G) \cap (U \cap G) \neq \emptyset$. In other words, $f \in (U \cap G)(U \cap G)^{-1}$, showing that

$$\{f \in G \mid fK \cap K \neq \emptyset\} \subseteq (U \cap G)(U \cap G)^{-1}.$$

As the latter set is coarsely bounded in G, the action $\lambda\colon G \curvearrowright X$ is coarsely proper. $\qquad\square$

Theorem 5.14 *The following conditions are equivalent for a Polish group G.*

(1) *G has bounded geometry,*
(2) *G is coarsely equivalent to a metric space of bounded geometry,*
(3) *G is coarsely equivalent to a proper metric space,*
(4) *G admits a continuous, coarsely proper, modest and cocompact action $G \curvearrowright X$ on a locally compact metrisable space X,*
(5) *G is locally bounded and every continuous, coarsely proper and modest action $G \curvearrowright X$ on a locally compact Hausdorff space X with a dense orbit is cocompact.*

Proof That (1) and (2) are equivalent follow immediately from the fact that bounded geometry is invariant under coarse equivalences and that every Polish group of bounded geometry is locally bounded, Lemma 5.6, and hence has metrisable coarse structure.

Suppose that G has bounded geometry and let d be a gauge metric for G. Let also X be a maximal 2-discrete subset of G, i.e., maximal so that $d(x, y) \geqslant 2$ for all $x \neq y$ in X. Note that, if $\beta > 0$ is a fixed diameter, we may pick a finite set $F \subseteq G$ so that the ball $B_d(3\beta) = \{x \in G \mid d(x, 1) < 3\beta\}$ is contained in $F \cdot B_d(1)$, whence every set of diameter at most β is contained in a single left-translate of $F B_d(1)$. However, because X is 2-discrete, two distinct points of X cannot belong to the same left-translate of $B_d(1)$, showing that subsets of X of diameter $\leqslant \beta$ have cardinality at most $|F|$.

It follows that (X, d) is a uniformly locally finite and thus proper metric space, which, being cobounded in G, is coarsely equivalent to G. Thus (1) implies (3).

Conversely, assume that G is coarsely equivalent to a proper metric space (X, d), which, by picking a discrete subset, we may assume to be locally finite, i.e., finite-diameter subsets are finite. Fix a coarse equivalence $(X, d) \xrightarrow{\phi} G$. Because $\phi[X]$ is cobounded in G, there is a symmetric coarsely bounded set $1 \in A \subseteq G$ so that $G = \phi[X] \cdot A$.

We claim that A is a gauge for G and thus that G is of bounded geometry. Indeed, suppose $B \subseteq G$ is coarsely bounded. Then, as ϕ is a coarse equivalence and (X, d) is locally finite, the set

$$\phi^{-1}(BA) = \{x \in X \mid \phi(x)A \cap B \neq \emptyset\}$$

is bounded and thus finite. But $B \subseteq \phi[X] \cdot A$, so actually $B \subseteq \phi[\phi^{-1}(BA)] \cdot A$, showing that B is covered by finitely many left-translates of A. Thus (3) implies (1).

The implications (4)$\Rightarrow$(1)$\Rightarrow$(5) follow directly from Proposition 5.12 and Lemma 5.6, so let us consider the implication (5)$\Rightarrow$(4). Assume that (5) holds. By Theorem 3.40, we may assume that G is a coarsely embedded closed subgroup of the isometry group $\mathrm{Isom}(\mathbb{U})$ of the Urysohn metric space. As the latter is locally Roelcke pre-compact by Theorem 3.10, Lemma 5.13 gives us a coarsely proper and modest continuous action

$$\mathrm{Isom}(\mathbb{U}) \curvearrowright X,$$

where X is the locally compact Roelcke completion of $\mathrm{Isom}(\mathbb{U})$. Note that the action $G \curvearrowright X$ is automatically modest and, because G is coarsely embedded in $\mathrm{Isom}(\mathbb{U})$, it is also coarsely proper. Letting $Y = \overline{G \cdot x}$ for some $x \in X$, we obtain a modest, coarsely proper continuous action $G \curvearrowright Y$ on a locally compact Hausdorff space with a dense orbit. It thus follows from (5) that the action is also cocompact, i.e., that (4) holds. $\square$

5.3 Examples

As noted earlier, the obvious example of a Polish group with bounded geometry is a locally compact group. Also, a coarsely bounded group such as S_∞ or $\mathrm{Homeo}(\mathbb{S}^n)$ is automatically of bounded geometry because the entire group is a gauge for itself. However, apart from these specimens and simple algebraic constructs over these, there are far more interesting examples. We begin with a simple sufficient criterion for having bounded geometry.

Proposition 5.15 *Suppose G is a Polish group admitting a conjugacy invariant Roelcke pre-compact identity neighbourhood U. Then G has bounded geometry.*

Proof By Theorem 3.11, in a locally Roelcke precompact Polish group, the coarsely bounded sets and the Roelcke precompact sets coincide. In particular, if A is any coarsely bounded subset of G, then A is Roelcke pre-compact and hence, for some finite subset $F \subseteq G$, we have

$$A \subseteq UFU \subseteq FUU = F \cdot U^2,$$

where $UF = FU$ follows from the conjugacy invariance of U. But this shows that every coarsely bounded set is covered by finitely many left-translates of the specific coarsely bounded set U^2 and hence that G has bounded geometry. $\square$

Example 5.16 ($\mathrm{Homeo}_{\mathbb{Z}}(\mathbb{R})$ has bounded geometry) Recall from Example 3.12 the group $\mathrm{Homeo}_{\mathbb{Z}}(\mathbb{R})$ of orientation-preserving homeomorphisms of the real line commuting with integral shifts. There, it is also shown that the identity neighbourhood

$$V = \{f \in \mathrm{Homeo}_{\mathbb{Z}}(\mathbb{R}) \mid -2 < f(0) < 2\}$$

is Roelcke pre-compact.

We now claim that $\mathrm{Homeo}_{\mathbb{Z}}(\mathbb{R})$ admits a conjugacy-invariant Roelcke pre-compact identity neighbourhood. To see this, let

$$W = \{f \in \mathrm{Homeo}_{\mathbb{Z}}(\mathbb{R}) \mid -1 < f(0) < 1\}.$$

Then, for any $g \in \mathrm{Homeo}_{\mathbb{Z}}(\mathbb{R})$ and $f \in W$, choose $k \in \mathbb{Z}$ so that $k \leqslant g(0) < k+1$, whereby

$$k-1 < k+f(0) = f(k) \leqslant fg(0) < f(k+1) = f(0)+k+1 < k+2$$

and so

$$-2 < g^{-1}(k+1) - 2 = g^{-1}(k-1) < g^{-1}fg(0) < g^{-1}(k+2)$$
$$= g^{-1}(k) + 2 \leqslant 2.$$

In other words, if $g \in \mathrm{Homeo}_{\mathbb{Z}}(\mathbb{R})$ and $f \in W$, then $g^{-1}fg \in V$, showing that the set

$$W^{\mathrm{Homeo}_{\mathbb{Z}}(\mathbb{R})} = \{xyx^{-1} \mid y \in W \text{ and } x \in \mathrm{Homeo}_{\mathbb{Z}}(\mathbb{R})\}$$

is contained in the Roelcke pre-compact set V. Thus, $W^{\mathrm{Homeo}_{\mathbb{Z}}(\mathbb{R})}$ is a Roelcke pre-compact conjugacy-invariant identity neigbourhood in $\mathrm{Homeo}_{\mathbb{Z}}(\mathbb{R})$ and hence, by Proposition 5.15, this latter group has bounded geometry.

One more noteworthy thing is that the left- and right-coarse structures agree on $\mathrm{Homeo}_{\mathbb{Z}}(\mathbb{R})$. To see this, note that, because $U = W^{\mathrm{Homeo}_{\mathbb{Z}}(\mathbb{R})}$ is conjugacy invariant, so are all its powers U^n. Now, because $\mathrm{Homeo}_{\mathbb{Z}}(\mathbb{R})$ is connected, any coarsely bounded subset A must be contained in some U^n, whereby also its set of conjugates $A^{\mathrm{Homeo}_{\mathbb{Z}}(\mathbb{R})}$ is contained in U^n and thus is coarsely bounded. By Proposition 3.30, we find that $\mathcal{E}_L = \mathcal{E}_R$ on $\mathrm{Homeo}_{\mathbb{Z}}(\mathbb{R})$.

Although Example 5.16 establishes that $\mathrm{Homeo}_{\mathbb{Z}}(\mathbb{R})$ has bounded geometry, it is equally important to identify the specific geometry.

Proposition 5.17 *The group* $\mathrm{Homeo}_{\mathbb{Z}}(\mathbb{R})$ *is quasi-isometric to* $\mathbb{Z}$ *and is thus of bounded geometry. Moreover, the action*

$$\mathrm{Homeo}_{\mathbb{Z}}(\mathbb{R}) \curvearrowright \mathbb{R}$$

is transitive, modest and coarsely proper.

Proof We show that the quotient map $\mathrm{Homeo}_{\mathbb{Z}}(\mathbb{R}) \xrightarrow{\pi} \mathrm{Homeo}_+(\mathbb{S}^1)$ admits a Borel measurable section $\mathrm{Homeo}_+(\mathbb{S}^1) \xrightarrow{\phi} \mathrm{Homeo}_{\mathbb{Z}}(\mathbb{R})$ with finite defect

$$\Delta = \{\phi(gh)^{-1}\phi(g)\phi(h) \mid g, h \in \mathrm{Homeo}_+(\mathbb{S}^1)\}^{\pm} \cup \{\mathrm{id}\}$$

in $\mathbb{Z}$. Indeed, for $h \in \mathrm{Homeo}_+(\mathbb{S}^1)$, simply let $\phi(h)$ be the unique lift of h to a homeomorphism of $\mathbb{R}$ so that $\phi(h)(0) \in [0, 1[$. Then $\phi(gh)^{-1}\phi(g)\phi(h)$ is an integral translation, i.e., belongs to the subgroup $\mathbb{Z}$, and, evaluating it at 0, we see that it has possible values 0 or 1. So $\Delta \subseteq \{-1, 0, 1\}$.

So ϕ is a Borel measurable section for the quotient map and, moreover, ϕ is a quasi-morphism. As $\mathrm{Homeo}_+(\mathbb{S}^1)$ is coarsely bounded and thus, a fortiori, locally bounded, it follows from Theorem 4.35 that the product

$\mathbb{Z} \times \mathrm{Homeo}_+(\mathbb{S}^1)$ is coarsely equivalent to $\mathrm{Homeo}_{\mathbb{Z}}(\mathbb{R})$ via the map $(n, h) \mapsto \tau_n \phi(h)$, where $\tau_n(x) = x + n$ is the translation by n. However, because $\mathrm{Homeo}_+(\mathbb{S}^1)$ is coarsely bounded, it is a trivial factor in $\mathbb{Z} \times \mathrm{Homeo}_+(\mathbb{S}^1)$ and hence the inclusion $n \mapsto \tau_n$ is a coarse equivalence between $\mathbb{Z}$ and $\mathrm{Homeo}_{\mathbb{Z}}(\mathbb{R})$.

Now, observe that the map $g \in \mathrm{Homeo}_{\mathbb{Z}}(\mathbb{R}) \overset{\sigma}{\longmapsto} \tau_n \in \mathbb{Z}$, where $g = \tau_n h$ is the unique decomposition of g into an integral translation τ_n and an element $h \in \mathrm{Homeo}_{\mathbb{Z}}(\mathbb{R})$ satisfying $h(0) \in [0, 1[$, is a coarse equivalence. In particular, a subset $A \subseteq \mathrm{Homeo}_{\mathbb{Z}}(\mathbb{R})$ is coarsely bounded if and only if the corresponding set $\sigma(A)$ of translations is finite. As elements $h \in \mathrm{Homeo}_{\mathbb{Z}}(\mathbb{R})$ satisfying $h(0) \in [0, 1[$ move every $x \in \mathbb{R}$ by at most distance 1, it easily follows that the action is both modest and coarsely proper. $\qquad \square$

A case of Polish groups of bounded geometry of special interest is when a gauge can be taken to be an open subgroup. Namely, suppose that V is a open subgroup of a Polish group G and let $\mathrm{Sym}(G/V)$ be the group of all permutations of the countable homogeneous space G/V of left V-cosets equipped with the *permutation group topology*, i.e., by declaring pointwise stabilisers to be open. Set $G \overset{\pi}{\longrightarrow} \mathrm{Sym}(G/V)$ to be the homomorphism corresponding to the left-translation action of G on G/V. Recall that the *commensurator* of V in G is the subgroup defined by

$$\mathrm{Comm}_G(V) = \{g \in G \mid [V : V \cap gVg^{-1}] < \infty \text{ and } [V : V \cap g^{-1}Vg] < \infty\}$$

and note that $V \leqslant \mathrm{Comm}_G(V) \leqslant G$.

Theorem 5.18 *The following conditions are equivalent for a coarsely bounded subgroup V of a Polish group G.*

(1) *V is a gauge for G and hence G has bounded geometry,*
(2) *$G = \mathrm{Comm}_G(V)$,*
(3) *every double coset VgV is a finite union of left V-cosets,*
(4) *the homomorphism $G \overset{\pi}{\longrightarrow} \mathrm{Sym}(G/V)$ induced by the left-translation action of G on the homogeneous space G/V is a coarse equivalence between G and a locally compact subgroup $\overline{\pi[G]} \leqslant \mathrm{Sym}(G/V)$.*

Proof (2)$\Leftrightarrow$(3) We begin by observing that the index $[V : V \cap gVg^{-1}]$ equals the number of left-cosets of V contained in VgV. Therefore, G equals the commensurator $\mathrm{Comm}_G(V)$ if and only if every double coset VgV is a union of finitely many left V-cosets.

(1)$\Leftrightarrow$(3) Now, every coarsely bounded set in G can be covered by some $(FV)^n$ with $F \subseteq G$ finite and $n \geqslant 1$. So, for V to be a gauge for G, it suffices by an inductive argument to check that each double coset VgV is a union of finitely many left V-cosets. Conversely, if V is a gauge, then as VgV is coarsely bounded it must be a union finitely many left V-cosets.

(2)$\Rightarrow$(4) Now, if $G = \mathrm{Comm}_G(V)$, then $\overline{\pi[G]}$ is a locally compact subgroup of $\mathrm{Sym}(G/V)$. Indeed, if $U \leqslant \mathrm{Sym}(G/V)$ denotes the pointwise stabiliser of the coset $1V$, then $\pi^{-1}(U) = V$ and the V-orbit of any coset gV is just the collection of V-cosets contained in VgV, which is finite by the assumption $G = \mathrm{Comm}_G(V)$. So this shows that $U \cap \overline{\pi[G]}$ has finite orbits on G/V and thus is a compact open subgroup of $\overline{\pi[G]}$. So $\overline{\pi[G]}$ is locally compact. Moreover, we claim that $G \xrightarrow{\pi} \overline{\pi[G]}$ is coarsely proper. For, if $B \subseteq G$ fails to be coarsely bounded, it must intersect infinitely many distinct left-cosets of V, whereby $B \cdot V$ is an infinite subset of G/V and $\pi[B]$ cannot be relatively compact. So $G \xrightarrow{\pi} \overline{\pi[G]}$ is a coarsely proper continuous homomorphism whose image $\pi[G]$ is cocompact and, as $\overline{\pi[G]}$ is locally compact, also cobounded in $\overline{\pi[G]}$. This shows that, if $G = \mathrm{Comm}_G(V)$, then $G \xrightarrow{\pi} \overline{\pi[G]}$ is a coarse equivalence of G with a totally disconnected locally compact group.

(4)$\Rightarrow$(1) Conversely, suppose that $\overline{\pi[G]}$ is locally compact. Then there is a finite number of left-cosets $g_1 V, \ldots, g_n V$ so that the stabiliser of these orbits, namely

$$W = g_1 V g_1^{-1} \cap \cdots \cap g_n V g_n^{-1},$$

induces only finite orbits on G/V. This means that, for every $f \in G$, WfV is a finite union of left V-cosets. It follows that, if $F \subseteq G$ is finite and $n \geqslant 1$, there are finite sets $F_1, \ldots, F_n \subseteq G$ so that

$$(WF)^n \subseteq (WF)^{n-1} WFV \subseteq (WF)^{n-1} F_1 V \subseteq (WF)^{n-2} F_2 V \subseteq \cdots \subseteq F_n V.$$

As every coarsely bounded set is covered by some $(WF)^n$, we conclude that V covers every coarsely bounded set by finitely many left-translates, i.e., that V is a gauge for G. $\qquad\square$

In general though, even in the case of non-Archimedean Polish groups of bounded geometry, we should not expect to have open subgroups as gauges and must therefore develop other tools to analyse the structure of groups of bounded geometry.

Now, suppose $\mathbf{M}$ is some countable first-order structure and $\mathrm{Aut}(\mathbf{M})$ its group of automorphisms. We say that the tautological action

$$\mathrm{Aut}(\mathbf{M}) \curvearrowright \mathbf{M}$$

is *oligomorphic* if, for all $n \in \mathbb{N}$, the diagonal action

$$\mathrm{Aut}(\mathbf{M}) \curvearrowright \mathbf{M}^n$$

has only finitely many orbits. One often abuses language and simply says that Aut($\mathbf{M}$) is oligomorphic. When Aut($\mathbf{M}$) is given the permutation group topology, i.e., where the pointwise stabilisers of finite sets are declared open, then an oligomorphic automorphism group Aut($\mathbf{M}$) will be Roelcke pre-compact. Furthermore, by a well-known theorem [58] of E. Engeler, C. Ryll-Nardzewski and L. Svenonius, the countable first-order structures whose automorphism groups are oligomorphic are exactly the $\aleph_0$-*categorical* structures, that is, those structures that are characterised up to isomorphism by their first-order theory.

Example 5.19 (Automorphisms of the dense linear order) Let $\mathrm{Aut}_{\mathbb{Z}}(\mathbb{Q})$ be the group of order-preserving permutations of $\mathbb{Q}$ commuting with integral translations. With the permutation group topology, this is a non-Archimedean Polish group that we claim is coarsely equivalent to $\mathbb{Z}$ and thus, in particular, has bounded geometry.

First, let us note that the countable linear order $(\mathbb{Q}, <)$ is $\aleph_0$-categorical, whereby its automorphism group Aut($\mathbb{Q}$) is oligomorphic and hence Roelcke precompact. Because the identity neighbourhood

$$V = \{g \in \mathrm{Aut}_{\mathbb{Z}}(\mathbb{Q}) \mid g(0) = 0\} = \{g \in \mathrm{Aut}_{\mathbb{Z}}(\mathbb{Q}) \mid g(k) = k, \ \forall k \in \mathbb{Z}\}$$

in $\mathrm{Aut}_{\mathbb{Z}}(\mathbb{Q})$ is an open subgroup isomorphic to Aut($\mathbb{Q}$), we see that $\mathrm{Aut}_{\mathbb{Z}}(\mathbb{Q})$ is locally Roelcke pre-compact. Thus, by Theorem 3.11, the coarsely bounded and the Roelcke pre-compact sets coincide in $\mathrm{Aut}_{\mathbb{Z}}(\mathbb{Q})$.

We now show that the sets

$$W_m = \{f \in \mathrm{Aut}_{\mathbb{Z}}(\mathbb{Q}) \mid -m < f(0) < m\}$$

with $m \in \mathbb{N}$ form a basis for the ideal $\mathcal{CB}$ of coarsely bounded sets. On the one hand, if A is coarsely bounded, then A is Roelcke pre-compact, so we may find a finite set F so that $A \subseteq VFV$. Note now that, if $g \in VfV$, then $g(0) \in Vf(0)$ and so

$$\lfloor f(0) \rfloor \leqslant g(0) \leqslant \lceil f(0) \rceil.$$

In particular, this shows that VFV and hence also A must be contained in some set W_m for m large enough.

On the other hand, to see that W_m is itself coarsely bounded, let, for $q \in \mathbb{Q}$, τ_q denote the translation $\tau_q(x) = x + q$. Now, given $f \in W_m$, write $f(0) = k+r$ with $k \in \mathbb{Z}$ and $0 \leqslant r < 1$. Choose also $g \in V$ so that $g(\frac{1}{2}) = r$, whereby $f(0) = \tau_k g \tau_{\frac{1}{2}}(0)$, i.e.,

$$f \in \tau_k g \tau_{\frac{1}{2}} V \subseteq \tau_k V \tau_{\frac{1}{2}} V.$$

It follows that

$$W_m \subseteq \{\tau_k \mid -m \leqslant k < m\} \cdot V\tau_{\frac{1}{2}}V,$$

showing W_m to be Roelcke pre-compact.

Now, extend every $g \in \mathrm{Aut}_{\mathbb{Z}}(\mathbb{Q})$ to a homeomorphism of $\mathbb{R}$ and consider the resulting cocompact action

$$\mathrm{Aut}_{\mathbb{Z}}(\mathbb{Q}) \curvearrowright \mathbb{R}.$$

As the W_m form a basis for the ideal of coarsely bounded sets, we see that this action is coarsely proper, modest and cocompact, whereas $\mathbb{Z}$ is a cobounded subgroup of $\mathrm{Aut}_{\mathbb{Z}}(\mathbb{Q})$. In particular, $\mathrm{Aut}_{\mathbb{Z}}(\mathbb{Q})$ has bounded geometry.

We now aim to show that no open subgroup can be a gauge for $\mathrm{Aut}_{\mathbb{Z}}(\mathbb{Q})$. For this, suppose W is a coarsely bounded open subgroup of $\mathrm{Aut}_{\mathbb{Z}}(\mathbb{Q})$ and let

$$x = \sup\{w(0) \mid w \in W\}.$$

Because the orbit of 0 under the coarsely bounded set W is bounded, we see that x is finite. We remark also that x is fixed by W under the action $\mathrm{Aut}_{\mathbb{Z}}(\mathbb{Q}) \curvearrowright \mathbb{R}$. It follows that, if $g, f \in \mathrm{Aut}_{\mathbb{Z}}(\mathbb{Q})$ with $g(x) \neq f(x)$, then $gW \neq fW$. Note also that $\mathrm{Aut}_{\mathbb{Z}}(\mathbb{Q})$ acts transitively on $\mathbb{R} \setminus \mathbb{Q}$, because any irrational Dedekind cut in $\mathbb{Q}$ can be moved to any other. So x cannot be irrational, as otherwise W would have an uncountable index in $\mathrm{Aut}_{\mathbb{Z}}(\mathbb{Q})$, contradicting that it is open. Thus, W fixes the rational point x and therefore, for every finite set $F \subseteq \mathrm{Aut}_{\mathbb{Z}}(\mathbb{Q})$, we have that $FW(x)$ is a finite subset of $\mathbb{Q}$. But this shows that W cannot be a gauge for $\mathrm{Aut}_{\mathbb{Z}}(\mathbb{Q})$, because, for example, $V\tau_{\frac{1}{2}-x}$ is coarsely bounded whereas $V\tau_{\frac{1}{2}-x}(x) = \mathbb{Q} \cap]0, 1[$ is infinite and hence $V\tau_{\frac{1}{2}-x}$ cannot be covered by finitely many left-translates of W.

Summing up, we have a locally Roelcke pre-compact non-Archimedean Polish group $\mathrm{Aut}_{\mathbb{Z}}(\mathbb{Q})$ of bounded geometry, so that no open subgroup is a gauge for $\mathrm{Aut}_{\mathbb{Z}}(\mathbb{Q})$.

We can also construct a coarsely proper transitive isometric action of $\mathrm{Aut}_{\mathbb{Z}}(\mathbb{Q})$ on a countable connected graph by letting $\mathbb{Q}$ be the vertex set and connecting two vertices $x, y \in \mathbb{Q}$ by an edge if $|x - y| < 1$. The resulting graph is quasi-isometric to $\mathbb{Z}$, showing again by the Milnor–Schwarz Lemma, Theorem 2.77, that $\mathrm{Aut}_{\mathbb{Z}}(\mathbb{Q})$ is quasi-isometric to $\mathbb{Z}$.

Example 5.20 (Absolutely continuous homeomorphisms) Recall that a map $f: [0, 1] \to \mathbb{R}$ is *absolutely continuous* if, for every $\epsilon > 0$, there is $\delta > 0$ so that

$$\sum_k |f(y_k) - f(x_k)| < \epsilon$$

whenever $0 \leqslant x_1 < y_1 < x_2 < y_2 < \cdots < x_n < y_n \leqslant 1$ satisfies $\sum_k (y_k - x_k) < \delta$. Equivalently, f is absolutely continuous if f' exists almost everywhere, f' is Lebesgue integrable and $f(x) = f(0) + \int_0^x f'(t)\, dt$. W. Herndon [38] has investigated the subgroup $\mathrm{AC}^*_{\mathbb{Z}}(\mathbb{R})$ of $\mathrm{Homeo}_{\mathbb{Z}}(\mathbb{R})$ consisting of all h so that h and h^{-1} are absolutely continuous when restricted to $[0,1]$. This can be shown to be a Polish group in a finer topology and Herndon proved that the inclusion of $\mathbb{Z}$ into $\mathrm{AC}^*_{\mathbb{Z}}(\mathbb{R})$ is a coarse equivalence.

Whereas the above three examples, $\mathrm{Homeo}_{\mathbb{Z}}(\mathbb{R})$, $\mathrm{Aut}_{\mathbb{Z}}(\mathbb{Q})$ and $\mathrm{AC}^*_{\mathbb{Z}}(\mathbb{R})$ are very different topologically, they are quite similar algebraically and in their coarse structure. Namely, they are all obtained as central extensions by $\mathbb{Z}$ of coarsely bounded Polish subgroups of the homeomophism group of the circle and, in particular, they are all coarsely equivalent to $\mathbb{Z}$. Thus, apart from locally compact groups, it is not quite clear in which contexts Polish groups of bounded geometry arise. It would be especially interesting to obtain new topologically simple non-locally compact examples.

Problem 5.21 Find topologically simple, non-locally compact, non-coarsely bounded, Polish groups of bounded geometry.

Example 5.22 (An augmented Heisenberg group) The discrete Heisenberg group $H_3(\mathbb{Z})$ is the group of upper triangular matrices

$$\begin{pmatrix} 1 & a & c \\ 0 & 1 & b \\ 0 & 0 & 1 \end{pmatrix}$$

with integral coefficients a, b, c. As the matrix product gives

$$\begin{pmatrix} 1 & a & c \\ 0 & 1 & b \\ 0 & 0 & 1 \end{pmatrix} \cdot \begin{pmatrix} 1 & x & z \\ 0 & 1 & y \\ 0 & 0 & 1 \end{pmatrix} = \begin{pmatrix} 1 & a+x & c+z+ay \\ 0 & 1 & b+y \\ 0 & 0 & 1 \end{pmatrix},$$

it follows that $H_3(\mathbb{Z})$ may be written as a central extension of $\mathbb{Z}^2$ by $\mathbb{Z}$. Namely, if the cocycle $\omega \colon \mathbb{Z}^2 \times \mathbb{Z}^2 \to \mathbb{Z}$ is defined by $\omega((a,b),(x,y)) = ay$, then the Heisenberg group is isomorphic to the cartesian product $\mathbb{Z} \times \mathbb{Z}^2$ with a product skewed by ω,

$$(c,(a,b)) \cdot (z,(x,y)) = (c+z+\omega((a,b),(x,y)),(a+x,b+y)).$$

Conversely, recall from Section 4.4 of Chapter 4 that, if H is an arbitrary group and $\omega \colon H \times H \to Z(G)$ is a map into the centre of a group G satisfying the normalised cocycle equations

(1) $\omega(1,x) = \omega(x,1) = 1$,
(2) $\omega(y,z)\omega(x,yz) = \omega(x,y)\omega(xy,z)$,

then the formula

$$(g,x) \cdot (f,y) = (gf\omega(x,y),xy)$$

defines a group operation on $G \times H$ giving rise to an extension

$$G \xrightarrow{\iota} G \times_\omega H \xrightarrow{\pi} H$$

with embedding $\iota(g) = (g,1)$ and epimorphism $\pi(g,x) = x$.

We may therefore construct an augmented Heisenberg group $\mathbb{H}_3(\mathbb{Z})$ by defining the cocycle

$$\omega \colon \mathbb{Z}^2 \times \mathbb{Z}^2 \to Z\big(\mathrm{Homeo}_\mathbb{Z}(\mathbb{R})\big) = \mathbb{Z}$$

as before, i.e., $\omega\big((a,b),(x,y)\big) = \tau_{ay}$, and then setting

$$\mathbb{H}_3(\mathbb{Z}) = \mathrm{Homeo}_\mathbb{Z}(\mathbb{R}) \times_\omega \mathbb{Z}^2.$$

Observe that, as ω is trivially continuous, $\mathbb{H}_3(\mathbb{Z}) = \mathrm{Homeo}_\mathbb{Z}(\mathbb{R}) \times_\omega \mathbb{Z}^2$ is a topological group in the product topology. Also, identifying $\mathbb{Z}$ with the subgroup of integral translations in $\mathrm{Homeo}_\mathbb{Z}(\mathbb{R})$, we may canonically identify $H_3(\mathbb{Z}) = \mathbb{Z} \times_\omega \mathbb{Z}^2$ with a subgroup of $\mathbb{H}_3(\mathbb{Z}) = \mathrm{Homeo}_\mathbb{Z}(\mathbb{R}) \times_\omega \mathbb{Z}^2$.

We claim that the inclusion map of $H_3(\mathbb{Z})$ into $\mathbb{H}_3(\mathbb{Z})$ is a quasi-isometry between these groups. To see this, let

$$B = \big\{h \in \mathrm{Homeo}_\mathbb{Z}(\mathbb{R}) \mid 0 \leqslant h(0) < 1\big\}$$

and observe that, because B is coarsely bounded in $\mathrm{Homeo}_\mathbb{Z}(\mathbb{R})$, its homomorphic image $B \times \{\mathbf{0}\}$ is coarsely bounded in $\mathrm{Homeo}_\mathbb{Z}(\mathbb{R}) \times_\omega \mathbb{Z}^2$. As, moreover, $\mathbb{Z} \cdot B = \mathrm{Homeo}_\mathbb{Z}(\mathbb{R})$, we see that $H_3(\mathbb{Z}) \cdot (B \times \{\mathbf{0}\}) = \mathbb{H}_3(\mathbb{Z})$ and thus that $H_3(\mathbb{Z})$ is cobounded in $\mathbb{H}_3(\mathbb{Z})$.

Now, by Corollary 4.25, a subset $A \subseteq H_3(\mathbb{Z}) = \mathbb{Z} \times_\omega \mathbb{Z}^2$ is coarsely bounded in $\mathbb{H}_3(\mathbb{Z}) = \mathrm{Homeo}_\mathbb{Z}(\mathbb{R}) \times_\omega \mathbb{Z}^2$ if and only if the projection $\mathrm{proj}_{\mathrm{Homeo}_\mathbb{Z}(\mathbb{R})}(A)$ is coarsely bounded in $\mathrm{Homeo}_\mathbb{Z}(\mathbb{R})$, whereas $\mathrm{proj}_{\mathbb{Z}^2}(A)$ is a finite subset of $\mathbb{Z}^2$. However, in this case, as $\mathrm{proj}_{\mathrm{Homeo}_\mathbb{Z}(\mathbb{R})}(A)$ is a subset of the coarsely embedded subgroup $\mathbb{Z}$ of $\mathrm{Homeo}_\mathbb{Z}(\mathbb{R})(A)$, also $\mathrm{proj}_{\mathrm{Homeo}_\mathbb{Z}(\mathbb{R})}(A)$ is finite. In other words, a subset $A \subseteq H_3(\mathbb{Z})$ is coarsely bounded in $\mathbb{H}_3(\mathbb{Z})$ if and only if it is finite, i.e., if and only if it is coarsely bounded in $H_3(\mathbb{Z})$.

It follows that $H_3(\mathbb{Z})$ is a coarsely embedded, cobounded subgroup of $\mathbb{H}_3(\mathbb{Z})$ and thus that the inclusion map is a coarse equivalence. Because $H_3(\mathbb{Z})$ is finitely generated, the map is, in fact, a quasi-isometry.

Thus, $\mathbb{H}_3(\mathbb{Z})$ is a Polish group of bounded geometry, quasi-isometric to the discrete Heisenberg group $H_3(\mathbb{Z})$.

As the above examples of Polish groups of bounded geometry are all coarsely equivalent to locally compact groups, it is natural to wonder whether this is necessarily the case.

Problem 5.23 Is every Polish group of bounded geometry coarsely equivalent to a locally compact second countable group?

Several questions of similar nature have been studied in the literature. Namely, A. Eskin, D. Fisher and K. Whyte [29], answering a question of W. Woess, established the first examples of locally compact second countable groups not coarsely equivalent to any countable discrete group. Moreover, their examples were of two kinds, Lie groups and totally disconnected locally compact groups. Thus, not every locally compact group G has a discrete model, but one could ask whether it always has a combinatorial model, i.e., whether it is coarsely equivalent to a vertex transitive, connected, locally finite graph $\mathbb{X}$ or, equivalently, to its automorphism group Aut($\mathbb{X}$). However, even this turns out not to be the case (examples can be found in Section 6.C of [21]).

Let us nevertheless note the following simple fact.

Proposition 5.24 *Let G be a monogenic Polish group. Then G is quasi-isometric to a vertex transitive countable graph.*

Proof Suppose $V \ni 1$ is a coarsely bounded, symmetric open generating set for G. Let also $\Gamma \leqslant G$ be a countable dense subgroup of G and define a graph $\mathbb{X}$ by setting

$$\text{Vert } \mathbb{X} = \Gamma$$

and, for $x, y \in \Gamma$, $x \neq y$,

$$(x, y) \in \text{Edge } \mathbb{X} \Leftrightarrow x \in yV.$$

Then Edge $\mathbb{X}$ is a symmetric relation on the vertex set Γ, which is invariant under the evidently transitive, left-shift action of Γ on itself. So $\mathbb{X}$ is a vertex-transitive graph.

Let also $\rho_{\mathbb{X}}$ denote the shortest-path metric on $\mathbb{X}$ and ρ_V denote the word metric on G given by the generating set V. We claim that $\rho_{\mathbb{X}}$ is simply the restriction of ρ_V to Γ. Indeed, the inequality

$$\rho_V \leqslant \rho_{\mathbb{X}}$$

is obvious. For the other direction, fix $x, y \in \Gamma$ and suppose that $\rho_V(x, y) \leqslant k$, i.e., $y \in xV^k$. Then there are $v_1, \ldots, v_k \in V$ so that $y = xv_1 \cdots v_k$. However, we may not have $xv_1 \cdots v_i \in \Gamma$, so this may not give us a path in $\mathbb{X}$. Instead, note that because $yV^{k-1} \cap xV \neq \emptyset$, using the density of Γ in G, there is some $z_1 \in \Gamma \cap yV^{k-1} \cap xV$. Similarly, choose $z_2 \in yV^{k-2} \cap z_1 V$, etc. This produces a path $x, z_1, z_2, \ldots, z_{k-1}, y$ in $\mathbb{X}$, showing that $\rho_{\mathbb{X}}(x, y) \leqslant k$, as required.

Since G is quasi-isometric to the metric space (G, ρ_V) and Γ is a cobounded subset of (G, ρ_V), it follows that G is quasi-isometric to $\mathbb{X}$. $\square$

5.4 Topological Couplings

We now arrive at the generalisation of Gromov's Theorem on topological couplings to groups of bounded geometry. Observe first that, by Proposition 5.10 and Theorem 5.14, the class of groups that allow cocompact actions on locally compact spaces faithfully representing their geometry are exactly those of bounded geometry. So the groups of bounded geometry are the widest possible setting for Gromov's Theorem.

5.4.1 Coarse Equivalence from Topological Couplings

Definition 5.25 A *topological coupling* of Polish groups G and F is a pair

$$G \curvearrowright X \curvearrowleft F$$

of commuting, coarsely proper, modest, cocompact continuous actions on a locally compact Hausdorff space X.

As mentioned earlier, the basic motivating example for this definition is the coupling

$$\mathbb{Z} \curvearrowright \mathbb{R} \curvearrowleft \mathrm{Homeo}_{\mathbb{Z}}(\mathbb{R}).$$

We are now ready to establish the easy direction of our theorem, namely, that topological couplings give coarse equivalences. However, we will actually prove a stronger result also incorporating coarse embeddings.

Proposition 5.26 *Let* $G \curvearrowright X \curvearrowleft F$ *be a pair of commuting, coarsely proper and modest continuous actions of Polish groups on a locally compact Hausdorff space. Then the following hold.*

(1) *If the action of F is cocompact, then G coarsely embeds into F.*
(2) *If both G and F act cocompactly, then G and F are coarsely equivalent.*

Proof Assume first that F acts cocompactly on X and fix a compact set $K \subseteq X$ with $X = F \cdot K$. Pick also a point $x \in X$. We choose a map $G \xrightarrow{\phi} F$ so that $g^{-1}x \in \phi(g)K$ for all $g \in G$. As the G and F actions commute, it follows that also $\phi(g)^{-1}x \in gK$. Recall that, by Proposition 5.10, the entourages

$$E_C = \{(f_1, f_2) \in F \times F \mid f_1 C \cap f_2 C \neq \emptyset\}$$

with C compact form a basis for the coarse structure on F. Similarly for G.

Let us first see that ϕ is bornologous. Indeed, suppose $B \ni 1$ is a coarsely bounded set in G. Then, if $g \in G$ and $b \in B$, we have $g^{-1}x \in \phi(g)K \subseteq \phi(g)\overline{BK}$ and

$$g^{-1}x = b(gb)^{-1}x \in b\phi(gb)K = \phi(gb)bK \subseteq \phi(gb)\overline{BK}.$$

As the G action is modest, $\overline{BK}$ is compact and so this shows that

$$(g, h) \in E_B \implies \phi(g)\overline{BK} \cap \phi(h)\overline{BK} \neq \emptyset \implies \big(\phi(g), \phi(h)\big) \in E_{\overline{BK}},$$

i.e., ϕ is bornologous. Conversely, to see that ϕ is expanding, assume that $A \ni 1$ is a coarsely bounded set in F. Suppose that $g, h \in G$ and that $\big(\phi(g), \phi(h)\big) \in E_A$, i.e., that $\phi(h) = \phi(g)a$ for some $a \in A$. Then $\phi(g)^{-1}x \in gK \subseteq g\overline{AK}$ and $a^{-1}\phi(g)^{-1} = \phi(h)^{-1} \in hK$, whereby

$$\phi(g)^{-1}x \in g\overline{AK} \cap ahK = g\overline{AK} \cap haK \subseteq g\overline{AK} \cap h\overline{AK}.$$

In other words,

$$\big(\phi(g), \phi(h)\big) \in E_A \implies (g, h) \in E_{\overline{AK}},$$

showing that ϕ is expanding and thus a coarse embedding of G into F.

Assume now that, in addition, the action by G is cocompact. Then the compact set K can be chosen so that both $X = F \cdot K$ and $X = G \cdot K$. We then define a map $F \xrightarrow{\psi} G$ so that $f^{-1}x \in \psi(g)K$, which makes ψ a coarse embedding of F into G, the argument being identical to the one for ϕ. To see that G and F are coarsely equivalent, it suffices to check that $\psi \circ \phi$ and $\phi \circ \psi$ are close to the identities on G and F respectively. But, for all $g \in G$,

$$\phi(g)^{-1}x \in gK \cap \psi(\phi(g))K,$$

and thus

$$\big(g, \psi(\phi(g))\big) \in E_K$$

which means that $\psi \circ \phi$ is close to the identity on G. The argument for $\phi \circ \psi$ is similar. $\square$

On the face of it, it is far from clear that the relation between Polish groups of having a topological coupling is an equivalence relation. The problem, of course, is transitivity. Namely, given two topological couplings

$$G \curvearrowright X \curvearrowleft F \curvearrowright Y \curvearrowleft H$$

of Polish groups G, F and H, how does one construct a topological coupling

$$G \curvearrowright Z \curvearrowleft H?$$

In the case of F being locally compact, one may let $Z = X \times_F Y$, i.e., the quotient of $X \times Y$ by the orbit equivalence relation given by the diagonal action of F. However, if F is only Polish, but not locally compact, this quotient Z is, in general, a pathological space and may not be Hausdorff. Still, as a consequence of Theorem 5.31, we will see that having a topological coupling is indeed an equivalence relation on the class of Polish groups of bounded geometry.

Problem 5.27 Suppose G, F and H are Polish groups of bounded geometry and

$$G \curvearrowright X \curvearrowleft F \curvearrowright Y \curvearrowleft H$$

are two pairs of topological couplings between G and F, respectively between F and H. Find a direct description of a topological coupling

$$G \curvearrowright Z \curvearrowleft H$$

between G and H.

5.4.2 Vietoris–Rips Complexes, Arens–Eells Spaces and Kantorovich Metrics

The construction of topological couplings will require us to introduce some new machinery, which is likely to be of use in many other contexts.

In the following, fix a set X and let $\mathbb{R}X$ denote the free $\mathbb{R}$-vector space over X. We may think of $\mathbb{R}X$ as the vector space of finitely supported functions $X \xrightarrow{\xi} \mathbb{R}$ or, alternatively, $\mathbb{R}X$ can be viewed as the space with basis $\{\mathbf{1}_x\}_{x \in X}$. Let also ΔX be the simplex in $\mathbb{R}X$ consisting of all finite convex combinations $\beta = \sum_{i=1}^{m} \lambda_i \mathbf{1}_{x_i}$ of basis vectors and $\mathbb{M}X$ be the linear subspace of $\mathbb{R}X$ consisting of *molecules*, that is, elements $m \in \mathbb{R}X$ with zero mean, $\sum_{x \in X} m(x) = 0$. Alternatively, $\mathbb{M}X$ can be described as the hyperplane in $\mathbb{R}X$ consisting of linear combinations of *atoms* $\mathbf{1}_x - \mathbf{1}_y$ with $x, y \in X$.

Assume now that X is given a metric d. Then the space $\mathbb{M}X$ of molecules can be equipped with the *Arens–Eells norm* $\|\cdot\|_{\text{Æ}}$ defined by

$$\|m\|_{\text{Æ}} = \inf\left(\sum_{i=1}^{n}|t_i|d(x_i,y_i)\;\middle|\;m = \sum_{i=1}^{n}t_i(\mathbf{1}_{x_i}-\mathbf{1}_{y_i})\right).$$

Observe that, by replacing the term $t_i(\mathbf{1}_{x_i}-\mathbf{1}_{y_i})$ with $-t_i(\mathbf{1}_{y_i}-\mathbf{1}_{x_i})$, one may suppose that all the t_i in the above expression are positive. Also, by a short argument based on the triangle inequality for d, one may restrict the attention to presentations $m = \sum_{i=1}^{n}t_i(\mathbf{1}_{x_i}-\mathbf{1}_{y_i})$, where all the x_i, y_i belong to the support of m. A fundamental fact (see [**93**]) is that this norm may alternatively be computed by

$$\|m\|_{\text{Æ}} = \sup\left(\sum_{x\in X}m(x)F(x)\;\middle|\;F:(X,d)\to\mathbb{R}\text{ is 1-Lipschitz}\right).$$

In particular, given $x,y\in X$, consider the 1-Lipschitz function $F(z)=d(z,y)$. Then, for $m = \mathbf{1}_x - \mathbf{1}_y$,

$$d(x,y)=\sum_{z\in X}m(z)F(z)\leqslant\|\mathbf{1}_x-\mathbf{1}_y\|_{\text{Æ}}.$$

On the other hand, $d(x,y)$ is clearly an upper bound on the norm $\|\mathbf{1}_x-\mathbf{1}_y\|_{\text{Æ}}$, so we find that

$$\|\mathbf{1}_x-\mathbf{1}_y\|_{\text{Æ}}=d(x,y)$$

for all $x,y\in X$.

If A and B are subsets of X, we write

$$\text{dist}_d(A,B)=\inf_{a\in A,b\in B}d(a,b).$$

If $A=\{a\}$ is just a singleton, we just write $\text{dist}_d(a,B)$ in place of $\text{dist}_d(A,B)$.

Finally, we define a metric $d_{\text{Æ}}$ on ΔX by letting $d_{\text{Æ}}(\alpha,\beta)=\|\alpha-\beta\|_{\text{Æ}}$ for $\alpha,\beta\in\Delta X$. The distance $d_{\text{Æ}}(\alpha,\beta)$ is also sometimes called the *Kantorovich* distance and measures the cost of an optimal transport between the manufacturers in $\text{supp}(\alpha)$ and the consumers in $\text{supp}(\beta)$, where the transport cost is proportional to the distance between manufacturers and consumers.

For example, given $\alpha,\beta\in\Delta X$, define a 1-Lipschitz function F on X by $F(x)=\text{dist}_d(x,\text{supp}(\beta))$. Then $F(x)=0$ for $x\in\text{supp}(\beta)$ and hence

$$\text{dist}_d\big(\text{supp}(\alpha),\text{supp}(\beta)\big) = \sum_{\alpha(x)>0} \alpha(x)\text{dist}_d\big(\text{supp}(\alpha),\text{supp}(\beta)\big)$$

$$\leqslant \sum_{\alpha(x)>0} \alpha(x)\text{dist}_d\big(x,\text{supp}(\beta)\big)$$

$$= \sum_{x\in X}(\alpha(x)-\beta(x))F(x)$$

$$\leqslant \|\alpha-\beta\|_{\text{Æ}}.$$

On the other hand, any representation $\alpha-\beta = \sum_{i=1}^{m}\lambda_i(\mathbf{1}_{x_i}-\mathbf{1}_{y_i})$ with $x_i \in \text{supp}(\alpha)$ and $y_i \in \text{supp}(\beta)$ shows that

$$\|\alpha-\beta\|_{\text{Æ}} \leqslant \text{diam}_d\big(\text{supp}(\alpha)\cup\text{supp}(\beta)\big).$$

Summing up, we have that

$$\text{dist}_d\big(\text{supp}(\alpha),\text{supp}(\beta)\big) \leqslant d_{\text{Æ}}(\alpha,\beta) \leqslant \text{diam}_d\big(\text{supp}(\alpha)\cup\text{supp}(\beta)\big)$$

for all $\alpha,\beta \in \Delta X$.

Suppose constants M and K are given. Then we define the *Vietoris–Rips complex* of X with constants M and K by

$$\text{VR}(X,M,K) = \big\{\beta \in \Delta X \mid \text{diam}_d\big(\text{supp}(\beta)\big) \leqslant K \text{ and } \big|\text{supp}(\beta)\big| \leqslant M\big\}.$$

In order to avoid any confusion about the metric d, we shall sometimes include this explicitly in the notation as $\text{VR}(X,d,M,K)$. Observe that $\text{VR}(X,M,K)$ is an M-dimensional complex contained in ΔX and thus it inherits the metric $d_{\text{Æ}}$ and the induced topology.

Now, suppose $G \curvearrowright X$ is a group action. Then we get an induced linear action $G \curvearrowright \mathbb{R}X$, namely, the regular representation, which for $\xi = \sum_{i=1}^{m} t_i\mathbf{1}_{x_i}$ and $g \in G$ is given by

$$(g\cdot\xi)(z) = \xi(g^{-1}z)$$

or alternatively

$$g\cdot\xi = \sum_{i=1}^{m} t_i\mathbf{1}_{gx_i}.$$

Clearly, the subsets $\mathbb{M}X$ and ΔX are both invariant under this action. Also, if G acts isometrically on X, then G acts linearly isometrically on $\big(\mathbb{M}X,\|\cdot\|_{\text{Æ}}\big)$ and affinely isometrically on $\big(\Delta X, d_{\text{Æ}}\big)$. Furthermore, in this case, the complex $\text{VR}(X,M,K)$ will be G-invariant.

In this context, our first lemma concerns continuous actions that are not necessarily by isometries.

Lemma 5.28 *Suppose $G \curvearrowright X$ is a continuous action by a topological group G on a proper metric space (X,d). Then, for all constants M, K, the induced map*

$$G \times \mathsf{VR}(X, M, K) \to \Delta X$$

is continuous.

Proof Fix some $g \in G$ and $\beta \in \mathsf{VR}(X, M, K)$. We show that the restricted action map $G \times \mathsf{VR}(X, M, K) \to \Delta X$ is continuous at (g, β). So let $\epsilon > 0$ be given. Then we must find a neighbourhood U of g and some $\delta > 0$ so that

$$d_{\!\mathit{Æ}}\big(g\beta, f\alpha\big) < \epsilon$$

for all $f \in U$ and $\alpha \in \mathsf{VR}(X, M, K)$ with $d_{\!\mathit{Æ}}(\beta, \alpha) < \delta$.

Because d is a proper metric, the set $C = \big(\mathsf{supp}(\beta)\big)_{K+1}$ is compact. Also, by continuity of the action $G \curvearrowright X$, for each $x \in C$, we can find some open $V_x \ni x$ and open $U_x \ni g$ so that

$$d(fy, gx) < \epsilon/4$$

for all $f \in U_x$ and $y \in V_x$. By compactness, we have $C \subseteq V_{x_1} \cup \cdots \cup V_{x_n}$ for some finite collection of $x_1, \ldots, x_n \in C$. Set $U = \bigcap_{i=1}^{n} U_{x_i}$. Then, if $y \in C$ and $f \in U$, we may pick i so that $y \in V_{x_i}$, whereby

$$d(fy, gy) \leqslant d(fy, gx_i) + d(gy, gx_i) < \frac{\epsilon}{4} + \frac{\epsilon}{4} = \frac{\epsilon}{2}.$$

Now, if $\alpha \in \mathsf{VR}(X, M, K)$ with $d_{\!\mathit{Æ}}(\alpha, \beta) < 1$, then also

$$\mathsf{dist}_d\big(\mathsf{supp}(\alpha), \mathsf{supp}(\beta)\big) \leqslant d_{\!\mathit{Æ}}(\alpha, \beta) < 1$$

and so, as $\mathsf{supp}(\alpha)$ has diameter at most K,

$$\mathsf{supp}(\alpha) \subseteq \big(\mathsf{supp}(\beta)\big)_{K+1} \subseteq C.$$

It thus follows from the above that $d_{\!\mathit{Æ}}\big(f\alpha, g\alpha\big) < \epsilon/2$ whenever $f \in U$ and $d_{\!\mathit{Æ}}(\alpha, \beta) < 1$.

As C is compact, the restriction of the transformation g to C is uniformly continuous and the maximal relative displacement

$$R = \sup_{x, y \in C} d(gx, gy)$$

is finite. So find $0 < \delta < \frac{4R}{\epsilon}$ small enough so that, for all $x, y \in C$,

$$d(x, y) < \delta \;\;\Rightarrow\;\; d(gx, gy) < \epsilon/4.$$

Assume now that $\alpha \in \mathsf{VR}(X, M, K)$ satisfies $d_{Æ}(\alpha, \beta) = \|\alpha - \beta\|_{Æ} < \frac{\epsilon\delta}{4R} < 1$. Then, by the definition of the norm, we can write α and β as finite convex combinations

$$\alpha = \sum_{i=1}^{m} \lambda_i \mathbf{1}_{x_i} \qquad \beta = \sum_{i=1}^{m} \lambda_i \mathbf{1}_{y_i},$$

where $\lambda_i > 0$, $x_i, y_i \in X$ (possibly with $m > M$ and with some terms repeated) and

$$\sum_{i=1}^{m} \lambda_i d(x_i, y_i) < \frac{\epsilon\delta}{4R}.$$

In particular,

$$\sum_{d(x_i, y_i) \geqslant \delta} \lambda_i \leqslant \frac{1}{\delta} \cdot \sum_{i=1}^{m} \lambda_i d(x_i, y_i) < \frac{\epsilon}{4R}.$$

It follows that

$$d_{Æ}(g\alpha, g\beta) \leqslant \sum_{i=1}^{m} \lambda_i d(gx_i, gy_i)$$

$$\leqslant \sum_{d(x_i, y_i) \geqslant \delta} \lambda_i d(gx_i, gy_i) + \sum_{d(x_i, y_i) < \delta} \lambda_i d(gx_i, gy_i)$$

$$\leqslant \sum_{d(x_i, y_i) \geqslant \delta} \lambda_i R + \sum_{d(x_i, y_i) < \delta} \lambda_i \cdot \epsilon/4$$

$$< \epsilon/4 + \epsilon/4.$$

All in all, we see that

$$d_{Æ}(g\beta, f\alpha) \leqslant d_{Æ}(g\beta, g\alpha) + d_{Æ}(g\alpha, f\alpha) < \frac{\epsilon}{2} + \frac{\epsilon}{2} = \epsilon$$

for all $f \in U$ and $\alpha \in \mathsf{VR}(X, M, K)$ for which $d_{Æ}(\alpha, \beta) < \delta$. $\qquad \square$

For the next lemma, observe that, because $d_{Æ}(\mathbf{1}_x, \mathbf{1}_y) = \|\mathbf{1}_x - \mathbf{1}_y\|_{Æ} = d(x, y)$ for all $x, y \in X$, we see that the mapping $x \mapsto \mathbf{1}_x$ is an isometric embedding of (X, d) into $\mathsf{VR}(X, M, K)$ for any constants $M \geqslant 1$ and K. We shall therefore often identify X with its image in $\mathsf{VR}(X, M, K)$ or in ΔX and, whenever convenient, write x in place of the more cumbersome $\mathbf{1}_x$.

Lemma 5.29 *Let* $X \xrightarrow{\phi} E$ *be a bornologous map from a metric space of bounded geometry* X *to a metric space* E. *Then there are constants* M *and* K *and a uniformly continuous bornologous map*

$$X \xrightarrow{\psi} \mathsf{VR}(E, M, K)$$

so that ψ *is close to* ϕ.

Proof By rescaling the metric on X, we may assume that $\frac{1}{2}$ is a gauge for X, meaning that, for every β, there is a bound m so that every set of diameter β can be covered by m open balls of radius $\frac{1}{2}$. Let $N \subseteq X$ be a maximally 1-discrete subset, i.e., maximal so that $d(x, y) \geqslant 1$ for distinct $x, y \in N$. By maximality, N is 1-dense in X. Observe that, if M is chosen so that every set of diameter 2 can be covered by M open balls of radius $\frac{1}{2}$, then M is also an upper bound on the cardinality of a 1-discrete set in a ball of radius 2.

For every $x \in N$, define $\theta_x : X \to [0,2]$ by $\theta_x(y) = \max\{0, 2 - d(y,x)\}$. Note that θ_x is 1-Lipschitz and $\theta_x \geqslant 1$ on the ball of radius 1 centred at x, while $\operatorname{supp}(\theta_x)$ is contained in the 2-ball around x. Thus, every $y \in X$ belongs to the support of at most M many distinct functions θ_x, while $\theta_x(y) \geqslant 1$ for at least one $x \in N$. From this, it follows that

$$\Theta(y) = \sum_{x \in N} \theta_x(y)$$

defines a Lipschitz function $X \xrightarrow{\Theta} [1, 2M]$. Therefore, setting $\lambda_x = \frac{\theta_x}{\Theta}$, we have a partition of unity $\{\lambda_x\}_{x \in N}$ by Lipschitz functions with some uniform Lipschitz constant C and each λ_x supported in the 2-ball centred at x. Also, for every $y \in X$, the set

$$S_y = \{x \in N \mid \lambda_x(y) > 0\}$$

has diameter at most 4 and cardinality at most M. Let $K = \sup_{d(x,y) \leqslant 4} d(\phi x, \phi y)$.

We now define $X \xrightarrow{\psi} \mathsf{VR}(E, M, K)$ by

$$\psi(y) = \sum_{x \in X} \lambda_x(y)\mathbf{1}_{\phi(x)} = \sum_{x \in S_y} \lambda_x(y)\mathbf{1}_{\phi(x)}$$

for all $y \in X$. To see that ψ is uniformly continuous, suppose that $y, z \in X$ are given. Then

$$d_{\mathbb{E}}\big(\psi(y), \psi(z)\big) = \left\| \sum_{x \in S_y \cup S_z} \lambda_x(y)\mathbf{1}_{\phi(x)} - \sum_{x \in S_y \cup S_z} \lambda_x(z)\mathbf{1}_{\phi(x)} \right\|_{\mathbb{E}}$$

$$= \left\| \sum_{x \in S_y \cup S_z} \big(\lambda_x(y) - \lambda_x(z)\big)\mathbf{1}_{\phi(x)} \right\|_{\mathbb{E}}$$

$$\leqslant \sum_{x \in S_y \cup S_z} \big|\lambda_x(y) - \lambda_x(z)\big| \cdot \operatorname{diam}(\phi[S_y \cup S_z])$$

$$\leqslant \sum_{x \in S_y \cup S_z} Cd(y,z) \cdot \mathsf{diam}(\phi[S_y \cup S_z])$$

$$\leqslant 2MCd(y,z) \cdot \mathsf{diam}(\phi[S_y \cup S_z]).$$

Because ϕ is bornologous, we find that

$$\inf_{\delta > 0} \sup_{d(y,z) < \delta} d(y,z) \cdot \mathsf{diam}(\phi[S_y \cup S_z]) = 0,$$

so the above bound implies uniform continuity.

Finally, to see that ψ is close to ϕ, just note that $\mathsf{supp}(\psi(y)) = \phi[S_y] \subseteq \phi\big[B(y,2)\big]$ and so

$$d_{\!\it\AE}(\phi(y), \psi(y)) = \left\| 1_{\phi(y)} - \sum_{x \in S_y} \lambda_x(y) 1_{\phi(x)} \right\|_{\it\AE}$$

$$\leqslant \sum_{x \in S_y} \lambda_x(y) d\big(\phi(y), \phi(x)\big) \leqslant K$$

for all $y \in X$. $\qquad\qquad\qquad\qquad\qquad\qquad\qquad\qquad\qquad\qquad\quad\square$

Lemma 5.30 *Suppose (X,d) is a proper metric space and M and K are constants. Then the Kantorovich metric $d_{\it\AE}$ is a proper metric when restricted to the Vietoris–Rips complex*

$$\mathsf{VR}(X, M, K).$$

Moreover, X is cobounded in $\mathsf{VR}(X, M, K)$.

Proof It suffices to show that any $d_{\it\AE}$-bounded sequence (β_n) in $\mathsf{VR}(X, M, K)$ has a convergent subsequence. To see this, write $\beta_n = \sum_{i=1}^{M} \lambda_{i,n} 1_{x_{i,n}}$ for some $\lambda_{i,n} \geqslant 0$ and $x_{i,n} \in X$ so that $\mathsf{diam}\big(\{x_{i,n}\}_{i=1}^{M}\big) \leqslant K$. For all n and $i = 1, \ldots, M$,

$$d(x_{i,n}, x_{i,1}) - 2K \leqslant \mathsf{dist}\big(\mathsf{supp}(\beta_n), \mathsf{supp}(\beta_1)\big) \leqslant d_{\it\AE}(\beta_n, \beta_1)$$

$$\leqslant \sup_m d_{\it\AE}(\beta_m, \beta_1) < \infty,$$

which shows that $\{x_{i,n}\}_{n=1}^{\infty}$ is bounded. Because d is a proper metric on X, by passing to a subsequence, we may thus suppose that $x_i = \lim_n x_{i,n}$ exists for all $i = 1, \ldots, M$. Similarly, we may also assume that the $\lambda_i = \lim_n \lambda_{i,n}$ exist. We let $\alpha = \sum_{i=1}^{M} \lambda_i 1_{x_i}$ and observe that $\alpha \in \mathsf{VR}(X, M, K)$.

Now, for all n, we have

$$
\begin{aligned}
d_{\text{Æ}}(\beta_n, \alpha) &= \left\| \sum_{i=1}^{M} \lambda_{i,n} \mathbf{1}_{x_{i,n}} - \sum_{i=1}^{M} \lambda_i \mathbf{1}_{x_i} \right\|_{\text{Æ}} \\
&= \left\| \sum_{i=1}^{M} \lambda_{i,n}(\mathbf{1}_{x_{i,n}} - \mathbf{1}_{x_i}) + \sum_{i=1}^{M} (\lambda_{i,n} - \lambda_i)\mathbf{1}_{x_i} \right\|_{\text{Æ}} \\
&\leqslant \left\| \sum_{i=1}^{M} \lambda_{i,n}(\mathbf{1}_{x_{i,n}} - \mathbf{1}_{x_i}) \right\|_{\text{Æ}} + \left\| \sum_{i=1}^{M} (\lambda_{i,n} - \lambda_i)\mathbf{1}_{x_i} \right\|_{\text{Æ}} \\
&\leqslant \sum_{i=1}^{M} \lambda_{i,n} \left\| \mathbf{1}_{x_{i,n}} - \mathbf{1}_{x_i} \right\|_{\text{Æ}} + \left\| \sum_{i=1}^{M} (\lambda_{i,n} - \lambda_i)\mathbf{1}_{x_i} \right\|_{\text{Æ}} \\
&\leqslant \sum_{i=1}^{M} \lambda_{i,n} d(x_{i,n}, x_i) + \sum_{i=1}^{M} \left| \lambda_{i,n} - \lambda_i \right| \cdot \text{diam}(\{x_i\}_{i=1}^{M}) \\
&\leqslant \max_i d(x_{i,n}, x_i) + K \cdot \sum_{i=1}^{M} \left| \lambda_{i,n} - \lambda_i \right|.
\end{aligned}
$$

Therefore, $d_{\text{Æ}}(\beta_n, \alpha) \underset{n}{\longrightarrow} 0$, which shows that $d_{\text{Æ}}$ is proper on $\text{VR}(X, M, K)$. Finally, to verify that X is cobounded in $\text{VR}(X, M, K)$, just note that

$$
\left\| \mathbf{1}_{z_1} - \sum_{i=1}^{M} \eta_i \mathbf{1}_{z_i} \right\|_{\text{Æ}} \leqslant \text{diam}(\{z_i\}_{i=1}^{M}) \leqslant K
$$

for any element $\sum_{i=1}^{M} \eta_i \mathbf{1}_{z_i} \in \text{VR}(X, M, K)$.   □

5.4.3 Gromov's Theorem for Polish Groups of Bounded Geometry

With the setup of the preceding section, we now have the tools and language to establish the converse to Proposition 5.26.

Theorem 5.31 *Two Polish groups of bounded geometry are coarsely equivalent if and only if they admit a topological coupling.*

Proof By Proposition 5.26, it remains to prove that any two coarsely equivalent groups of bounded geometry admit a topological coupling. So suppose $G \xrightarrow{\phi} H$ is a coarse equivalence between Polish groups of bounded geometry and fix a compatible left-invariant gauge metric ∂ on G. Without loss of generality, by Theorems 3.10 and 3.40, we may also suppose that H is a coarsely embedded closed subgroup of a locally Roelcke precompact Polish group $\mathbb{H}$. Let d be a compatible left-invariant coarsely proper metric on $\mathbb{H}$, whence the restriction of d to H is also coarsely proper. Now, by Lemma 5.29, there are constants M and K along with a uniformly continuous map

$$G \xrightarrow{\psi} \mathsf{VR}(H,d,M,K)$$

so that ψ is close to ϕ. Observe that, since d is left-invariant, the Vietoris–Rips complex $\mathsf{VR}(H,d,M,K)$ is invariant under the induced action $H \curvearrowright \Delta H$.

Let $d_\wedge$ be the compatible metric for the Roelcke uniformity on $\mathbb{H}$ given by

$$d_\wedge(h,f) = \inf_{k \in \mathbb{H}} d(h,k) + d(k^{-1}, f^{-1})$$

and note that

$$d_\wedge(h,1) = \inf_{k \in \mathbb{H}} d(h,k) + d(k^{-1},1) = \inf_{k \in \mathbb{H}} d(h,k) + d(k,1) = d(h,1)$$

for all $h \in \mathbb{H}$. In particular, the open balls

$$B_{d_\wedge}(1,\eta) = \{h \in \mathbb{H} \mid d_\wedge(h,1) < \eta\} = \{h \in \mathbb{H} \mid d(h,1) < \eta\}$$

are coarsely bounded. Because, by Zielinski's Theorem, Theorem 3.11, the coarsely bounded sets in $\mathbb{H}$ are also Roelcke precompact, we see that $B_{d_\wedge}(1,\eta)$ is Roelcke pre-compact. By compatibility with the uniformity, $d_\wedge$ extends to a compatible metric on the Roelcke completion $\widehat{\mathbb{H}}$ of $\mathbb{H}$, which we shall still denote $d_\wedge$. Furthermore, by the preceding comments, we see that $d_\wedge$ is a proper metric on $\widehat{\mathbb{H}}$.

Let X denote the closure of H inside $\widehat{\mathbb{H}}$. Then X is locally compact and the restriction of $d_\wedge$ to X remains proper. Also, as H has bounded geometry, by Lemma 5.13 and the proof of Theorem 5.14, the left-shift action of H on itself extends to a coarsely proper, modest, cocompact continuous action $H \curvearrowright X$. Therefore, by Lemma 5.28, the induced map

$$H \times \mathsf{VR}(X,d_\wedge,M,K) \to \Delta X$$

is continuous.

Henceforth, we let $d_{\text{Æ}}$ and $d_{\text{Æ}}^\wedge$ denote the Kantorovich metrics on the complexes $\mathsf{VR}(H,d,M,K)$ and $\mathsf{VR}(X,d_\wedge,M,K)$ associated with the metric spaces (H,d) and $(X,d_\wedge)$ respectively. Observe that since $d_\wedge \leqslant d$ on $\mathbb{H}$, the natural inclusion of (H,d) into $(X,d_\wedge)$ is 1-Lipschitz and hence so is the induced inclusion

$$\mathsf{VR}(H,d,M,K) \to \mathsf{VR}(X,d_\wedge,M,K).$$

Now, $\mathsf{VR}(H, d, M, K)$ is invariant under the action $H \curvearrowright \Delta X$, so, because

$$\overline{\mathsf{VR}(H,d,M,K)}^{d_{\mathbb{E}}^{\wedge}} \subseteq \mathsf{VR}(X, d_{\wedge}, M, K)$$

and the action map $H \times \mathsf{VR}(X, d_{\wedge}, M, K) \to \Delta X$ is continuous, it follows that also

$$Y = \overline{\mathsf{VR}(H,d,M,K)}^{d_{\mathbb{E}}^{\wedge}}$$

is invariant under the H-action. Moreover, the action $H \curvearrowright Y$ is continuous.

Consider now the product space Y^G equipped with the Tychonoff topology. Alternatively, Y^G can be viewed as the space of all maps from G to Y equipped with topology of pointwise convergence. In particular, the map ψ may be viewed as an element of Y^G. We define commuting left and right actions $H \curvearrowright Y^G \curvearrowleft G$ by setting, for $g, f \in G$, $h \in H$ and $\xi \in Y^G$,

$$\left(h \cdot \xi\right)_f = h \cdot \xi_f \quad \text{and} \quad \left(\xi \cdot g\right)_f = \xi_{gf}.$$

Because H acts continuously on Y and H acts separately on every factor of Y^G, we see that $H \curvearrowright Y^G$ is continuous. On the other hand, whereas the action of G is clearly by homeomorphisms, it may not be continuous. Nevertheless, because both H and G act by homeomorphisms on Y^G, the subspace

$$\Omega = \overline{H \cdot \psi \cdot G}$$

is both H- and G-invariant.

Now, because $(G, \partial) \xrightarrow{\psi} \mathsf{VR}(H, d, M, K)$ is close to the bornologous map ϕ, we see that the corresponding modulus of uniform continuity

$$\omega_\psi(t) = \sup_{\partial(g,f) \leqslant t} d_{\mathbb{E}}(\psi_g, \psi_f)$$

is finite for $t \in \mathbb{R}_+$. Also, as ψ is uniformly continuous, we have $\lim_{t \to 0+} \omega_\psi(t) = 0$.

Claim 5.32 For all $\xi \in \Omega$ and $f_1, f_2 \in G$, we have

$$d_{\mathbb{E}}^{\wedge}(\xi_{f_1}, \xi_{f_2}) \leqslant \omega_\psi\left(\partial(f_1, f_2)\right).$$

By consequence, the action $\Omega \curvearrowleft G$ is continuous.

Proof Suppose first that $h \in H$ and $g, f_1, f_2 \in G$. Then

$$
\begin{aligned}
d_{\mathbb{E}}^{\wedge}(h\psi_{gf_1}, h\psi_{gf_2}) &\leqslant d_{\mathbb{E}}(h\psi_{gf_1}, h\psi_{gf_2}) \\
&= d_{\mathbb{E}}(\psi_{gf_1}, \psi_{gf_2}) \\
&\leqslant \omega_\psi\left(\partial(gf_1, gf_2)\right) \\
&= \omega_\psi\left(\partial(f_1, f_2)\right).
\end{aligned}
$$

Now, for a general element $\xi \in \Omega$, given $\epsilon > 0$, find $h \in H$ and $g \in G$ so that

$$d_{\cancel{E}}^{\wedge}(\xi_{f_i}, h\psi_{gf_i}) < \epsilon$$

for $i = 1, 2$. Then

$$\begin{aligned}
d_{\cancel{E}}^{\wedge}(\xi_{f_1}, \xi_{f_2}) &\leqslant d_{\cancel{E}}^{\wedge}(\xi_{f_1}, h\psi_{gf_1}) + d_{\cancel{E}}^{\wedge}(h\psi_{gf_1}, h\psi_{gf_2}) + d_{\cancel{E}}^{\wedge}(h\psi_{gf_2}, \xi_{f_2}) \\
&\leqslant \omega_\psi\big(\partial(f_1, f_2)\big) + 2\epsilon.
\end{aligned}$$

We thus find that $d_{\cancel{E}}^{\wedge}(\xi_{f_1}, \xi_{f_2}) \leqslant \omega_\psi(\partial(f_1, f_2))$ for all $\xi \in \Omega$.

Now, to see that the action $\Omega \curvearrowleft G$ is continuous, observe that, for all $\xi, \zeta \in \Omega$ and $g, f, k \in G$, we have

$$\begin{aligned}
d_{\cancel{E}}^{\wedge}\big((\xi \cdot g)_k, (\zeta \cdot f)_k\big) &= d_{\cancel{E}}^{\wedge}\big(\xi_{gk}, \zeta_{fk}\big) \\
&\leqslant d_{\cancel{E}}^{\wedge}\big(\xi_{gk}, \zeta_{gk}\big) + d_{\cancel{E}}^{\wedge}\big(\zeta_{gk}, \zeta_{fk}\big) \\
&\leqslant d_{\cancel{E}}^{\wedge}\big(\xi_{gk}, \zeta_{gk}\big) + \omega_\psi\big(\partial(gk, fk)\big).
\end{aligned}$$

Now, suppose ξ and g, k are fixed and $\epsilon > 0$. Choose $\eta > 0$ so that $\omega_\psi(\eta) < \epsilon$. Then, when $\zeta \in \Omega$ is sufficiently close to ξ, we have $d_{\cancel{E}}^{\wedge}\big(\xi_{gk}, \zeta_{gk}\big) < \epsilon$ and, when f is sufficiently close to g, we have $\partial(gk, fk) < \eta$, whereby $\omega_\psi\big(\partial(gk, fk)\big) < \epsilon$ and thus ultimately

$$d_{\cancel{E}}^{\wedge}\big((\xi \cdot g)_k, (\zeta \cdot f)_k\big) < 2\epsilon.$$

This show that, for a fixed coordinate $k \in G$, the map $(\xi, g) \mapsto (\xi \cdot g)_k$ is continuous and hence that $\Omega \curvearrowleft G$ is a continuous action. $\square$

Claim 5.33 The space Ω is locally compact. In fact, for every $r > 0$ and $\xi \in \Omega$, the set $\{\zeta \in \Omega \mid d_{\cancel{E}}^{\wedge}(\xi_1, \zeta_1) \leqslant r\}$ is a compact neighbourhood of ξ.

Proof Suppose that $\zeta \in \Omega$ satisfies $d_{\cancel{E}}^{\wedge}(\xi_1, \zeta_1) \leqslant r$. Then by Claim 5.32 we have

$$\begin{aligned}
d_{\cancel{E}}^{\wedge}(\xi_f, \zeta_f) &\leqslant d_{\cancel{E}}^{\wedge}(\xi_f, \xi_1) + d_{\cancel{E}}^{\wedge}(\xi_1, \zeta_1) + d_{\cancel{E}}^{\wedge}(\zeta_1, \zeta_f) \\
&\leqslant 2\omega_\psi\big(\partial(f, 1)\big) + r
\end{aligned}$$

for all $f \in G$. It therefore follows that

$$\{\zeta \in \Omega \mid d_{\cancel{E}}^{\wedge}(\xi_1, \zeta_1) \leqslant r\} \subseteq \prod_{f \in G} \{\beta \in Y \mid d_{\cancel{E}}^{\wedge}(\beta, \xi_f) \leqslant 2\omega_\psi\big(\partial(f, 1)\big) + r\}.$$

Because the metric $d_{\cancel{E}}^{\wedge}$ is proper on Y, the latter product is compact. $\square$

Claim 5.34 The action $\Omega \curvearrowleft G$ is modest.

Proof To see that the action is modest, suppose B is a coarsely bounded subset of G and that $\xi \in \Omega$ and $r > 0$ are given. By Claim 5.33, we must show that

$$\{\zeta \in \Omega \mid d_{\cancel{E}}^{\wedge}(\xi_1, \zeta_1) \leqslant r\} \cdot B$$

is relatively compact in Ω. But, if $d_{\bar{E}}^{\wedge}(\xi_1, \zeta_1) \leqslant r$ and $g \in B$, then

$$
\begin{aligned}
d_{\bar{E}}^{\wedge}\big(\xi_1, (\zeta \cdot g)_1\big) &= d_{\bar{E}}^{\wedge}(\xi_1, \zeta_g) \\
&\leqslant d_{\bar{E}}^{\wedge}(\xi_1, \zeta_1) + d_{\bar{E}}^{\wedge}(\zeta_1, \zeta_g) \\
&\leqslant r + \omega_\psi\big(\partial(g, 1)\big) \\
&\leqslant r + \omega_\psi\left(\sup_{f \in B} \partial(f, 1)\right).
\end{aligned}
$$

So

$$
\{\zeta \in \Omega \mid d_{\bar{E}}^{\wedge}(\xi_1, \zeta_1) \leqslant r\} \cdot B \subseteq \left\{\zeta \in \Omega \;\middle|\; d_{\bar{E}}^{\wedge}(\xi_1, \zeta_1) \leqslant r + \omega_\psi\left(\sup_{f \in B} \partial(f, 1)\right)\right\},
$$

showing that the action is modest.   □

For the next step, we let κ_ψ denote the compression modulus of the mapping $(G, \partial) \xrightarrow{\psi} \mathsf{VR}(H, d, M, K)$, that is,

$$
\kappa_\psi(t) = \inf_{\partial(g, f) \geqslant t} d_{\bar{E}}(\psi_g, \psi_f).
$$

As ψ is close to the coarse embedding ϕ, we see that $\lim_{t \to \infty} \kappa_\psi(t) = \infty$.

Claim 5.35 The action $\Omega \curvearrowleft G$ is coarsely proper.

Proof Suppose $C \subseteq \Omega$ is compact. Then we must show that the set

$$
\{f \in G \mid Cf \cap C \neq \emptyset\}
$$

is coarsely bounded in G. So pick $r > 0$ large enough so that $d_{\bar{E}}^{\wedge}(\zeta_1, \psi_1) < r$ for all $\zeta \in C$ and let $U = \{x \in X \mid \mathrm{dist}_{d_\wedge}(x, \mathrm{supp}(\psi_1)) < r\}$. Note that $U \subseteq X$ is open and relatively compact, whereby $U \cap H$ is Roelcke pre-compact and has finite d-diameter.

Assume that $h \in H$ and $g, f \in G$ satisfy

$$
\mathrm{supp}(h\psi_g) \cap U \neq \emptyset \quad \text{and} \quad \mathrm{supp}(h\psi_{gf}) \cap U \neq \emptyset.
$$

Then, as $h\psi_g, h\psi_{gf} \in \mathsf{VR}(H, d, M, K)$, we have

$$
\begin{aligned}
\kappa_\psi\big(\partial(1, f)\big) = \kappa_\psi\big(\partial(g, gf)\big) &\leqslant d_{\bar{E}}(\psi_g, \psi_{gf}) \\
&= d_{\bar{E}}(h\psi_g, h\psi_{gf}) \\
&\leqslant \mathrm{diam}_d\big(\mathrm{supp}(h\psi_g) \cup \mathrm{supp}(h\psi_{gf})\big) \\
&\leqslant 2K + \mathrm{diam}_d(U \cap H).
\end{aligned}
$$

As $\lim_{t\to\infty} \kappa_\psi(t) = \infty$, it follows that there is some constant c so that, for all g, $f \in G$ and $h \in H$,

$$\mathsf{supp}(h\psi_g) \cap U \neq \emptyset \text{ and } \mathsf{supp}(h\psi_{gf}) \cap U \neq \emptyset \implies \partial(1, f) \leqslant c.$$

Also, if $\zeta \in \Omega$ and $f \in G$ satisfy $\mathsf{supp}(\zeta_1) \cap U \neq \emptyset$ and $\mathsf{supp}(\zeta_f) \cap U \neq \emptyset$, then we can find some $h\psi g$ sufficiently close to ζ so that also $\mathsf{supp}(h\psi_g) \cap U \neq \emptyset$ and $\mathsf{supp}(h\psi_{gf}) \cap U \neq \emptyset$, which implies that $\partial(1, f) \leqslant c$.

Assume finally that $\zeta \in C$ and $f \in G$ satisfy $\zeta \cdot f \in C$. Then $d_{\textit{Æ}}(\zeta_1, \psi_1) < r$ and $d_{\textit{Æ}}(\zeta_f, \psi_1) < r$, whereby

$$\mathsf{dist}_{d_\wedge}(\mathsf{supp}(\zeta_1), \mathsf{supp}(\psi_1)) < r, \quad \mathsf{dist}_{d_\wedge}(\mathsf{supp}(\zeta_f), \mathsf{supp}(\psi_1)) < r.$$

Therefore, $\mathsf{supp}(\zeta_1) \cap U \neq \emptyset$ and $\mathsf{supp}(\zeta_f) \cap U \neq \emptyset$ and we conclude that $\partial(1, f) \leqslant c$. This shows that the set $\{f \in G \mid Cf \cap C \neq \emptyset\}$ is coarsely bounded in G. $\qquad\square$

Claim 5.36 The action $\Omega \curvearrowleft G$ is cocompact.

Proof Recall that $G \xrightarrow{\phi} H$ is a coarse equivalence and that the mapping $G \xrightarrow{\psi} \mathsf{VR}(H, d, M, K)$ is close to ϕ. Since $\phi[G]$ is cobounded in H, it follows that there is some R so that

$$R > \sup_{h \in H} \inf_{f \in G} d_{\textit{Æ}}(\psi_f, h).$$

In particular, this means that for all $h \in H$ and $g \in G$, there is some $f \in G$ so that

$$d_{\textit{Æ}}\big((h\psi g)_f, 1_H\big) = d_{\textit{Æ}}(h\psi_{gf}, 1_H) = d_{\textit{Æ}}(\psi_{gf}, h^{-1}) < R.$$

We will show that, for any fixed $\xi \in \Omega$, there is $f \in G$ so that

$$d_{\textit{Æ}}^\wedge\big((\xi \cdot f)_1, 1_H\big) = d_{\textit{Æ}}^\wedge(\xi_f, 1_H) \leqslant \omega_\psi(1) + R,$$

which, by Claim 5.33, shows that the set

$$\big\{\zeta \in \Omega \mid d_{\textit{Æ}}^\wedge(\zeta_1, 1_H) \leqslant \omega_\psi(1) + R\big\}$$

is a compact fundamental domain for the action $\Omega \curvearrowleft G$.

To see this, suppose $\xi \in \Omega$ is given and let U be the relatively compact open set in X defined by

$$U = \{x \in X \mid \mathsf{dist}_{d_\wedge}(x, \mathsf{supp}(\xi_1)) < 1\}.$$

Then $U \cap H$ is Roelcke pre-compact and thus $\mathsf{diam}_d(U \cap H) < \infty$. Observe also that, if $\beta \in \mathsf{VR}(H, d, M, K)$ satisfies $\mathsf{supp}(\beta) \cap U \neq \emptyset$, then $\mathsf{supp}(\beta)$ lies within the d-ball of radius

$$Q = \text{dist}_d(1_H, U \cap H) + \text{diam}_d(U \cap H) + K$$

centred at 1_H. In particular, it follows that $d_{\mathbb{Æ}}(\beta, 1_H) < Q$. Finally, let

$$r > \sup\{t \mid \kappa_\psi(t) \leqslant R + Q\}$$

and pick elements $f_1, \ldots, f_n$ that are 1-dense in $B_\partial(1_G, r)$.

Assume towards a contradiction that

$$d_{\mathbb{Æ}}^\wedge(\xi_f, 1_H) > \omega_\psi(1) + R$$

for all $f \in G$. Choose $h\psi g \in H \cdot \psi \cdot G$ close enough to ξ so that $d_{\mathbb{Æ}}^\wedge((h\psi g)_1, \xi_1) < 1$ and

$$d_{\mathbb{Æ}}^\wedge((h\psi g)_{f_i}, 1_H) > \omega_\psi(1) + R$$

for all i. As

$$\text{dist}_{d_\wedge}\big(\text{supp}((h\psi g)_1), \text{supp}(\xi_1)\big) \leqslant d_{\mathbb{Æ}}^\wedge((h\psi g)_1, \xi_1) < 1,$$

we find that $\text{supp}((h\psi g)_1) \cap U \neq \emptyset$ and so $d_{\mathbb{Æ}}((h\psi g)_1, 1_H) < Q$. Now pick $f \in G$ satisfying $d_{\mathbb{Æ}}((h\psi g)_f, 1_H) < R$. Then

$$
\begin{aligned}
\kappa_\psi\big(\partial(f, 1_G)\big) = \kappa_\psi\big(\partial(gf, g)\big) \\
\leqslant d_{\mathbb{Æ}}\big(\psi_{gf}, \psi_g\big) \\
= d_{\mathbb{Æ}}\big(h\psi_{gf}, h\psi_g\big) \\
\leqslant d_{\mathbb{Æ}}\big((h\psi g)_f, 1_H\big) + d_{\mathbb{Æ}}\big(1_H, (h\psi g)_1\big) \\
< R + Q
\end{aligned}
$$

and so $\partial(f, 1_G) < r$, i.e., $f \in B_\partial(1_G, r)$. So find some f_i with $\partial(f, f_i) \leqslant 1$. Then, by Claim 5.32, we have

$$
\begin{aligned}
d_{\mathbb{Æ}}^\wedge((h\psi g)_{f_i}, 1_H) \leqslant d_{\mathbb{Æ}}((h\psi g)_{f_i}, 1_H) \\
\leqslant d_{\mathbb{Æ}}\big((h\psi g)_{f_i}, (h\psi g)_f\big) + d_{\mathbb{Æ}}\big((h\psi g)_f, 1_H\big) \\
< \omega_\psi(1) + R,
\end{aligned}
$$

which contradicts the assumption on f_i. $\square$

We are finally left with the task of verifying that the action $H \curvearrowright \Omega$ is coarsely proper, modest and cocompact.

For coarse properness, note that, by Claim 5.33, every compact subset of Ω is contained in some set of the form $D = \{\zeta \in \Omega \mid d_{\mathbb{Æ}}^\wedge(\zeta_1, 1_H) \leqslant r\}$. So let

$$Z = \{x \in X \mid d_\wedge(x, 1_H) \leqslant r + K\}$$

and observe that, as the action of H on X is coarsely proper, there is a coarsely bounded set $B \subseteq H$ so that $h \cdot Z \cap Z = \emptyset$ for all $h \in H \setminus B$. Thus, for any $h \in H \setminus B$ and $\zeta \in \Omega$ with

$$\text{dist}_{d_\wedge}\left(\text{supp}(\zeta_1), 1_H\right) \leqslant d_{\cancel{E}}^\wedge(\zeta_1, 1_{1_H}) \leqslant r,$$

we have $\text{supp}(\zeta_1) \subseteq Z$ and thus

$$d_{\cancel{E}}^\wedge\left((h\zeta)_1, 1_H\right) \geqslant \text{dist}_{d_\wedge}\left(h \cdot \text{supp}(\zeta_1), 1_H\right) > r + K.$$

Thus, $h \cdot D \cap D = \emptyset$, showing that the action is coarsely proper.

Similarly, to verify modesty, suppose a compact set

$$D = \{\zeta \in \Omega \mid d_{\cancel{E}}^\wedge(\zeta_1, 1_H) \leqslant r\}$$

and a coarsely bounded set $B \subseteq H$ are given. Again, let

$$Z = \{x \in X \mid d_\wedge(x, 1_H) \leqslant r + K\}$$

and note that, if $\zeta \in D$, then $\text{supp}(\zeta_1) \subseteq Z$. Thus, if $\zeta \in D$ and $h \in B$, then $\text{supp}((h\zeta)_1) \subseteq B \cdot Z$. Letting $s = \sup_{x \in B \cdot Z} d_\wedge(x, 1_H)$, we see that

$$B \cdot D \subseteq \{\eta \in \Omega \mid d_{\cancel{E}}^\wedge(\eta_1, 1_H) \leqslant s\},$$

establishing modesty.

Because the action of H on X is cocompact, pick a compact set $Z \subseteq X$ so that $X = H \cdot Z$. Then, for every $\xi \in \Omega$, there is some $h \in H$ for which $\text{supp}((h\xi)_1) = h \cdot \text{supp}(\xi_1)$ intersects Z. Picking some $z \in Z$, we see that, for any $\xi \in \Omega$, there is $h \in H$ so that

$$d_{\cancel{E}}^\wedge\left((h\xi)_1, 1_z\right) \leqslant K + \text{diam}_{d_\wedge}(Z).$$

Setting $D = \{\zeta \in \Omega \mid d_{\cancel{E}}(\zeta_1, 1_z) \leqslant K + \text{diam}_{d_\wedge}(Z)\}$, which is compact, we have $H \cdot D = \Omega$, thus verifying cocompactness. $\square$

In the proof of Theorem 5.31, we use only the assumption that $G \xrightarrow{\phi} H$ is cobounded in one place, namely, for the proof of Claim 5.36 stating that the G-action on Ω is cocompact. This means that, with an unchanged proof except for using condition (1) of Proposition 5.26, the following result is established.

Theorem 5.37 *Suppose G and H are Polish groups of bounded geometry. Then G coarsely embeds into H if and only if there are commuting, coarsely proper and modest continuous actions of Polish groups on a locally compact Hausdorff space*

$$G \curvearrowright X \curvearrowleft H$$

so that the H-action is cocompact.

5.5 Coarse Couplings

We shall now consider a more algebraic notion of similarity between Polish groups strengthening coarse equivalence and reminiscent of the Gromov–Hausdorff distance between metric spaces. Recall first that the *Hausdorff distance* between two subsets A and B of a metric space (X, d) is given by

$$d_{\mathrm{Haus}}(A, B) = \max\{\sup_{a \in A} \inf_{b \in B} d(a, b), \sup_{b \in B} \inf_{a \in A} d(a, b)\}.$$

For example, if $G \curvearrowright (X, d)$ is an isometric group action, then the Hausdorff distance between any two orbits $G \cdot x$ and $G \cdot y$ is simply

$$d_{\mathrm{Haus}}(Gx, Gy) = \inf_{g \in G} d\big(x, g(y)\big),$$

which is finite. Thus, if $G, F \curvearrowright (X, d)$ are isometric group actions having two orbits $G \cdot x$ and $F \cdot y$ with finite Hausdorff distance, then any two orbits of G and F have finite Hausdorff distance. We define two isometric actions $G, F \curvearrowright (X, d)$ to be *proximal* if the Hausdorff distance between some two or, equivalently, any two G- and F-orbits is finite.

More abstractly, the *Gromov–Hausdorff distance* between two metric spaces X and Y is given by the infimum of the Hausdorff distance $d_{\mathrm{Haus}}\big(i[X], j[Y]\big)$ where i and j vary over all isometric embeddings $i \colon X \to Z$ and $j \colon Y \to Z$ into a common metric space (Z, d).

In case of topological groups, we would like also to preserve the algebraic structure of the groups and are led to consider two levels of proximity. Thus, suppose $G \hookrightarrow H \hookleftarrow F$ is a pair of coarsely proper isomorphic embeddings of Polish groups G and F into a Polish group H and identify G and F with their images in H. Consider the following two conditions.

(1) Both G and F are cobounded in $G \cup F$, i.e., $F \subseteq GB$ and $G \subseteq FB$ for some coarsely bounded set $B \subseteq H$,
(2) both G and F are cobounded in H, i.e., $H = GB = FB$ for some coarsely bounded set $B \subseteq H$.

Evidently, (2) implies (1), whereas, on the other hand, (1) implies that G and F are coarsely equivalent, as they are both coarsely equivalent to $G \cup F$ with the coarse structure induced from H. Moreover, if H has a coarsely proper compatible left-invariant metric d, then (1) is equivalent to requiring that G and F have finite Hausdorff distance in H. We spell this out in the following definition.

Definition 5.38 A *coarse coupling* of Polish groups G and F is a pair of coarsely proper isomorphic embeddings $G \hookrightarrow H \hookleftarrow F$ into a locally bounded Polish group H so that, when identifying G and F with their images in H,

$$d_{\mathrm{Haus}}(G, F) < \infty$$

for any coarsely proper metric d on H.

We now have the following basic equivalence characterising coarse couplings.

Theorem 5.39 *The following are equivalent for Polish groups G and F.*

(1) *G and F admit a coarse coupling $G \hookrightarrow H \hookleftarrow F$,*
(2) *G and F have proximal, coarsely proper, continuous, isometric actions on a metric space.*

If, furthermore, G and F have bounded geometry, these conditions imply that G and F admit a topological coupling.

Proof Suppose $G, F \curvearrowright (X, d_X)$ are two proximal, coarsely proper, continuous, isometric actions on a metric space (X, d_X). Observe then that, for $x \in X$ fixed, the subspace $Y = \bigcup_n (FG)^n \cdot x$ is separable and simultaneously G- and F-invariant. Fix also compatible left-invariant metrics $d_G \leqslant 1$ on G and $d_F \leqslant 1$ on F and equip $Y \times G \times F$ with the metric

$$\partial\big((y_1, g_1, f_1), (y_2, g_2, f_2)\big) = d_X(y_1, y_2) + d_G(g_1, g_2) + d_F(f_1, f_2).$$

Then G and F act continuously and isometrically on $Y \times G \times F$ by

$$g \cdot (y, g', f) = (gy, gg', f), \qquad f \cdot (y, g, f') = (fy, g, ff').$$

Observe that, because d_G and d_F are bounded, these two actions remain proximal.

Now, by a construction of Uspenskiĭ [**91**], we may find an isometric embedding ι of $(Y \times G \times F, \partial)$ into the Urysohn metric space $\mathbb{U}$ and an isomorphic embedding

$$\mathrm{Isom}(Y \times G \times F, \partial) \xrightarrow{\theta} \mathrm{Isom}(\mathbb{U})$$

so that, for each $h \in \mathrm{Isom}(Y \times G \times F, \partial)$, the following diagram commutes.

$$Y \times G \times F \xrightarrow{\ h\ } Y \times G \times F$$

$$\downarrow{\iota} \qquad\qquad \downarrow{\iota}$$

$$\mathbb{U} \xrightarrow{\ \ \theta(h)\ \ } \mathbb{U}$$

Now, because the metrics d_G and d_F are compatible with the group topologies, it follows that G and F embed isomorphically into $\mathrm{Isom}(Y \times G \times F, \partial)$ via their respective actions when the latter group is equipped with the topology of pointwise convergence. Composing with θ, we thus get isomorphic embeddings of G and F into $\mathrm{Isom}(\mathbb{U})$. Moreover, by the equivariance of ι, we find that the resulting actions of G and F on $\mathbb{U}$ are coarsely proper and proximal.

We may thus identify G and F with closed subgroups of $\mathrm{Isom}(\mathbb{U})$ whose actions on $\mathbb{U}$ are proximal and both coarsely proper. Now fix some $z \in \mathbb{U}$. Then the orbit map $h \in \mathrm{Isom}(\mathbb{U}) \mapsto h(z) \in \mathbb{U}$ is a coarse equivalence. It follows from the properness of the actions that G and F are coarsely embedded in $\mathrm{Isom}(\mathbb{U})$ and by the proximality of the actions that G and F are cobounded in $G \cup F$.

Conversely, if $G \hookrightarrow H \hookleftarrow F$ is a coarse coupling, fix a compatible left-invariant coarsely proper metric d on H. Then, as G and F are coarsely embedded, d is coarsely proper on each of them, so the left-shift actions $G, F \curvearrowright (H, d)$ are coarsely proper. Moreover, as $G \subseteq FB$ and $F \subseteq GB$ for some coarsely bounded set $B \subseteq H$, the orbits $G \cdot 1$ and $F \cdot 1$ have finite Hausdorff distance and so the actions are proximal.   □

By virtue of Theorem 5.31, the relation of having a topological coupling defines an equivalence relation on the class of Polish groups of bounded geometry. However, this is of course far from clear from the definition of topological couplings itself. Similarly, it is unclear if admitting a coarse coupling is a transitive relation.

Problem 5.40 Is the relation of having a coarse coupling an equivalence relation on the class of Polish groups?

5.6 Representations on Reflexive Spaces

For the following result, we should recall some facts about proper affine isometric actions. Namely, suppose that G is a locally compact Polish group. Then, if G is amenable, it admits a continuous proper affine isometric action on a Hilbert space [4] (see also [17]). On the other hand, if G is non-amenable, it can in general only be shown to have a continuous proper affine isometric action on a reflexive Banach space [14].

For general Polish groups, these results are no longer valid and the obstructions seem to be of two different kinds; harmonic analytic and large-scale geometric. For the harmonic analytic obstructions, note there are Polish groups such as $\text{Homeo}_+([0,1])$ [59] and $\text{Isom}(\mathbb{U})$ with no non-trivial continuous linear or affine isometric representations on reflexive spaces whatsoever. Moreover, these examples are, in fact, even amenable.

Eschewing such groups, one can restrict the attention to non-Archimedean groups, which all have faithful unitary representations. But even in this setting, large-scale geometric obstructions appear. Indeed, there are amenable non-Archimedean Polish groups such as $\text{Isom}(\mathbb{Z}\mathbb{U})$ all of whose affine isometric actions on reflexive Banach spaces have fixed points [80] and so are not coarsely proper.

However, by [14], every metric space of bounded geometry is coarsely embeddable into a reflexive space and, by combining that construction with an averaging result of F. M. Schneider and A. Thom [82] improving an earlier result in [80], we obtain the following.

Theorem 5.41 *Let G be an amenable Polish group of bounded geometry. Then G admits a coarsely proper continuous affine isometric action on a reflexive Banach space.*

Proof Fix a gauge metric d on G and let $X \subseteq G$ be maximal so that $d(x,y) \geqslant 2$ for all distinct $x, y \in X$. It follows that, for every diameter n, there is an upper bound k_n on the cardinality of subsets of X of diameter n.

For $g \in G$, we now define a function $\phi_g^n \colon X \to [0,1]$ by

$$\phi_g^n(x) = \begin{cases} 1 - \frac{d(x,g)}{n} & \text{if } d(x,g) \leqslant n, \\ 0 & \text{otherwise.} \end{cases}$$

By construction of ϕ_g^n, we find that, for all $g, f \in G$,

$$\|\phi_g^n - \phi_f^n\|_\infty \leqslant \frac{d(g,f)}{n},$$

while

$$\|\phi_g^n - \phi_f^n\|_\infty \geqslant \frac{n-4}{n},$$

whenever $d(g, f) \geqslant n$. Because the supports of ϕ_g^n and ϕ_f^n have diameter $\leqslant 2n$ and thus cardinality $\leqslant k_{2n}$, there is some sufficiently large coefficient $p_n < \infty$ so that

$$\|\phi_g^n - \phi_f^n\|_\infty \leqslant \|\phi_g^n - \phi_f^n\|_{p_n} \leqslant 2\|\phi_g^n - \phi_f^n\|_\infty \leqslant \frac{2}{n} \cdot d(g, f)$$

for all $g, f \in G$.

Thus, $g \in G \mapsto \phi_g^n \in \ell^{p_n}(X)$ is a $\frac{2}{n}$-Lipschitz map satisfying $\|\phi_g^n - \phi_f^n\|_{p_n} \geqslant \frac{n-4}{n}$ whenever $d(g, f) \geqslant n$. As G is amenable, it follows from Theorem 6.1 in [82] that G admits a continuous isometric linear representation $\pi_n : G \curvearrowright V_n$ on a separable Banach space V_n finitely representable in $L^{p_n}(\ell^{p_n}(X)) = L^{p_n}$ along with a $\frac{2}{n}$-Lipschitz cocycle $b_n : G \to V_n$ for π_n satisfying $\|b_n(g) - b_n(f)\|_{V_n} \geqslant \frac{n-4}{n}$ whenever $d(g, f) \geqslant n$. As V_n is finitely representable in L^{p_n}, we have that V_n is reflexive.

Now let $E = \left(\bigoplus_n V_n\right)_2$ be the ℓ^2-sum of the V_n and let $\pi : G \curvearrowright E$ be the product of the linear representations π_n. We claim that $b(g) = (b_1(g), b_2(g), \ldots)$ defines a continuous cocycle $b : G \to E$ for π. To see this, note that

$$\|b(g) - b(f)\|_E^2 = \sum_n \|b_n(g) - b_n(f)\|_{V_n}^2 \leqslant \left(\sum_n \frac{4}{n^2}\right) \cdot d(g, f)^2.$$

Similarly, if $d(g, f) \geqslant n$, then $\|b_m(g) - b_m(f)\|_{V_m} \geqslant \frac{m-4}{m} \geqslant \frac{1}{2}$ for all $m = 8, \ldots, n$ and so

$$\|b(g) - b(f)\|^2 \geqslant \sum_{m=8}^n \|b_m(g) - b_m(f)\|_{V_m}^2 \geqslant \frac{n-7}{4},$$

which shows that b is coarsely proper. $\qquad\square$

Let us observe that bounded geometry in itself is not sufficient for Theorem 5.41. For this, consider the group $\mathrm{Homeo}_\mathbb{Z}(\mathbb{R})$ and let H_0 and $H_{\frac{1}{2}}$ denote the isotropy subgroups of 0 and $\frac{1}{2}$ respectively. Then it is fairly easy to see that every translation τ_α of $\mathbb{R}$ of amplitude $|\alpha| < \frac{1}{6}$ can be written as $\tau_\alpha = gf$ for $g \in H_0$ and $f \in H_{\frac{1}{2}}$. As also every element of $\mathrm{Homeo}_\mathbb{Z}(\mathbb{R})$ is a product of some translation τ_β and an element of H_0, we see that $\mathrm{Homeo}_\mathbb{Z}(\mathbb{R})$ is generated by the two subgroups H_0 and $H_{\frac{1}{2}}$. As each of these is isomorphic to the group $\mathrm{Homeo}_+([0, 1])$ that has no non-trivial continuous isometric linear and thus also affine representations on reflexive Banach spaces [59], we conclude that the same holds for $\mathrm{Homeo}_\mathbb{Z}(\mathbb{R})$.

5.7 Compact *G*-Flows and Unitary Representations

In this section, we shall consider the impact of bounded geometry on the topological dynamics of a Polish group. Let us recall that a *compact G-flow* of a topological group G is a continuous action $G \curvearrowright K$ on a compact Hausdorff space K. Such a flow is *minimal* provided that all orbits are dense. Among all minimal compact G-flows, there is a flow $G \curvearrowright M$ of which all other flows are a factor, i.e., so that for any other minimal compact G-flow $G \curvearrowright K$ there is a continuous G-equivariant map $\phi: M \to K$. Moreover, up to isomorphism of G-flows, the flow $G \curvearrowright M$ is unique and is therefore denoted the *universal minimal flow* of G.

For a topological group G, let $\mathrm{LUC}(G)$ denote the commutative $\mathbb{C}^*$-algebra of all bounded left-uniformly continuous complex valued functions on G with the supremum norm. Then the right-regular representation $\rho: G \curvearrowright \mathrm{LUC}(G)$ is continuous and thus induces a continuous action ρ^* on the Gelfand spectrum $\mathfrak{A}(G) = \mathrm{spec}(\mathrm{LUC}(G))$ via

$$\left(\rho^*(g)\omega\right)\phi = \omega\left(\rho(g)\phi\right) = \omega\left(\phi(\,\cdot\,g)\right)$$

for $\omega \in \mathfrak{A}(G)$ and $\phi \in \mathrm{LUC}(G)$.

Although the compact flow $G \curvearrowright \mathfrak{A}(G)$ is not necessarily minimal, it does have a dense orbit $G \cdot \omega$. Indeed, observe that each $g \in G$ defines an element $\omega_g \in \mathfrak{A}(G)$ by $\omega_g(\phi) = \phi(g)$ for $\phi \in \mathrm{LUC}(G)$. The map $g \mapsto \omega_g$ is a homeomorphic embedding of G into $\mathfrak{A}(G)$ with dense image and, as $\rho^*(g)\omega_f = \omega_{fg}$, we see that ω_{1_G} has a dense G-orbit in $\mathfrak{A}(G)$. Moreover, if $G \curvearrowright K$ is any compact G-flow with a dense orbit $G \cdot x$, then there is a unique continuous G-equivariant map $\phi: \mathfrak{A}(G) \to K$ so that $\phi(\omega_{1_G}) = x$. The space $\mathfrak{A}(G)$ is called the *greatest ambit* of G and every minimal subflow of $G \curvearrowright \mathfrak{A}(G)$ is a realisation of the universal minimal flow of G.

While the greatest ambit is in general a very large non-metrisable compact space, a much investigated and fairly common phenomenon among Polish groups G is that the universal minimal flow is metrisable or even reduces to a single point. In the latter case, we say that G is *extremely amenable*, since this is equivalent to every compact G-flow having a fixed point and thus implies amenability. On the other hand, as shown by R. Ellis [28] for discrete groups and later W. Veech [92] in all generality, every locally compact group acts freely on its greatest ambit and thus fails to be extremely amenable unless trivial. Our first goal is to indicate that this result is really geometric, rather than topological, by giving an appropriate generalisation to Polish groups of bounded geometry.

Proposition 5.42 *Suppose G is a Polish group of bounded geometry with gauge metric d. Then, for every constant α, there is a uniform entourage U_α on the greatest ambit $\mathfrak{A}(G)$ of G, so that*

$$\left(\rho^*(g)\omega, \rho^*(f)\omega\right) \notin U_\alpha$$

whenever $\omega \in \mathfrak{A}(G)$ and $g, f \in G$ with $9 \leqslant d(g, f) \leqslant \alpha$.

In particular, every $g \in G$ with $d(g, 1) \geqslant 9$ acts freely on $\mathfrak{A}(G)$.

Proof Let $X \subseteq G$ be a maximal 2-discrete subset and suppose $\alpha \geqslant 9$ is given. Then (X, d) is uniformly locally finite and can therefore be partitioned into finitely many $2(\alpha + 4)$-discrete sets $X = \bigcup_{i=1}^m X_i$. To see this, let m be the maximal cardinality of a ball of radius $2(\alpha + 4)$ in X. We let X_1 be a maximal $2(\alpha + 4)$-discrete set in X, X_2 be a maximal $2(\alpha + 4)$-discrete set in $X \setminus X_1$, X_3 a maximal $2(\alpha + 4)$-discrete set in $X \setminus (X_1 \cup X_2)$, etc. So, we see that, if x is a point outside $X_1 \cup \cdots \cup X_m$, then there must be some $x_1 \in X_1$ within distance $2(\alpha + 4)$ from x, a point $x_2 \in X_2$ within distance $2(\alpha + 4)$ from x, etc. As the X_i are disjoint, it follows that x and $x_1, \ldots, x_m$ are distinct points of the ball $B(x, 2(\alpha + 4))$, contradicting that $|B(x, 2(\alpha + 4))| \leqslant m$.

Define bounded 1-Lipschitz functions $(G, d) \xrightarrow{\phi_i} \mathbb{R}$ by

$$\phi_i(x) = \min\{\alpha + 4, d(x, X_i)\}$$

and let

$$U_\alpha = \{(\omega, \eta) \in \mathfrak{A}(G) \times \mathfrak{A}(G) \mid |\omega(\phi_i) - \eta(\phi_i)| < 1, \forall i\}.$$

To see that U_α is as required, assume towards a contradiction that $\omega \in \mathfrak{A}(G)$ and $g, f \in G$ with $9 \leqslant d(g, f) \leqslant \alpha$ satisfy $(\rho^*(g)\omega, \rho^*(f)\omega) \in U_\alpha$, i.e., that

$$\max_i \left|\left(\rho^*(g)\omega\right)(\phi_i) - \left(\rho^*(f)\omega\right)(\phi_i)\right| < 1.$$

Because the ω_x with $x \in G$ are dense in $\mathfrak{A}(G)$, there is some $x \in G$ so that also

$$\max_i \left|\phi_i(xg) - \phi_i(xf)\right| = \max_i \left|\left(\rho^*(g)\omega_x\right)(\phi_i) - \left(\rho^*(f)\omega_x\right)(\phi_i)\right| < 1.$$

Now find $a, b \in X$ with $d(xg, a) < 2, d(xf, b) < 2$, whence

$$|d(a, b) - d(g, f)| = |d(a, b) - d(xg, xf)| < 4$$

and thus $d(a, b) < d(g, f) + 4 \leqslant \alpha + 4$. Assume that $a \in X_i$ and observe that, because X_i is $2(\alpha + 4)$-discrete, a is the unique point in X_i within distance

$< \alpha + 4$ of b, whereby $\phi_i(b) = d(b, X_i) = d(b, a)$. Thus, as ϕ_i is 1-Lipschitz and $\phi_i(a) = 0$, we find that

$$|\phi_i(xg) - \phi_i(xf)| \geqslant |\phi_i(a) - \phi_i(b)| - 4 = d(a, b) - 4 \geqslant d(g, f) - 8 \geqslant 1,$$

contradicting our assumption on x. □

To put this result into perspective, recall that $\text{Homeo}_{\mathbb{Z}}(\mathbb{R})$ is generated by two isomorphic copies of the extremely amenable group $\text{Homeo}_+([0, 1])$. Both of these copies are thus coarsely bounded in $\text{Homeo}_{\mathbb{Z}}(\mathbb{R})$ and, in fact, every extremely amenable subgroup must be contained in some fixed coarsely bounded set $B \subseteq \text{Homeo}_{\mathbb{Z}}(\mathbb{R})$.

Elaborating on the result of Ellis and Veech, in [48] it was shown that every non-compact locally compact group has non-metrisable universal minimal flow. Although the proof of this breaks down for Polish groups of bounded geometry, we are able instead to rely on a recent analysis due to I. Ben Yaacov, J. Melleray and T. Tsankov [7], who analysed metrisable universal minimal flows of Polish groups.

Proposition 5.43 *Let G be Polish group of bounded geometry with metrisable universal minimal flow. Then G is coarsely bounded.*

Proof By the main result of [7], if G has metrisable universal minimal flow, then G contains an extremely amenable, closed, co-pre-compact subgroup H. Here be *co-pre-compact* means that, for every identity neighbourhood V in G, there is a finite set $F \subseteq G$ so that $G = VFH$. However, observe that, by Proposition 5.42, the extremely amenable subgroup H is coarsely bounded in G. Thus, if we let V be a coarsely bounded identity neighbourhood and choose F finite so that $G = VFH$, then G is a product of three coarsely bounded sets and therefore coarsely bounded itself. □

As an example of a non-compact Polish group with metrisable universal minimal flow, one can of course take any extremely amenable group. For a more interesting case, consider $\text{Homeo}_+(\mathbb{S}^1)$ whose tautological action on $\mathbb{S}^1$ is its universal minimal flow. While $\text{Homeo}_+(\mathbb{S}^1)$ is coarsely bounded, the central extension $\text{Homeo}_{\mathbb{Z}}(\mathbb{R})$, on the other hand, is only of bounded geometry and thus has non-metrisable universal minimal flow.

A second question, related to extreme amenability, is the richness of the unitary representations of amenable Polish groups. To briefly summarise the link, suppose G is an amenable Polish group that fails to be extremely amenable. Then, one can show that G admits a non-trivial metrisable compact flow $G \curvearrowright K$, which, by the amenability of G, carries an invariant probability measure μ. It thus follows that the associated unitary representation

$$G \curvearrowright L^2(K, \mu)$$

is non-trivial too.

In the case of Polish groups of bounded geometry, we have a more-detailed picture.

Proposition 5.44 *Let G be an amenable Polish group of bounded geometry with gauge metric d. Then there are a strongly continuous unitary representation $\pi: G \curvearrowright \mathcal{H}$ and unit vectors $\xi_n \in \mathcal{H}$ so that*

$$\left\| \pi(g)\xi_n - \xi_n \right\| = \sqrt{2},$$

whenever $9 \leqslant d(g, 1) \leqslant n$.

Proof For each n, let U_n be the uniform entourage in $\mathfrak{A}(G)$ given by Proposition 5.42. Then, as $\mathfrak{A}(G)$ is compact, there is a finite covering $\mathfrak{A}(G) = \bigcup_{i=1}^{k} W_i$ by open sets satisfying $W_i \times W_i \subseteq U_n$. Observe that, if $\omega \in W_i$ and $9 \leqslant d(g, 1) \leqslant n$, then $(\rho^*(g)\omega, \omega) \notin U_n \supseteq W_i \times W_i$, i.e., $\rho^*(g)\omega \notin W_i$ and so $\rho^*(g)[W_i] \cap W_i = \emptyset$.

Because G is amenable, it fixes a regular Borel probability measure μ on $\mathfrak{A}(G)$. So pick some i so that W_i has positive μ-measure and let $\xi_n = \frac{1}{\sqrt{\mu(W_i)}}\chi_{W_i} \in L^2(\mathfrak{A}(G), \mu)$. Then, whenever $9 \leqslant d(g, 1) \leqslant n$, we see that $\pi(g)\xi_n$ and ξ_n are disjointly supported normalised L^2-functions and so $\left\| \pi(g)\xi_n - \xi_n \right\| = \sqrt{2}$. $\qquad\square$

5.8 Efficiently Contractible Groups

The complexities of the proof of Theorem 5.31 indicate the utility of when an arbitrary bornologous map may be replaced by a uniformly continuous map. A standard trick using partitions of unity (see, for example, Lemma A.1 in [23]), shows that this may be accomplished provided the domain is locally compact and the range is a Banach space. We will extend this further by using generalised convex combinations in contractible groups.

Definition 5.45 A Polish group G is said to be *efficiently contractible* if it admits a continuous contraction $(R_\alpha)_{\alpha \in [0,1]}: G \to G$ onto 1_G so that, for every coarsely bounded set A, the restriction $R: [0, 1] \times A \to G$ is uniformly continuous and has coarsely bounded image.

Example 5.46 Suppose G is a contractible locally compact Polish group. Then the contraction $R: [0,1] \times G \to G$ will be uniformly continuous and have compact image whenever restricted to $[0,1] \times A$, where $A \subseteq G$ is compact.

Example 5.47 If X is a Banach space, then the usual contraction $R_\alpha(x) = (1 - \alpha)x$ is uniformly continuous and have norm-bounded image when restricted to the bounded sets $[0,1] \times nB_X$.

Example 5.48 (Efficient contractibility of $\mathrm{Homeo}_{\mathbb{Z}}(\mathbb{R})$) Recall that the formula

$$d(g, f) = \sup_{x \in \mathbb{R}} |g^{-1}(x) - f^{-1}(x)|$$

defines a compatible left-invariant coarsely proper metric on the Polish group $\mathrm{Homeo}_{\mathbb{Z}}(\mathbb{R})$. Let also H be the closed subgroup

$$H = \{g \in \mathrm{Homeo}_{\mathbb{Z}}(\mathbb{R}) \mid g(n) = n, \forall n \in \mathbb{Z}\}.$$

We first define a continuous contraction $[0,1] \times \mathrm{Homeo}_{\mathbb{Z}}(\mathbb{R}) \xrightarrow{R} \mathrm{Homeo}_{\mathbb{Z}}(\mathbb{R})$ onto H by setting

$$R_\alpha(g) = g\tau_{\alpha g^{-1}(0)},$$

where τ_β denotes the translation by β, and note that, for $g, f \in \mathrm{Homeo}_{\mathbb{Z}}(\mathbb{R})$, $\alpha, \beta \in [0,1]$ and $x \in \mathbb{R}$,

$$\begin{aligned}
&\left| R_\alpha(g)^{-1}(x) - R_\beta(f)^{-1}(x) \right| \\
&= \left| \tau_{\alpha g^{-1}(0)}^{-1} g^{-1}(x) - \tau_{\beta f^{-1}(0)}^{-1} f^{-1}(x) \right| \\
&= \left| g^{-1}(x) - \alpha g^{-1}(0) - f^{-1}(x) + \beta f^{-1}(0) \right| \\
&\leqslant \left| g^{-1}(x) - f^{-1}(x) \right| + \left| \alpha g^{-1}(0) - \beta f^{-1}(0) \right| \\
&\leqslant d(g, f) + \left| \alpha g^{-1}(0) - \alpha f^{-1}(0) \right| + \left| \alpha f^{-1}(0) - \beta f^{-1}(0) \right| \\
&\leqslant 2 \cdot d(g, f) + |\alpha - \beta| \cdot |f^{-1}(0)|.
\end{aligned}$$

Thus,

$$d\big(R_\alpha(g), R_\beta(f) \big) \leqslant 2 \cdot d(g, f) + |\alpha - \beta| \cdot |f^{-1}(0)|$$

and so, in particular, if $A \subseteq \mathrm{Homeo}_{\mathbb{Z}}(\mathbb{R})$ is a coarsely bounded set, the restriction $[0,1] \times A \xrightarrow{R} G$ is uniformly continuous and has coarsely bounded image.

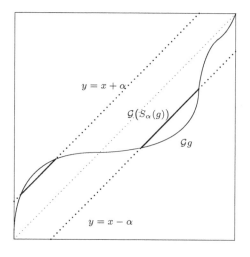

Figure 5.1 The α-truncation of g; $\mathcal{G}g$ = graph of g.

Now, H is isomorphic to the coarsely bounded group $\mathrm{Homeo}_+([0,1])$ and thus is coarsely bounded itself. As in Example 1.19 of [**78**], we now define a uniformly continuous contraction $[0,1] \times H \xrightarrow{S} H$ onto id by letting

$$S_\alpha(g)(x) = \max\left\{\min\{g(x), x+1-\alpha\}, x-1+\alpha\right\}$$

for $g \in H$ and $\alpha \in [0,1]$. In other words, $S_\alpha(g)$ is just the truncation of g between the two translations $\tau_{1-\alpha}$ and $\tau_{\alpha-1}$. In particular, as $\tau_{1-\alpha}^{-1} = \tau_{\alpha-1}$, we see that $S_\alpha(g^{-1}) = S_\alpha(g)^{-1}$. A graphical depiction of this truncation is given in Figure 5.1.

More explicitly, observe that the group $\mathrm{Homeo}_+(\mathbb{R})$ of order-preserving homeomorphisms of $\mathbb{R}$ is a distributive lattice under the two operations $g \vee f = \max\{g, f\}$ and $g \wedge f = \min\{g, f\}$. Furthermore, these satisfy $(g \vee f)^{-1} = g^{-1} \wedge f^{-1}$, so

$$\begin{aligned}
S_\alpha(g)^{-1} &= \left((g \wedge \tau_{1-\alpha}) \vee \tau_{\alpha-1}\right)^{-1} \\
&= (g \wedge \tau_{1-\alpha})^{-1} \wedge \tau_{1-\alpha} \\
&= (g^{-1} \vee \tau_{\alpha-1}) \wedge \tau_{1-\alpha} \\
&= (g^{-1} \wedge \tau_{1-\alpha}) \vee \tau_{\alpha-1} \\
&= S_\alpha(g^{-1}).
\end{aligned}$$

Note also that $|S_\alpha(g)(x) - S_\alpha(f)(x)| \leqslant |g(x) - f(x)|$ and that $|S_\alpha(g)(x) - S_\beta(g)(x)| \leqslant |\alpha - \beta|$. Therefore, for $g, f \in H$, $\alpha, \beta \in [0, 1]$ and $x \in \mathbb{R}$,

$$
\begin{aligned}
&\left|S_\alpha(g)^{-1}(x) - S_\beta(f)^{-1}(x)\right| \\
&= \left|S_\alpha(g^{-1})(x) - S_\beta(f^{-1})(x)\right| \\
&\leqslant \left|S_\alpha(g^{-1})(x) - S_\alpha(f^{-1})(x)\right| + \left|S_\alpha(f^{-1})(x) - S_\beta(f^{-1})(x)\right| \\
&\leqslant \left|g^{-1}(x) - f^{-1}(x)\right| + |\alpha - \beta|.
\end{aligned}
$$

So $d\big(S_\alpha(g), S_\alpha(g)\big) \leqslant d(g, f) + |\alpha - \beta|$ and $[0, 1] \times H \xrightarrow{S} H$ is therefore uniformly continuous.

Now define a contraction $[0, 1] \times \mathrm{Homeo}_\mathbb{Z}(\mathbb{R}) \xrightarrow{T} \mathrm{Homeo}_\mathbb{Z}(\mathbb{R})$ onto id by

$$
T_\alpha(g) = \begin{cases} R_{2\alpha}(g) & \text{for } 0 \leqslant \alpha \leqslant \frac{1}{2}, \\ S_{2\alpha-1}R_1(g) & \text{for } \frac{1}{2} \leqslant \alpha \leqslant 1. \end{cases}
$$

Then, for $A \subseteq \mathrm{Homeo}_\mathbb{Z}(\mathbb{R})$, the restriction $[0, 1] \times A \xrightarrow{T} \mathrm{Homeo}_\mathbb{Z}(\mathbb{R})$ is uniformly continuous and has a coarsely bounded image. In other words, T is an efficient contraction of $\mathrm{Homeo}_\mathbb{Z}(\mathbb{R})$ to id.

Although trivially verified in all cases above, it is not clear if the assumption that $R\big[[0, 1] \times A\big]$ is coarsely bounded for coarsely bounded A is not already implied by uniform continuity.

Problem 5.49 Suppose G is a locally bounded Polish group and $[0, 1] \times G \xrightarrow{R} G$ is a continuous contraction onto 1_G so that, for every coarsely bounded set A, the restriction $[0, 1] \times A \xrightarrow{R} G$ is uniformly continuous. Is R then an efficient contraction, that is, is $R\big[[0, 1] \times A\big]$ coarsely bounded in G for every coarsely bounded set A?

We shall now use the contraction R to define generalised convex combinations in the group G. First, for $x \in G$, let $S(1, x) = x$. We set

$$
\Delta_n = \left\{ (\lambda_i) \in [0, 1]^n \;\middle|\; \sum_{i=1}^n \lambda_i = 1 \right\}
$$

and, for $(\lambda_i) \in \Delta_n$ and $x_1, \ldots, x_n \in G$, define

$$
S(\lambda_1, x_1, \ldots, \lambda_n, x_n) = \begin{cases} x_n & \text{if } \lambda_n = 1, \\ x_n R_{\lambda_n}\left(x_n^{-1} S\left(\frac{\lambda_1}{1-\lambda_n}, x_1, \ldots, \frac{\lambda_{n-1}}{1-\lambda_n}, x_{n-1}\right)\right) & \text{otherwise.} \end{cases}
$$

We think of $S(\lambda_1, x_1, \ldots, \lambda_n, x_n)$ as being akin to the 'convex combination' of the points $x_1, \ldots, x_n$ with coefficients $\lambda_1, \ldots, \lambda_n$. However, because the construction is not associative, S will in general not be symmetric in its variables, i.e., if $\pi \in \mathrm{Sym}(n)$, then we may have

$$S(\lambda_{\pi(1)}, x_{\pi(1)}, \ldots, \lambda_{\pi(n)}, x_{\pi(n)}) \neq S(\lambda_1, x_1, \ldots, \lambda_n, x_n).$$

Nevertheless, variables x_k with coefficient $\lambda_k = 0$ are redundant, in the sense that

$$S(\lambda_1, x_1, \ldots, 0, x_k, \ldots \lambda_n, x_n) = S(\lambda_1, x_1, \ldots, \widehat{0}, \widehat{x_k}, \ldots, \lambda_n, x_n),$$

where $\widehat{}$ indicates that the term has been omitted. This is easily seen by induction on n.

Lemma 5.50 *Let G be an efficiently contractible Polish group with a coarsely proper compatible left-invariant metric d. Then, for all K and $n \geqslant 1$, there is some constant $\theta(K, n)$ so that*

$$d\big(S(\lambda_1, x_1, \ldots, \lambda_n, x_n), x_n\big) < \theta(K, n)$$

whenever $x_1, \ldots, x_n \in G$ and $(\lambda_i) \in \Delta_n$ satisfy $\mathrm{diam}_d(\{x_1, x_2, \ldots, x_n\}) \leqslant K$.

Proof We prove this by induction on $n \geqslant 1$ uniformly for all K. When $n = 1$, we simply have $S(1, x_1) = x_1$ and the result is trivial.

So assume by induction that the result holds for some n and all K. Fix a value of K. Then, because R is an efficient contraction and d is coarsely proper, there is some radius $\theta(K, n+1)$ so that

$$R\big[[0,1] \times B_d\big(\theta(K, n) + K\big)\big] \subseteq B_d\big(\theta(K, n+1)\big),$$

where $B_d(\alpha)$ denotes the open ball of radius α centred at the identity.

Now, suppose that $x_1, \ldots, x_{n+1} \in G$ and $(\lambda_i) \in \Delta_{n+1}$ are given and satisfy

$$\mathrm{diam}_d(\{x_1, x_2, \ldots, x_{n+1}\}) \leqslant K.$$

If $\lambda_{n+1} = 1$, then $S(\lambda_1, x_1, \ldots, \lambda_{n+1}, x_{n+1}) = x_{n+1}$ and hence

$$d\big(S(\lambda_1, x_1, \ldots, \lambda_{n+1}, x_{n+1}), x_{n+1}\big) < \theta(K, n+1).$$

So assume instead that $\lambda_{n+1} < 1$. Then, by the inductive hypothesis,

$$d\left(x_{n+1}^{-1} S\left(\frac{\lambda_1}{1 - \lambda_{n+1}}, x_1, \ldots, \frac{\lambda_n}{1 - \lambda_{n+1}}, x_n\right), 1\right)$$

$$= d\left(S\left(\frac{\lambda_1}{1 - \lambda_{n+1}}, x_1, \ldots, \frac{\lambda_n}{1 - \lambda_{n+1}}, x_n\right), x_{n+1}\right)$$

$$\leqslant d\left(S\left(\frac{\lambda_1}{1-\lambda_{n+1}},x_1,\ldots,\frac{\lambda_n}{1-\lambda_{n+1}},x_n\right),x_n\right)+d(x_n,x_{n+1})$$

$$< \theta(K,n)+K$$

and so

$$R_{\lambda_{n+1}}\left(x_{n+1}^{-1}S\left(\frac{\lambda_1}{1-\lambda_{n+1}},x_1,\ldots,\frac{\lambda_n}{1-\lambda_{n+1}},x_n\right)\right)\in B_d\big(\theta(K,n+1)\big).$$

It thus follows that

$$d\big(S(\lambda_1,x_1,\ldots,\lambda_{n+1},x_{n+1}),x_{n+1}\big)$$

$$= d\left(x_{n+1}R_{\lambda_{n+1}}\left(x_{n+1}^{-1}S\left(\frac{\lambda_1}{1-\lambda_{n+1}},x_1,\ldots,\frac{\lambda_n}{1-\lambda_{n+1}},x_n\right)\right),x_{n+1}\right)$$

$$= d\left(R_{\lambda_{n+1}}\left(x_{n+1}^{-1}S\left(\frac{\lambda_1}{1-\lambda_{n+1}},x_1,\ldots,\frac{\lambda_n}{1-\lambda_{n+1}},x_n\right)\right),1\right)$$

$$< \theta(K,n+1),$$

which finishes the inductive step from n to $n+1$.   □

Lemma 5.51 *Let G be an efficiently contractible Polish group with a coarsely proper compatible left-invariant metric d. Also let K and $n \geqslant 1$ be fixed. Then, for all $\epsilon > 0$, there is $\delta > 0$ so that*

$$d(S(\lambda_1,x_1,\ldots,\lambda_n,x_n),S(\sigma_1,x_1,\ldots,\sigma_n,x_n)) < \epsilon$$

whenever $\mathsf{diam}_d\big(\{x_1,x_2,\ldots,x_n\}\big) \leqslant K$ *and* $\max_i |\lambda_i - \sigma_i| < \delta$.
In other words, assuming that $\mathsf{diam}_d\big(\{x_1,x_2,\ldots,x_n\}\big) \leqslant K$, *the mapping*

$$(\lambda_i) \in \Delta_n \mapsto S(\lambda_1,x_1,\ldots,\lambda_n,x_n) \in G$$

is uniformly continuous and, moreover, the modulus of uniform continuity is independent of the (x_i).

Proof The proof is by induction on $n \geqslant 1$ uniformly for all K. As $S(1,x_1)=x_1$, the base case of $n = 1$ is trivial. So suppose that the result holds for some $n \geqslant 1$ and all K. Let K and $\epsilon > 0$ be given. Then, by Lemma 5.50, there is some constant $\theta = \theta(K,n)$ so that

$$d\big(S(\lambda_1,x_1,\ldots,\lambda_n,x_n),x_n\big) < \theta$$

whenever $\mathsf{diam}_d\big(\{x_1,x_2,\ldots,x_n\}\big) \leqslant K$. Because R is uniformly continuous on bounded sets, we may pick $\delta > 0$ so that $d(R_\lambda(y),R_\sigma(z)) < \epsilon$, whenever $d(y,1),d(z,1) \leqslant K+\theta$, $d(y,z) < \delta$ and $|\lambda - \sigma| < 3\delta$. In particular,

$$d(R_\lambda(y),1) = d(R_\lambda(y),R_1(y)) < \epsilon$$

assuming $d(y,1) \leqslant K + \theta$ and $\lambda > 1 - 3\delta$.

By the induction hypothesis, choose also some $\eta > 0$ so that

$$d\big(S(\lambda_1, x_1, \ldots, \lambda_n, x_n), S(\sigma_1, x_1, \ldots, \sigma_n, x_n)\big) < \delta,$$

whenever $\mathsf{diam}_d(\{x_1, x_2, \ldots, x_n\}) \leqslant K$ and $|\lambda_i - \sigma_i| < \frac{3\eta}{\delta^2}$ for all i.

Assume now that $x_1, \ldots, x_{n+1} \in G$ and $(\lambda_i), (\sigma_i) \in \Delta_{n+1}$ are fixed and satisfy $\mathsf{diam}_d(\{x_1, x_2, \ldots, x_{n+1}\}) \leqslant K$ and $|\lambda_i - \sigma_i| < \min\{\delta, \eta\}$ for all i. For the induction step, it suffices to show that

$$d\big(S(\lambda_1, x_1, \ldots, \lambda_{n+1}, x_{n+1}), S(\sigma_1, x_1, \ldots, \sigma_{n+1}, x_{n+1})\big) < 2\epsilon.$$

We begin by observing that, when $\lambda_{n+1} < 1$, we have

$$d\left(x_{n+1}^{-1} S\left(\frac{\lambda_1}{1 - \lambda_{n+1}}, x_1, \ldots, \frac{\lambda_n}{1 - \lambda_{n+1}}, x_n\right), 1\right)$$

$$= d\left(S\left(\frac{\lambda_1}{1 - \lambda_{n+1}}, x_1, \ldots, \frac{\lambda_n}{1 - \lambda_{n+1}}, x_n\right), x_{n+1}\right)$$

$$\leqslant d\left(S\left(\frac{\lambda_1}{1 - \lambda_{n+1}}, x_1, \ldots, \frac{\lambda_n}{1 - \lambda_{n+1}}, x_n\right), x_n\right) + d(x_n, x_{n+1})$$

$$< \theta + K$$

and similarly when $\sigma_{n+1} < 1$ we have

$$d\left(x_{n+1}^{-1} S\left(\frac{\sigma_1}{1 - \sigma_{n+1}}, x_1, \ldots, \frac{\sigma_n}{1 - \sigma_{n+1}}, x_n\right), 1\right) < \theta + K.$$

The proof now splits into two cases. First suppose that $\lambda_{n+1} > 1 - 2\delta$. Then either $\lambda_{n+1} = 1$ and thus $S(\lambda_1, x_1, \ldots, \lambda_{n+1}, x_{n+1}) = x_{n+1}$ or, on the other hand, $1 - 2\delta < \lambda_{n+1} < 1$. In the second case, we have

$$d\big(S(\lambda_1, x_1, \ldots, \lambda_{n+1}, x_{n+1}), x_{n+1}\big)$$

$$= d\left(x_{n+1} R_{\lambda_{n+1}}\left(x_{n+1}^{-1} S\left(\frac{\lambda_1}{1 - \lambda_{n+1}}, x_1, \ldots, \frac{\lambda_n}{1 - \lambda_{n+1}}, x_n\right)\right), x_{n+1}\right)$$

$$= d\left(R_{\lambda_{n+1}}\left(x_{n+1}^{-1} S\left(\frac{\lambda_1}{1 - \lambda_{n+1}}, x_1, \ldots, \frac{\lambda_n}{1 - \lambda_{n+1}}, x_n\right)\right), 1\right)$$

$$< \epsilon.$$

Moreover, we also have $\sigma_{n+1} > \lambda_{n+1} - \delta > 1 - 3\delta$, so by the same reasoning

$$d\big(S(\sigma_1, x_1, \ldots, \sigma_{n+1}, x_{n+1}), x_{n+1}\big) < \epsilon.$$

We thus find that

$$d\big(S(\lambda_1, x_1, \ldots, \lambda_{n+1}, x_{n+1}), S(\sigma_1, x_1, \ldots, \sigma_{n+1}, x_{n+1})\big)$$
$$\leqslant d\big(S(\lambda_1, x_1, \ldots, \lambda_{n+1}, x_{n+1}), x_{n+1}\big) + d\big(S(\sigma_1, x_1, \ldots, \sigma_{n+1}, x_{n+1}), x_{n+1}\big)$$
$$< 2\epsilon,$$

as required.

For the second case, suppose instead that $\lambda_{n+1} \leqslant 1 - 2\delta$. Then $\sigma_{n+1} \leqslant 1 - \delta$ and so both $\frac{1}{1-\lambda_{n+1}} \leqslant \frac{1}{\delta}$ and $\frac{1}{1-\sigma_{n+1}} \leqslant \frac{1}{\delta}$. Therefore, for any $i = 1, \ldots, n$, we have

$$\left| \frac{\lambda_i}{1 - \lambda_{n+1}} - \frac{\sigma_i}{1 - \sigma_{n+1}} \right|$$
$$= \left| \frac{\lambda_i(1 - \sigma_{n+1}) - \sigma_i(1 - \lambda_{n+1})}{(1 - \lambda_{n+1})(1 - \sigma_{n+1})} \right|$$
$$\leqslant \frac{1}{\delta^2} \left| (\lambda_i - \sigma_i) + (\sigma_i \lambda_{n+1} - \lambda_i \sigma_{n+1}) \right|$$
$$\leqslant \frac{|\lambda_i - \sigma_i|}{\delta^2} + \frac{|\sigma_i \lambda_{n+1} - \lambda_i \lambda_{n+1}|}{\delta^2} + \frac{|\lambda_i \lambda_{n+1} - \lambda_i \sigma_{n+1}|}{\delta^2}$$
$$< \frac{3\eta}{\delta^2}.$$

By the choice of η, it follows that

$$d\left(x_{n+1}^{-1} S\left(\frac{\lambda_1}{1 - \lambda_{n+1}}, x_1, \ldots, \frac{\lambda_n}{1 - \lambda_{n+1}}, x_n \right), \right.$$
$$\left. x_{n+1}^{-1} S\left(\frac{\sigma_1}{1 - \sigma_{n+1}}, x_1, \ldots, \frac{\sigma_n}{1 - \sigma_{n+1}}, x_n \right) \right)$$
$$= d\left(S\left(\frac{\lambda_1}{1 - \lambda_{n+1}}, x_1, \ldots, \frac{\lambda_n}{1 - \lambda_{n+1}}, x_n \right), \right.$$
$$\left. S\left(\frac{\sigma_1}{1 - \sigma_{n+1}}, x_1, \ldots, \frac{\sigma_n}{1 - \sigma_{n+1}}, x_n \right) \right)$$
$$< \delta.$$

Therefore,

$$d\Big(S\big(\lambda_1, x_1, \ldots, \lambda_{n+1}, x_{n+1}\big), S\big(\sigma_1, x_1, \ldots, \sigma_{n+1}, x_{n+1}\big)\Big)$$

$$= d\left[x_{n+1} R_{\lambda_{n+1}} \left(x_{n+1}^{-1} S\left(\frac{\lambda_1}{1-\lambda_{n+1}}, x_1, \ldots, \frac{\lambda_n}{1-\lambda_{n+1}}, x_n \right) \right), \right.$$
$$\left. x_{n+1} R_{\sigma_{n+1}} \left(x_{n+1}^{-1} S\left(\frac{\sigma_1}{1-\sigma_{n+1}}, x_1, \ldots, \frac{\sigma_n}{1-\sigma_{n+1}}, x_n \right) \right) \right]$$

$$= d\left[R_{\lambda_{n+1}} \left(x_{n+1}^{-1} S\left(\frac{\lambda_1}{1-\lambda_{n+1}}, x_1, \ldots, \frac{\lambda_n}{1-\lambda_{n+1}}, x_n \right) \right), \right.$$
$$\left. R_{\sigma_{n+1}} \left(x_{n+1}^{-1} S\left(\frac{\sigma_1}{1-\sigma_{n+1}}, x_1, \ldots, \frac{\sigma_n}{1-\sigma_{n+1}}, x_n \right) \right) \right]$$

$$< \epsilon.$$

This finishes the inductive step and thus the proof of the lemma. $\qquad\square$

Theorem 5.52 *Suppose $H \xrightarrow{\phi} G$ is a bornologous map from a Polish group H of bounded geometry to an efficiently contractible locally bounded Polish group G. Then ϕ is close to a uniformly continuous map $H \xrightarrow{\psi} G$.*

Before passing to the proof, we should mention that some assumption on H akin to bounded geometry is necessary. Indeed, A. Naor [66] constructs a bornologous map $X \xrightarrow{\phi} Y$ between two separable Banach spaces that is not close to any uniformly continuous map. The problem is, of course, that X is infinite-dimensional and hence does not have bounded geometry.

Proof Fix a gauge metric ∂ on H, a coarsely proper compatible left-invariant metric d on G and let $X \subseteq H$ be a maximally 2-discrete subset, i.e., $\partial(x, y) \geqslant 2$ for distinct $x, y \in X$. Then X is 2-dense in H and (X, ∂) is uniformly locally finite. Fix also a linear ordering $\prec$ of X, set

$$K = \sup \big(d(\phi(x), \phi(y)) \mid \partial(x, y) < 8\big)$$

and let M be the maximum cardinality of a diameter 8 subset of X.

For every $x \in X$, define a map $\theta_x \colon H \to [0, 3]$ by $\theta_x(h) = \max\{0, 3 - \partial(h, x)\}$. Note that θ_x is 1-Lipschitz and $\theta_x \geqslant 1$ on a ball of radius 2 centred at x, while $\mathsf{supp}(\theta_x)$ is contained in the 3-ball around x. Since X is uniformly locally finite, it follows that

$$\sup_{h \in H} \big|\{x \in X \mid \theta_x(h) > 0\}\big| < \infty.$$

and so, as X is also 2-dense in H, the mapping

$$\Theta(h) = \sum_{x \in X} \theta_x(h)$$

is a bounded Lipschitz function with $\Theta \geqslant 1$. It follows that setting $\lambda_x = \frac{\theta_x}{\Theta}$, we have a partition of unity $\{\lambda_x\}_{x \in X}$ by Lipschitz functions with some uniform Lipschitz constant C and each λ_x supported in the 3-ball centred at x.

Let now $h \in H$ be given and suppose $x_1 \prec \cdots \prec x_n$ are the elements $x \in X$ so that $\lambda_x(h) \neq 0$. We define

$$\psi(h) = S\big(\lambda_{x_1}(h), \phi(x_1), \ldots, \lambda_{x_n}(h), \phi(x_n)\big).$$

As the λ_x are only supported in the 3-ball centred at x, the x_i are within distance 3 of h, whereby $\mathsf{diam}_d(\{\phi(x_1), \ldots, \phi(x_n)\}) \leqslant K$ and $n \leqslant M$. Thus

$$d(\psi(h), \phi(h)) \leqslant d\big(S\big(\lambda_{x_1}(h), \phi(x_1), \ldots, \lambda_{x_n}(h), \phi(x_n)\big), \phi(x_n)\big)$$
$$+ d\big(\phi(x_n), \phi(h)\big),$$

which, by Lemma 5.50 and the fact that ϕ is bornologous, is bounded independently of h. Therefore, ψ is close to ϕ.

Let us now verify that $\psi \colon H \to G$ is uniformly continuous. So let $\epsilon > 0$ be given and choose, by Lemma 5.51, some $\delta > 0$ small enough so that

$$d\big(S(\lambda_1, a_1, \ldots, \lambda_n, a_n), S(\sigma_1, a_1, \ldots, \sigma_n, a_n)\big) < \epsilon,$$

whenever $n \leqslant M$, $\mathsf{diam}_d(\{a_1, \ldots, a_n\}) \leqslant K$ and $|\lambda_i - \sigma_i| < \delta$.

Suppose that $h, g \in H$ with $\partial(h, g) < \min\{1, \frac{\delta}{C}\}$, whence $|\lambda_x(h) - \lambda_x(g)| < \delta$ for all $x \in X$. Let also $x_1 \prec \cdots \prec x_n$ list $B_\partial(h, 4) \cap X$, whence $n \leqslant M$ and $\mathsf{diam}_d(\{\phi(x_1), \ldots, \phi(x_n)\}) \leqslant K$.

Since $\partial(h, g) < 1$, we see that, if $\lambda_x(h) \neq 0$ or $\lambda_x(g) \neq 0$, then x is among the $x_1, \ldots, x_n$. From this and by adding redundant variables, we see that, for some subsequences $1 \leqslant i_1 < \cdots < i_p \leqslant n$ and $1 \leqslant j_1 < \cdots < j_q \leqslant n$,

$$\psi(h) = S\big(\lambda_{x_{i_1}}(h), \phi(x_{i_1}), \ldots, \lambda_{x_{i_p}}(h), \phi(x_{i_p})\big)$$
$$= S\big(\lambda_{x_1}(h), \phi(x_1), \ldots, \lambda_{x_n}(h), \phi(x_n)\big)$$

and

$$\psi(g) = S\big(\lambda_{x_{j_1}}(g), \phi(x_{j_1}), \ldots, \lambda_{x_{j_q}}(g), \phi(x_{j_q})\big)$$
$$= S\big(\lambda_{x_1}(g), \phi(x_1), \ldots, \lambda_{x_n}(g), \phi(x_n)\big).$$

In particular, $d(\psi(h), \psi(g)) < \epsilon$ as desired. $\qquad\square$

Corollary 5.53 *Suppose H is a Polish group of bounded geometry and G is an efficiently contractible, locally bounded Polish group. If H is coarsely embeddable into G, then there is a uniformly continuous coarse embedding of H into G.*

As an instance of this, note that, by a result of N. Brown and E. Guentner [14], every uniformly locally finite metric space admits a coarse embedding into a reflexive space. This in turn implies that every Polish group of bounded geometry embeds coarsely into a reflexive space and then that this embedding can simultaneously be taken to be uniformly continuous.

Problem 5.54 Is the group $\mathrm{Isom}(\mathbb{U})$ efficiently contractible?

As an indication for this problem, let us note that J. Melleray [61] has shown that $\mathrm{Isom}(\mathbb{U})$ is homeomorphic to separable Hilbert space and thus is contractible. However, this contraction need of course not have anything to do with the coarse structure on the group.

5.9 Entropy, Growth Rates and Metric Amenability

The class of Polish groups of bounded geometry is very special in the sense that it will allow us to transfer a number of concepts of finitude from locally compact groups. Some of this can be done even in the larger category of coarse spaces of bounded geometry, e.g., [75], but we shall restrict our attention to Polish groups.

5.9.1 Entropy

The concept of a gauge of course allows us to do some very basic counting and measuring. For this, we recall A. N. Kolmogorov's notions of metric entropy and capacity [51, 52] in metric spaces.

Suppose (X,d) is a metric space of bounded geometry. Then $\alpha > 0$ is said to be a *gauge* for (X,d) if, for every $\beta < \infty$, there is a constant K_β so that every set of diameter $\leqslant \beta$ can be covered by K_β many open balls $B_d(x,\alpha) = \{y \in X \mid d(y,x) < \alpha\}$ of radius α. In this case, we define the α-*entropy*, $\mathrm{ent}_\alpha(A)$, of a bounded set $A \subseteq X$ to be the minimal number of open balls of radius α covering A.

Similarly, the α-*capacity*, $\mathrm{cap}_\alpha(A)$, is the largest size of a α-discrete subset D of A, i.e., so that $d(x,y) \geqslant \alpha$ for distinct points in D.

Observe that, if $D \subseteq A$ realises the α-capacity of A, then every point of A is within distance $< \alpha$ of some point of D and hence A is contained in the union of the $|D|$ open balls of radius α centred at the points of D. Thus, $\mathrm{ent}_\alpha(A) \leqslant \mathrm{cap}_\alpha(A)$. Conversely, no two points of distance $\geqslant 2\alpha$ can belong to the same open ball of radius α. Taken together, this shows that

$$\mathrm{cap}_{2\alpha}(A) \leqslant \mathrm{ent}_\alpha(A) \leqslant \mathrm{cap}_\alpha(A)$$

for any gauge $\alpha > 0$ and any bounded set $A \subseteq X$.

Note also, that if $\beta > \alpha$, then β is still a gauge for (X, d) and, as any open ball of radius β can be covered by K open balls of radius α for some K, we have

$$\mathrm{ent}_\beta \leqslant \mathrm{ent}_\alpha \leqslant K \cdot \mathrm{ent}_\beta.$$

Also, if d is a gauge metric on a Polish group G of bounded geometry, then, because the open unit ball $B_d(1_G, 1)$ centred at 1_G is a gauge for G, we see that the distance $\alpha = 1$ is a gauge for the metric space (G, d). For suggestiveness of notation, we shall write ent_d and cap_d for ent_1 and cap_1 respectively.

Yet another notation will be useful. Namely, suppose $A \subseteq G$ is a gauge and $B \subseteq G$ is a coarsely bounded set. Then we let $\mathrm{ent}_A(B)$ be the smallest number of left-translates of A covering B. That is,

$$\mathrm{ent}_A(B) = \mathrm{ent}_{E_A}(B) = \min(|F| \mid B \subseteq FA).$$

As above, we see that, for two gauges A and A' on a Polish group G of bounded geometry, we have

$$\frac{1}{N}\mathrm{ent}_A \leqslant \mathrm{ent}_{A'} \leqslant N \cdot \mathrm{ent}_A,$$

where $N = \max\{\mathrm{ent}_A(A'), \mathrm{ent}_{A'}(A)\}$. That is, the entropy functions associated to any two gauges are bi-Lipschitz equivalent. This also has the consequence that, if d and d' are two gauge metrics on G, then

$$\frac{1}{N}\mathrm{ent}_d \leqslant \mathrm{ent}_{d'} \leqslant N \cdot \mathrm{ent}_d$$

for some N.

Finally, if A is a subset of a metric space (X, d) and $\beta > 0$, we let

$$(A)_\beta = \{x \in X \mid d(x, A) < \beta\}$$

and

$$\partial_\beta A = (A)_\beta \setminus A.$$

For future reference, let us observe the following basic fact.

Lemma 5.55 *Suppose that G is a locally compact second countable group with left Haar measure λ and compatible left-invariant proper metric d. Then, for any subset $A \subseteq G$,*

$$\frac{\lambda(A)}{\lambda\big(B_d(1_G, 1)\big)} \leqslant \mathrm{ent}_d(A) \leqslant \mathrm{cap}_d(A) \leqslant \frac{\lambda\big((A)_{\frac{1}{2}}\big)}{\lambda\big(B_d(1_G, \frac{1}{2})\big)}.$$

Proof For the first inequality, observe that A is covered by $\mathrm{ent}_d(A)$ many left-translates of $B_d(1_G, 1)$ and hence, by left-translation invariance of λ,

$$\lambda(A) \leqslant \mathrm{ent}_d(A) \cdot \lambda\big(B_d(1_G, 1)\big).$$

For the last inequality, we similarly note that $(A)_{\frac{1}{2}}$ contains $\mathrm{cap}_d(A)$ many disjoint translates of $B_d(1_G, 1/2)$ and hence

$$\mathrm{cap}_d(A) \cdot \lambda\big(B_d(1_G, 1/2)\big) \leqslant \lambda\big((A)_{\frac{1}{2}}\big)$$

as claimed. $\qquad\qquad\qquad\qquad\qquad\qquad\qquad\qquad\qquad\qquad\qquad\qquad\qquad\quad \square$

5.9.2 Growth Rates

Assume now that G is a Polish group of bounded geometry generated by a coarsely bounded set A. Then, by increasing A, we may suppose that A is also a gauge for G and thus consider the corresponding increasing *growth function* for G given by

$$\mathfrak{g}_A(n) = \mathrm{ent}_A(A^n).$$

Assume now that B is a different gauge also generating G. Then, as observed above, the entropy function ent_B is bi-Lipschitz equivalent to that of A, say $\frac{1}{N}\mathrm{ent}_A \leqslant \mathrm{ent}_B \leqslant N \cdot \mathrm{ent}_A$ for some N. Moreover, as both A and B are symmetric, coarsely bounded generating sets containing 1, there is some sufficiently large M so that $A \subseteq B^M$ and $B \subseteq A^M$. It thus follows that

$$\mathfrak{g}_A(n) = \mathrm{ent}_A(A^n) \leqslant N \cdot \mathrm{ent}_B(A^n) \leqslant N \cdot \mathrm{ent}_B(B^{Mn}) = N \cdot \mathfrak{g}_B(Mn)$$

and similarly $\mathfrak{g}_B(n) \leqslant N \cdot \mathfrak{g}_A(Mn)$.

Definition 5.56 Two increasing functions $\mathfrak{g}, \mathfrak{f} \colon \mathbb{N} \to \mathbb{R}_+$ are said to have *equivalent growth rates* if there is some constant λ so that

$$\mathfrak{g}(n) \leqslant \lambda \cdot \mathfrak{f}(\lambda n + \lambda) + \lambda \quad \text{and} \quad \mathfrak{f}(n) \leqslant \lambda \cdot \mathfrak{g}(\lambda n + \lambda) + \lambda$$

for all n.

Thus, in view of the above, we see that the functions $\mathfrak{g}_A$ and $\mathfrak{g}_B$ have equivalent growth rates and thus define a natural invariant of G.

Moreover, the growth rate is quasi-isometry invariant. To see this, suppose that A and B are gauges generating Polish groups G and H respectively and let ρ_A and ρ_B be the associated word metrics. Suppose also that $G \xrightarrow{\phi} H$ is a quasi-isometry, say

$$\frac{1}{K}\rho_A(x,y) - K \leqslant \rho_B(\phi x, \phi y) \leqslant K\rho_A(x,y) + K$$

for all $x, y \in G$. Without loss of generality, we may assume that $\phi(1_G) = 1_H$, whence $\phi[A^n] \subseteq B^{Kn+K}$ for all n. Then, as the ρ_A-diameters of the inverse images $\phi^{-1}(hB)$ of left-translates of B are uniformly bounded, so are their A-entropies, i.e., $N = \sup_{h \in H} \operatorname{ent}_A\big(\phi^{-1}(hB)\big) < \infty$. It follows that

$$\operatorname{ent}_A(\phi^{-1}(C)) \leqslant N \cdot \operatorname{ent}_B(C)$$

for all $C \subseteq H$ and hence that

$$\mathfrak{g}_A(n) = \operatorname{ent}_A(A^n) \leqslant \operatorname{ent}_A\big(\phi^{-1}(B^{Kn+K})\big) \leqslant N \cdot \operatorname{ent}_B(B^{Kn+K})$$
$$= N \cdot \mathfrak{g}_B(Kn + K).$$

By symmetry, we find that $\mathfrak{g}_A$ and $\mathfrak{g}_B$ have equivalent growth rates.

Example 5.57 (Growth rates of locally compact groups) For a compactly generated, locally compact second countable group G, the growth rate is usually expressed as the volume growth, i.e., $\mathfrak{g}(n) = \lambda(A^n)$, where A is a compact generating set and λ is a left Haar measure. However, by Lemma 5.55, one sees that this gives an equivalent growth rate to that given by the entropy function ent_A.

5.9.3 Metric Amenability

We now consider the concept of metric amenability due to J. Block and S. Weinberger [12], see also [68]. This was originally introduced as a notion of amenability for metric spaces, but was expanded to coarse spaces of bounded geometry in [75]. To avoid confusion with the usual notion of amenability of topological groups, to which it is not equivalent, we shall use the more descriptive terminology of metric amenability.

In direct analogy with E. Følner's isoperimetric reformulation of amenability of discrete groups [32], we have the following concept of metric amenability [12].

Definition 5.58 Let (X, d) be a metric space of bounded geometry with gauge α. Then (X, d) is said to be *metrically amenable* if, for all $\epsilon > 0$ and $\beta < \infty$, there is a bounded set A with

$$\text{ent}_\alpha(\partial_\beta A) < \epsilon \cdot \text{ent}_\alpha(A).$$

A few words are in order with respect to this definition. Namely, observe first that it is independent of the specific choice of gauge α for (X, d). That is, (X, d) is metrically amenable with respect one α if and only if it is with respect to another gauge α'.

A second observation is that, because $(\overline{A})_\beta = (A)_\beta$ and thus

$$\partial_\beta \overline{A} = (\overline{A})_\beta \setminus \overline{A} \subseteq (A)_\beta \setminus A = \partial_\beta A,$$

we see that if (X, d) is metrically amenable one can always require the set A in the definition to be closed in X.

Thirdly, unless X is homogeneous, metric amenability may rely only on a part of the space. For example, suppose T_3 is the 3-regular tree and $t \in \mathsf{T}_3$ is any vertex. Then we may let $X = \mathsf{T}_3 \sqcup \mathbb{R}$ be the free amalgam where the distance between $s \in \mathsf{T}_3$ and $a \in \mathbb{R}$ is just

$$d(s, a) = d_{\mathsf{T}_3}(s, t) + |a|.$$

Then X has bounded geometry and, because of the sufficiently isolated copy of $\mathbb{R}$ in X, it is also metrically amenable despite the presence of the metrically non-amenable space T_3. However, as we are dealing with groups equipped with left-invariant metrics, mostly our examples will be homogenous.

Lemma 5.59 *Let (X, d) be a metric space of bounded geometry with gauge α. Then X fails to be metrically amenable if and only if, for all K, there is some σ so that*

$$\text{ent}_\alpha\big((A)_\sigma\big) \geqslant K \cdot \text{ent}_\alpha(A)$$

for all bounded sets A.

Proof That this condition implies that X cannot be metrically amenable is clear because

$$\text{ent}_\alpha\big((A)_\sigma\big) \leqslant \text{ent}_\alpha(\partial_\sigma A) + \text{ent}_\alpha(A).$$

On the other hand, suppose X fails to be metrically amenable and choose $\epsilon > 0$ and $\beta > 2\alpha$ so that

$$\text{ent}_\alpha(\partial_\beta A) \geqslant \epsilon \cdot \text{ent}_\alpha(A)$$

for all bounded $A \subseteq X$. Also fix K and let $A \subseteq X$ be any bounded non-empty set. Let $n > K/\epsilon$ and observe that

$$\mathrm{ent}_\alpha\left((A)_{(2m+1)\beta} \setminus (A)_{2m\beta}\right) \geq \mathrm{ent}_\alpha\left(\partial_\beta\left((A)_{2m\beta}\right)\right)$$
$$\geq \epsilon \cdot \mathrm{ent}_\alpha\left((A)_{2m\beta}\right)$$
$$\geq \epsilon \cdot \mathrm{ent}_\alpha(A)$$

for all m. Moreover, for $m < m'$, the minimal distance between $(A)_{(2m+1)\beta} \setminus (A)_{2m\beta}$ and $(A)_{(2m'+1)\beta} \setminus (A)_{2m'\beta}$ is at least

$$2m'\beta - (2m + 1)\beta \geq \beta > 2\alpha,$$

so no ball of radius α can intersect both of these sets. Therefore,

$$\mathrm{ent}_\alpha\left((A)_{(2n+1)\beta}\right) \geq \sum_{m=1}^{n} \mathrm{ent}_\alpha\left((A)_{(2m+1)\beta} \setminus (A)_{2m\beta}\right)$$
$$\geq n\epsilon \cdot \mathrm{ent}_\alpha(A)$$
$$> K \cdot \mathrm{ent}_\alpha(A).$$

So $\sigma = (2n + 1)\beta$ works for K.   □

The fifth and more substantial observation is that metric amenability is a coarse invariant of metric spaces of bounded geometry. Although this is certainly well known, for the purpose of exposition, we give a proof.

Proposition 5.60 *Suppose X and Y are coarsely equivalent metric spaces of bounded geometry. Then X is metrically amenable if and only if Y is.*

Proof Assume that X fails to be metrically amenable. We show that so does Y. So assume that $X \xrightarrow{\phi} Y$ is a coarse equivalence and pick δ so that

$$\sup_{y \in Y} d(y, \phi[X]) < \delta.$$

Suppose some constant K and gauge η for Y are given. Let α be a gauge for X large enough so that $d(\phi x, \phi x') \geq 2\eta$ whenever $d(x, x') \geq \alpha$. Then

$$\mathrm{ent}_\alpha(A) \leq \mathrm{cap}_\alpha(A) \leq \mathrm{cap}_{2\eta}(\phi[A]) \leq \mathrm{ent}_\eta(\phi[A])$$

for all bounded $A \subseteq X$.

Let $\eta' > \eta$ be chosen so that $d(\phi x, \phi x') < \eta'$ whenever $d(x, x') < \alpha$. Then ϕ maps open balls of radius α into open balls of radius η' and hence

$$\mathrm{ent}_{\eta'}\left(\phi[A]\right) \leq \mathrm{ent}_\alpha(A)$$

for all bounded $A \subseteq X$. Choose N so that, for bounded $B \subseteq Y$,

$$\mathrm{ent}_\eta(B) \leqslant N \cdot \mathrm{ent}_{\eta'+\delta}(B)$$

and let σ witness the failure of metric amenability of X, that is, so that

$$\mathrm{ent}_\alpha\big((A)_\sigma\big) \geqslant KN \cdot \mathrm{ent}_\alpha(A)$$

for all bounded $A \subseteq X$. Pick σ' so that $d(\phi x, \phi x') < \sigma'$ whenever $d(x, x') < \sigma$, whereby

$$\phi[(A)_\sigma] \subseteq \big(\phi[A]\big)_{\sigma'}$$

for $A \subseteq X$.

Then, for all bounded $A \subseteq X$, we have

$$
\begin{aligned}
KN \cdot \mathrm{ent}_{\eta'}\big(\phi[A]\big) &\leqslant KN \cdot \mathrm{ent}_\alpha(A) \\
&\leqslant \mathrm{ent}_\alpha\big((A)_\sigma\big) \\
&\leqslant \mathrm{ent}_\eta\big(\phi[(A)_\sigma]\big) \\
&\leqslant \mathrm{ent}_\eta\big((\phi[A])_{\sigma'}\big).
\end{aligned}
$$

On the other hand, if $B \subseteq Y$ is bounded, let $A = \phi^{-1}\big((B)_\delta\big)$, whereby the Hausdorff distance between B and $\phi[A]$ is at most δ. In particular,

$$
\begin{aligned}
K \cdot \mathrm{ent}_\eta(B) &\leqslant KN \cdot \mathrm{ent}_{\eta'+\delta}(B) \\
&\leqslant KN \cdot \mathrm{ent}_{\eta'}(\phi[A]) \\
&\leqslant \mathrm{ent}_\eta\big((\phi[A])_{\sigma'}\big) \\
&\leqslant \mathrm{ent}_\eta\big((B)_{\sigma'+\delta}\big).
\end{aligned}
$$

Thus, the constant $\sigma' + \delta$ witnesses the failure of metric amenability for Y.

$\square$

If G is a Polish group of bounded geometry and d is a gauge metric, we say that G is *metrically amenable* when (G, d) is. As different gauge metrics are always coarsely equivalent, by Proposition 5.60, we see that this definition is independent of the choice of gauge metric d on G. In fact, suppose A is any choice of gauge for G. Then G is metrically amenable exactly when, for all $\epsilon > 0$ and coarsely bounded set C, there is a coarsely bounded set B so that

$$\mathrm{ent}_A(BC \setminus B) < \epsilon \cdot \mathrm{ent}_A(B).$$

Similarly, as gauge metrics are compatible metrics for the coarse structure on G, we note that metric amenability is a coarse invariant of Polish groups of bounded geometry.

5.9.4 Amenability versus Metric Amenability
of Locally Compact Groups

Although locally compact Polish groups are not the main object of our study, all the examples of Polish groups of bounded geometry found so far are coarsely equivalent to locally compact Polish groups. For this reason, it is therefore useful to have a good picture of what metric amenability means in this context.

So suppose that G is a locally compact second countable group with left Haar measure λ and compatible left-invariant proper metric d. By the observations of Section 5.9.3, we have that G is metrically amenable if and only if, for all $\epsilon > 0$ and compact sets C, there is a compact set K with

$$\mathrm{ent}_d(KC \setminus K) < \epsilon \cdot \mathrm{ent}_d(K).$$

On the other hand, G being amenable is equivalent to Følner's condition [32] that, for all $\epsilon > 0$ and compact C, there is a compact set K with

$$\lambda(CK \setminus K) < \epsilon \cdot \lambda(K).$$

The superficial similarity between these two conditions hides some delicate issues having to do with the fact that the entropy ent_d and the measure λ are invariant under *left-translations*, while on the other hand KC is a union of *right-translates* of K. In fact, as shown by R. Tessera [87], metric amenability is equivalent to the conjunction of amenability and unimodularity. The next proposition expands on this fact.

Proposition 5.61 *Let G be a locally compact second countable group with left Haar measure λ and compatible left-invariant proper metric d. Then the following conditions are equivalent.*

(1) *G is metrically amenable, i.e., for all $\epsilon > 0$ and compact sets C, there is a compact set K with*

$$\mathrm{ent}_d(KC \setminus K) < \epsilon \cdot \mathrm{ent}_d(K),$$

(2) *for all $\epsilon > 0$ and compact sets C, there is a compact set K with*

$$\lambda(KC \setminus K) < \epsilon \cdot \lambda(K),$$

(3) *G is amenable and unimodular,*
(4) *for all $\epsilon > 0$ and compact sets C, there is a compact set K with*

$$\lambda(CKC \setminus K) < \epsilon \cdot \lambda(K),$$

(5) *for all $\epsilon > 0$ and compact sets C, there is a compact set K with*

$$\text{ent}_d(CKC \setminus K) < \epsilon \cdot \text{ent}_d(K).$$

Proof (1)$\Rightarrow$(2) Assume that G is metrically amenable and let $\epsilon > 0$ and C be given. Let $B = B_d(1_G, \frac{1}{2})$ and find a compact set K so that

$$\text{ent}_d(KBC \setminus K) < \frac{\epsilon\lambda(B)}{\lambda\big(B_d(1_G,1)\big)} \cdot \text{ent}_d(K).$$

Setting $K' = KB = (K)_{\frac{1}{2}}$ and using the Lipschitz bounds of Lemma 5.55, we then see that

$$\begin{aligned}
\lambda(K'C \setminus K') &\leqslant \lambda(KBC \setminus K) \\
&\leqslant \text{ent}_d(KBC \setminus K) \cdot \lambda(B_d(1_G,1)) \\
&< \epsilon\lambda(B) \cdot \text{ent}_d(K) \\
&\leqslant \epsilon \cdot \lambda\big((K)_{\frac{1}{2}}\big) \\
&= \epsilon \cdot \lambda(K').
\end{aligned}$$

So K' is the set sought in (2).

(2)$\Rightarrow$(3) Assume that (2) holds and let Δ be the modular function on G, i.e., Δ satisfies

$$\Delta(g) = \frac{\lambda(Kg)}{\lambda(K)}$$

for all compact non-null sets $K \subseteq G$. By (2) we have that, for any fixed $g \in G$ and $\epsilon > 0$, there is some compact K with $\lambda(Kg \setminus K) < \epsilon\lambda(K)$ and so

$$\lambda(Kg) \leqslant \lambda(K) + \lambda(Kg \setminus K) < (1+\epsilon)\lambda(K),$$

whence $\Delta(g) < 1 + \epsilon$. In other words, $\Delta(g) = 1$ for all $g \in G$ and so G is unimodular.

To see that also G is amenable, we note that, for all $\epsilon > 0$ and compact C, there is a compact set K so that

$$\lambda(KC^{-1} \setminus K) < \epsilon \cdot \lambda(K)$$

and hence, by taking inverses and applying unimodularity, we have

$$\lambda(CK^{-1} \setminus K^{-1}) = \lambda(KC^{-1} \setminus K) < \epsilon \cdot \lambda(K) = \epsilon \cdot \lambda(K^{-1}).$$

Thus, K^{-1} witnesses the Følner condition for G.

(3)$\Rightarrow$(4) This implication uses convolution of L^1-functions and can be found in Proposition 2 in [**69**].

(4)$\Rightarrow$(5) Assume (4) holds and let $\epsilon > 0$ and a compact set C be given. Let $B = B_d(1_G, \frac{1}{2})$ and find a compact set K so that

$$\lambda(CKBCB \setminus K) < \frac{\epsilon \lambda(B)}{\lambda(B_d(1_G, 1))} \cdot \lambda(K).$$

Setting again $K' = KB$ and employing the Lipschitz bounds of Lemma 5.55, we have

$$\text{ent}_d(CK'C \setminus K') \leqslant \frac{\lambda\big((CK'C \setminus K')_{\frac{1}{2}}\big)}{\lambda(B)}$$

$$= \frac{\lambda\big((CKBC \setminus KB) \cdot B\big)}{\lambda(B)}$$

$$\leqslant \frac{\lambda(CKBCB \setminus K)}{\lambda(B)}$$

$$< \frac{\epsilon \cdot \lambda(K)}{\lambda(B_d(1_G, 1))}$$

$$\leqslant \epsilon \cdot \text{ent}_d(K)$$

$$\leqslant \epsilon \cdot \text{ent}_d(K').$$

(5)$\Rightarrow$(1) This is trivial. $\qquad\qquad\qquad\qquad\qquad\qquad\qquad\qquad\square$

In particular, for countable discrete groups, the two notions of amenability coincide, because these groups are automatically unimodular. This is also part of Følner's theorem. However, for example, the metabelian locally compact group $\mathbb{R}_+ \ltimes \mathbb{R}$ is amenable, but fails to be unimodular and thus is not metrically amenable either. One may also note that $\mathbb{R}_+ \ltimes \mathbb{R}$ is coarsely equivalent to the metrically non-amenable hyperbolic plane $\mathbb{H}^2$.

In the context of Polish groups, metric amenability no longer implies amenability. Indeed, $\text{Homeo}_{\mathbb{Z}}(\mathbb{R})$ is a Polish group of bounded geometry coarsely equivalent to $\mathbb{Z}$ and therefore metrically amenable. However, $\text{Homeo}_{\mathbb{Z}}(\mathbb{R})$ is not amenable because, for example, it acts continuously on the compact space $\mathbb{S}^1$ without preserving a measure.

The equivalence of (1) and (5) in Proposition 5.61 of course begs the question whether this also holds in the context of general Polish groups of bounded geometry.

Problem 5.62 Let G be a Polish group of bounded geometry with a gauge metric d and assume that G is metrically amenable. Does it follow that, for all $\epsilon > 0$ and coarsely bounded C, there is a coarsely bounded set B so that

$$\text{ent}_d(CBC \setminus B) < \epsilon \cdot \text{ent}_d(B) ?$$

Observe that, if the left- and right-coarse structures coincide on G, then every coarsely bounded set C is contained in a coarsely bounded conjugacy-invariant set C', whereby $CB \subseteq C'B = BC'$ for all B. Thus, if G is also metrically amenable, then we can find B so that $\mathrm{ent}_d(BC'C \setminus B) < \epsilon \cdot \mathrm{ent}_d(B)$ and hence

$$\mathrm{ent}_d(CBC \setminus B) \leqslant \mathrm{ent}_d(BC'C \setminus B) < \epsilon \cdot \mathrm{ent}_d(B).$$

This reasoning applies, for example, to $\mathrm{Homeo}_{\mathbb{Z}}(\mathbb{R})$, which is metrically amenable because it is coarsely equivalent to the unimodular amenable group $\mathbb{Z}$. In other words, Problem 5.62 has a positive answer for $\mathrm{Homeo}_{\mathbb{Z}}(\mathbb{R})$.

5.10 Nets in Polish Groups

Recall that a subset X of a topological group G is *cobounded* in G if $G = XB$ for some coarsely bounded set $B \subseteq G$. Also, X is *uniformly discrete* if uniformly discrete with respect to the left-uniformity on G, i.e., if there is an identity neighbourhood U so that $xU \cap yU = \emptyset$ for all distinct $x, y \in X$.

Definition 5.63 A subset X of a topological group G is said to be a *net* in G if it is simultaneously cobounded and uniformly discrete.

We claim that a Polish group G admits a net if and only if it is locally bounded. Indeed, observe that, if X is a net in G, then, being uniformly discrete, X is countable and, by coboundedness, G is covered by countably many coarsely bounded sets, whereby G is locally bounded. Conversely, if G is locally bounded, we choose a net in G by letting X be a maximal 1-discrete set with respect to some coarsely proper metric on G.

In the case where a Polish group G has bounded geometry, we would like nets to reflect this fact. So we define a net $X \subseteq G$ to be *proper* if there is an open gauge $U \subseteq G$ so that $xU \cap yU = \emptyset$ for distinct $x, y \in X$. In this case, no two distinct points of X can belong to the same left-translate of U, which means that the entropy function ent_U, when restricted to subsets of X, is simply the counting measure.

As above, every Polish group of bounded geometry admits a proper net. The main quality of a proper net X that we shall be using is that, if d is a coarsely proper metric on G, then (X, d) is uniformly locally finite. Indeed, for every diameter r, there is a k so that every set $B \subseteq X$ of diameter r is covered by k left-translates of U. Because no two distinct points of X can belong to the same left-translate of U, this implies that such $B \subseteq X$ have cardinality at most k.

Also, if X is a net in a locally bounded Polish group G and d is a coarsely proper metric on G, then the isometric inclusion $(X, d) \hookrightarrow (G, d)$ is cobounded and thus a coarse equivalence. In other words, G is coarsely equivalent to the discrete metric space (X, d).

It is well known that all nets in an infinite-dimensional Banach space are bi-Lipschitz equivalent. We generalise this to Polish groups of unbounded geometry.

Proposition 5.64 *Let G and H be Polish groups of unbounded geometry with coarsely proper compatible left-invariant metrics d and ∂ and nets X and Y. Then G and H are coarsely equivalent if and only if there is a bijective coarse equivalence ϕ between (X, d) and (Y, ∂).*

Proof Clearly, if (X, d) and (Y, ∂) are coarsely equivalent, so are G and H.

Now, suppose conversely that G and H are coarsely equivalent, whereby also (X, d) and (Y, ∂) are coarsely equivalent by some map $X \xrightarrow{\psi} Y$. Assume that X is σ-cobounded in (G, d), i.e., that $G = \bigcup_{x \in X} B_d(x, \sigma)$. Then, because G has unbounded geometry, there is some $\alpha > 0$ so that no ball $B_d(g, \alpha - \sigma)$ of diameter $\alpha - \sigma$ can be covered by finitely many balls of radius σ. It follows that every ball $B_d(g, \alpha)$ of radius α and centre $g \in G$ has infinite intersection with X. Similarly, we may suppose that α is large enough so that, for every $h \in H$, the intersection $B_\partial(h, \alpha) \cap Y$ is infinite.

Pick $\beta > 2\alpha$ large enough so that

$$d(x, x') \geqslant \beta \;\; \Rightarrow \;\; \partial(\psi(x), \psi(x')) \geqslant 2\alpha$$

and let $Z \subseteq X$ be a maximal β-discrete subset. Then $\psi \colon Z \to Y$ is injective and $\psi[Z]$ is cobounded and 2α-discrete in Y.

Pick a partition $\{P_z\}_{z \in Z}$ of X so that, for all $z \in Z$,

$$X \cap B_d(z, \alpha) \subseteq P_z \subseteq B_d(z, \beta)$$

and, similarly, a partition $\{Q_z\}_{z \in Z}$ of Y so that, for some σ and all $z \in Z$,

$$Y \cap B_\partial(\psi(z), \alpha) \subseteq Q_z \subseteq B_\partial(\psi(z), \sigma).$$

Then each P_z and Q_z will be infinite, whereby we may extend $\psi|_Z$ to a bijection ϕ between X and Y so that $\phi[P_z] = Q_z$ for all $z \in Z$. As ϕ and ψ are easily seen to be close maps, i.e., $\sup_{x \in X} \partial(\phi(x), \psi(x)) < \infty$, it follows that ϕ is a coarse equivalence between (X, d) and (Y, ∂). $\qquad\square$

A similar statement is also true for metrically non-amenable groups provided we restrict our attention to proper nets in G and H. The reasoning that follows underlies the investigations of K. Whyte in [**95**].

Proposition 5.65 *Let G and H be metrically non-amenable Polish groups of bounded geometry with coarsely proper compatible left-invariant metrics d and ∂ and proper nets X and Y. Then G and H are coarsely equivalent if and only if there is a bijective coarse equivalence ϕ between (X,d) and (Y,∂).*

Proof Suppose that $G \xrightarrow{\phi} H$ is a coarse equivalence, whence

$$\sup_{h \in H} \operatorname{diam}_d(\phi^{-1}(h)) < \infty.$$

As (X,d) is uniformly locally finite, we see that $k = \sup_{h \in H} |\phi^{-1}(h) \cap X| < \infty$. In other words, $X \xrightarrow{\phi} H$ is at most k-to-1.

Assume that Y is α-cobounded in H and that U is an open gauge for H so that $yU \cap y'U = \emptyset$ for distinct y, y' in Y, whence ent_U is just the counting measure when restricted to subsets of Y. Pick N so that $\operatorname{ent}_U \leq N \cdot \operatorname{ent}_{(U)_\alpha}$ and, because H is metrically non-amenable, some σ so that $\operatorname{ent}_U((A)_\sigma) \geq Nk \cdot \operatorname{ent}_U(A)$ for all $A \subseteq H$. Because Y is α-cobounded in H, we see that, if $(A)_\alpha \cap Y \subseteq \bigcup_{i=1}^n h_i U$, then $A \subseteq \bigcup_{i=1}^n h_i \cdot (U)_\alpha$. In other words,

$$\operatorname{ent}_{(U)_\alpha}(A) \leq \operatorname{ent}_U((A)_\alpha \cap Y)$$

for all $A \subseteq H$.

Putting this together, we find that for any finite subset $A \subseteq H$

$$|A| \leq \operatorname{ent}_U(A)$$
$$\leq \frac{1}{Nk}\operatorname{ent}_U((A)_\sigma)$$
$$\leq \frac{1}{k}\operatorname{ent}_{(U)_\alpha}((A)_\sigma)$$
$$\leq \frac{1}{k}\operatorname{ent}_U((A)_{\sigma+\alpha} \cap Y)$$
$$= \frac{1}{k}|(A)_{\sigma+\alpha} \cap Y|.$$

In particular, if $D \subseteq X$ is finite, then, as ϕ is $\leq k$-to-1,

$$|D| \leq k \cdot |\phi[D]| \leq |(\phi[D])_{\sigma+\alpha} \cap Y|.$$

Define now a relation $\mathcal{R} \subseteq X \times Y$ by letting $x\mathcal{R}y \Leftrightarrow \partial(y,\phi(x)) < \sigma+\alpha$. By the above, any finite subset $D \subseteq X$ is $\mathcal{R}$-related to at least $|D|$ many elements of Y. So, by Hall's marriage lemma, this implies that there is an injection $X \xrightarrow{\zeta} Y$ whose graph is contained in $\mathcal{R}$, i.e., so that $\partial(\zeta(x),\phi(x)) < \sigma+\alpha$ for all $x \in X$.

Similarly, suppose $H \xrightarrow{\phi'} G$ is a coarse inverse to ϕ, that is, ϕ' is a coarse embedding so that $\phi' \circ \phi$ is close to id_G, i.e.,

$$v = \sup_{x \in G} d\big(x, \phi'(\phi(x))\big) < \infty.$$

Then one may produce an injection $Y \xrightarrow{\zeta'} X$ with $d(\zeta'(y), \phi'(y)) < \sigma' + \alpha'$ for all $y \in Y$, where α' and σ' correspond to α and σ in the construction above.

Finally, by the Schröder–Bernstein Theorem, there is a bijection $X \xrightarrow{\eta} Y$ whose graph $\mathcal{G}\eta$ in contained in the union $\mathcal{G}\zeta \cup (\mathcal{G}\zeta')^\mathsf{T}$. In other words, for all $x \in X$, either $\eta(x) = \zeta(x)$ or $x = \zeta'(\eta(x))$. Observe now that, if $\eta(x) = \zeta(x)$, then

$$\partial(\eta(x), \phi(x)) = \partial(\zeta(x), \phi(x)) < \sigma + \alpha.$$

On the other hand, suppose $x = \zeta'(\eta(x))$. Then

$$d\big(\phi'(\phi(x)), \phi'(\eta(x))\big) \leqslant d\big(\phi'(\phi(x)), x\big) + d\big(x, \phi'(\eta(x))\big)$$
$$\leqslant v + d\big(\zeta'(\eta(x)), \phi'(\eta(x))\big)$$
$$< v + \sigma' + \alpha'.$$

As ϕ' is expanding, this implies that $d\big(\phi(x), \eta(x)\big) < \mu$ for some constant μ independent of x. All in all, we therefore find that

$$\partial(\eta(x), \phi(x)) < \max\{\sigma + \alpha, \mu\}$$

for all $x \in X$ and thus η is close to the coarse equivalence ϕ and hence is itself a coarse equivalence.   □

Let us note that in the proofs of Propositions 5.64 and 5.65, from a coarse equivalence $G \xrightarrow{\psi} H$, we produce a bijective coarse equivalence $X \xrightarrow{\phi} Y$ that is close to ψ. In fact, setting $H = G$ and keeping track of the constants in the proof of Proposition 5.65, we can extract the following statement.

Lemma 5.66 *Suppose G is a metrically non-amenable Polish group of bounded geometry, d is a coarsely proper compatible left-invariant metric and X is a proper net. Then there is a constant K so that, for every isometry $(G, d) \xrightarrow{\psi} (G, d)$, there is a permutation ϕ of X with $\sup_{x \in X} d(\psi(x), \phi(x)) \leqslant K$.*

Suppose G is a metrically non-amenable Polish group of bounded geometry and pick a coarsely proper metric d and a proper net X. Let also K be the constant given by Lemma 5.66. Then every element $g \in G$ acts isometrically on (G, d) by left-translation and Lemma 5.66 thus provides an element $\phi(g)$ of the group $\text{Sym}(X)$ of all permutations of X so that $\sup_{x \in X} d(gx, \phi(g)x) \leqslant K$. In particular, for all $g, h \in G$ and $x, y \in X$,

(1) $\left| d(\phi(g)x, \phi(g)y) - d(x, y) \right| \leqslant 2K$, whereas
(2) $d(\phi(g)\phi(h)x, \phi(gh)x) \leqslant 3K$.

Because (X, d) is uniformly locally finite, it follows from condition (2) that the defect of $G \xrightarrow{\phi} \text{Sym}(X)$,

$$\Delta_\phi = \{1\} \cup \{\phi(gh)^{-1}\phi(g)\phi(h) \mid g, h \in G\}^{\pm},$$

is relatively compact when $\text{Sym}(X)$ is equipped with the permutation group topology.

In other words, if G is a metrically non-amenable Polish group of bounded geometry with a proper net X, then there is a quasi-morphism with relatively compact defect

$$G \xrightarrow{\phi} \text{Sym}(X)$$

so that, for each $g \in G$, the left-translation λ_g is close to $\phi(g)$ on X.

5.11 Two-Ended Polish Groups

In the following, we aim to determine the structure of Polish groups coarsely equivalent to $\mathbb{R}$. In the case of finitely generated groups, these are, of course, simply the two-ended groups classified via a result of H. Hopf [40]; namely, every finitely generated two-ended group contains a finite-index infinite cyclic subgroup (see also [60] for a proof). We will establish a generalisation of this to all Polish groups stating, in particular, that any Polish group coarsely equivalent to $\mathbb{R}$ contains a cobounded undistorted copy of $\mathbb{Z}$.

It will be useful to keep a few examples in mind that will indicate the possible behaviours. Apart from simple examples $\mathbb{Z}$ and $\mathbb{R}$, we of course have groups such as $\text{Homeo}_\mathbb{Z}(\mathbb{R})$ and $\text{Aut}_\mathbb{Z}(\mathbb{Q})$ that, though acting on $\mathbb{R}$, do not admit homomorphisms to $\mathbb{R}$. But one should also note the infinite dihedral group D_∞ of all isometries of $\mathbb{Z}$, which contains $\mathbb{Z}$ as an index-2 subgroup.

Since we will be passing to a subgroup of finite index, let us begin by observing that these are always coarsely embedded.

Proposition 5.67 *Suppose H is a closed subgroup of a Polish group G admitting a compact transversal, that is, a compact subset $T \subseteq G$ containing exactly one point from every left coset of H. Then H is coarsely embedded in G.*

Observe that, for example, this applies to the situation when H has finite index in G, in which case T is finite, or when G is a Zappa–Szép products between the closed subgroup H and a compact group T, i.e., $G = TH$ and $H \cap T = \{1\}$.

Note also that, if T is a transversal for the left cosets of H, then T^{-1} is a transversal for the right cosets and vice versa.

Proof Replacing T by T^{-1}, we may assume it is a transversal for the right cosets of H instead. This means that every element $g \in G$ has a unique representation as a product $g = ht$ with $h \in H$ and $t \in T$. We first show that the group multiplication map $H \times T \to G$ is a homeomorphism. As it is obviously continuous, we must verify that also the inverse is continuous, i.e., that, if a sequence $g_i = h_i t_i$ converges to some element $g = ht$ with $h, h_i \in H$ and $t, t_i \in T$, then also $h_i \to h$ and $t_i \to t$.

Assume first that $t_i \not\to t$. We pass to a subsequence of the t_i staying outside some fixed neighbourhood of t and then, by compactness of T, to a further subsequence t_i' that converges to some $s \in T$, $s \neq t$. Letting h_i' be the corresponding subsequence of the h_i, we still have $h_i' t_i' \to ht$ and so

$$\lim_i h_i' = \lim_i h_i' t_i' (t_i')^{-1} = (\lim_i h_i' t_i') \cdot (\lim(t_i')^{-1}) = ht \cdot s^{-1}.$$

As H is closed, we find that $f = hts^{-1} \in H$. But $ht = hts^{-1} \cdot s = fs$ and so, by the uniqueness of representations, we see that $h = f$ and $t = s$, contradicting our assumptions.

Thus, $t_i \to t$ and so, as $h_i t_i \to ht$, we have $h_i t \to ht$ and thus also $h_i \to h$. This shows that the group multiplication map $H \times T \to G$ is a homeomorphism.

We now proceed to show that H is coarsely embedded in G. So assume that a set $A \subseteq H$ is coarsely unbounded in H. We must show that it is coarsely unbounded in G too. For this, fix a continuous isometric action $H \curvearrowright X$ on a metric space (X, d) so that $A \cdot x$ has infinite diameter for some (and thus all) $x \in X$.

We let Ω be the space of H-equivariant continuous maps $G \xrightarrow{\xi} X$, i.e., so that, for all $g \in G$ and $h \in H$,

$$h \cdot \xi(g) = \xi(hg).$$

Equip Ω with the metric

$$d_\infty(\xi, \zeta) = \sup_{g \in G} d\big(\xi(g), \zeta(g)\big).$$

To see that d_∞ is indeed finite, note that, because T is compact and H acts isometrically on X, we have

$$d_\infty(\xi,\zeta) = \sup_{g \in G} d\big(\xi(g),\zeta(g)\big)$$

$$= \sup_{h \in H} \sup_{t \in T} d\big(\xi(ht),\zeta(ht)\big)$$

$$= \sup_{h \in H} \sup_{t \in T} d\big(h \cdot \xi(t), h \cdot \zeta(t)\big)$$

$$= \sup_{h \in H} \sup_{t \in T} d\big(\xi(t),\zeta(t)\big)$$

$$= \sup_{t \in T} d\big(\xi(t),\zeta(t)\big)$$

$$< \infty.$$

We can define a continuous isometric action of G on Ω via $g \cdot \xi = \xi(\cdot\, g)$. First, to see that the action is by isometries, note that, for $g \in G$ and $\xi, \zeta \in \Omega$,

$$d_\infty(g \cdot \xi, g \cdot \zeta) = \sup_{f \in G} d\big(\xi(fg),\zeta(fg)\big) = \sup_{k \in G} d\big(\xi(k),\zeta(k)\big) = d_\infty(\xi,\zeta).$$

Secondly, for continuity, as the action is by isometries, it suffices to show that, if $\xi \in \Omega$ and $\epsilon > 0$ are given, then there is some identity neighbourhood W in G so that

$$d_\infty(w \cdot \xi, \xi) \leqslant \epsilon$$

for all $w \in W$. So let ξ and $\epsilon > 0$ be given. By continuity of ξ, we find for every $t \in T$ some open identity neighbourhood V_t so that $\mathsf{diam}\big(\xi[t V_t^2]\big) < \epsilon$. Also, by compactness of T, pick a finite subset $F \subseteq T$ so that

$$T \subseteq \bigcup_{t \in F} t V_t$$

and let $W = \bigcap_{t \in F} V_t$. Now, if $w \in W$ and $s \in T$, there is some $t \in T$ so that $s \in t V_t \subseteq t V_t^2$, whereby also $sw \in t V_t W \subseteq t V_t^2$ and thus

$$d(\xi(sw),\xi(s)) < \epsilon.$$

This show that, for all $w \in W$,

$$d_\infty(w \cdot \xi, \xi) = \sup_{s \in T} \big((w \cdot \xi)(s), \xi(s)\big) = \sup_{s \in T} \big(\xi(sw),\xi(s)\big) \leqslant \epsilon$$

as required.

Let us now return to the subset $A \subseteq H$ and $x \in X$ so that $A \cdot x$ has infinite diameter in X. We let $\xi \in \Omega$ be defined by $\xi(ht) = h \cdot x$ for all $h \in H$ and $t \in T$.

That ξ is well defined follows from the unique representation of elements in G as products ht with $h \in H$ and $t \in T$ and the continuity follows from the fact that the group multiplication is a homeomorphism from $H \times T$ to G. Finally, to check H-equivariance, assume $h, f \in H$ and $t \in T$. Then

$$\xi(h \cdot ft) = \xi(hf \cdot t) = (hf) \cdot x = h \cdot (f \cdot x) = h \cdot \xi(ft),$$

as required. Finally, fix any $t \in T$ and note that

$$\sup_{h \in A} d_\infty(h \cdot \xi, \xi) \geqslant \sup_{h \in A} d((h \cdot \xi)(t), \xi(t))$$

$$\geqslant \sup_{h \in A} d(\xi(ht), \xi(t))$$

$$\geqslant \sup_{h \in A} d(h \cdot x, x)$$

$$= \infty.$$

In other words, $A \cdot \xi$ has infinite diameter in Ω and so A must be coarsely unbounded in G too. $\qquad\square$

In Proposition 5.67 above, H is automatically cobounded in G and so the inclusion mapping will not only be a coarse embedding but a coarse equivalence. In particular, if G is monogenic, then so is H and the inclusion mapping is a quasi-isometry between H and G.

A positive answer to the following problem would provide a conceptually simpler statement.

Problem 5.68 Suppose H is a cocompact closed subgroup of a Polish group G, i.e., $G = HK$ for some compact set $K \subseteq G$. Is H coarsely embedded in G?

Again, as every Polish group isomorphically embeds into a coarsely bounded group, for example, the isometry group of the Urysohn sphere, the above problem has a negative answer if we replace cocompact by cobounded. Even worse is the example of the semi-direct product $S_\infty \ltimes \mathbb{F}_\infty$, where S_∞ acts by permuting the free generators of the free non-abelian group $\mathbb{F}_\infty$ of denumerable rank. Here S_∞ is coarsely bounded, in fact even Roelcke precompact, but still $\mathbb{F}_\infty$ is not coarsely embedded in $S_\infty \ltimes \mathbb{F}_\infty$. So $\mathbb{F}_\infty$ is co-Roelcke pre-compact in $S_\infty \ltimes \mathbb{F}_\infty$, but fails to be coarsely embedded.

A weaker and perhaps more accessible set of problems than Problem 5.68 is the question of whether being locally bounded or monogenic passes to cocompact subgroups.

Problem 5.69 Suppose H is a cocompact closed subgroup of a locally bounded Polish group. Is H also locally bounded?

Problem 5.70 Suppose H is a cocompact closed subgroup of a monogenic Polish group. Is H also monogenic?

Let us now return to the problem of describing the two-ended Polish groups, i.e., those coarsely equivalent to $\mathbb{R}$. So, in the following, let G be a Polish group coarsely equivalent to $\mathbb{R}$, whence G is monogenic and thus admits a maximal metric d. Let X be a maximal 1-discrete subset of G. Suppose also that $\psi \colon G \to \mathbb{R}$ is a quasi-isometry, whereby, as X is 1-discrete, the restriction $\psi \!\restriction_X$ is K-Lipschitz for some $K > 0$. Thus, by the MacShane–Whitney extension theorem, there is a K-Lipschitz extension $\tilde{\psi} \colon (G,d) \to \mathbb{R}$ of $\psi \!\restriction_X$, namely,

$$\tilde{\psi}(y) = \inf_{x \in X} \psi(x) + K \cdot d(y,x).$$

Also, as X is maximally 1-discrete, we see that $\tilde{\psi}$ and ψ are close on G and hence $\tilde{\psi}$ remains a quasi-isometry. Therefore,

$$\frac{1}{K'}d(x,y) - K' \leqslant |\tilde{\psi}(x) - \tilde{\psi}(y)| \leqslant K'd(x,y)$$

for all $x, y \in G$ and some $K' > 0$. Setting $\phi = \frac{1}{K'}\tilde{\psi}$, we see that

$$\frac{1}{C}d(x,y) - C \leqslant |\phi x - \phi y| \leqslant d(x,y)$$

for some constant $C > 3$ so that $\phi[G]$ is C-cobounded in $\mathbb{R}$.

Lemma 5.71 *Suppose that $x, y, z \in G$ satisfy*

$$\phi x + C^4 \leqslant \phi y \leqslant \phi z - C^4.$$

Then, for every $g \in G$, either

$$\phi(gx) < \phi(gy) < \phi(gz)$$

or

$$\phi(gx) > \phi(gy) > \phi(gz).$$

Proof Note first that, for all $u, v \in G$,

$$|\phi(gu) - \phi(gv)| \geqslant \frac{1}{C}d(gu, gv) - C = \frac{1}{C}d(u,v) - C \geqslant \frac{1}{C}|\phi u - \phi v| - C,$$

so $\phi(gu) \neq \phi(gv)$ provided $|\phi u - \phi u| \geqslant C^4$. In particular, $\phi(gx), \phi(gy), \phi(gz)$ are all distinct.

Fix $g \in G$ and assume for a contradiction that $\phi(gz)$ lies between $\phi(gx)$ and $\phi(gy)$, say

$$\phi(gx) < \phi(gz) < \phi(gy).$$

Then, as $\phi[G]$ is C-cobounded in $\mathbb{R}$ and $\phi x < \phi y$, we may find a sequence $x_0 = x, x_1, x_2, \ldots, x_n = y \in G$ so that

$$\phi x_0 < \phi x_1 < \cdots < \phi x_n = \phi y < \phi z - C^4$$

and $|\phi x_i - \phi x_{i+1}| \leqslant 3C$ for all i. In particular,

$$|\phi(gx_i) - \phi(gx_{i+1})| \leqslant d(gx_i, gx_{i+1}) = d(x_i, x_{i+1})$$
$$\leqslant C|\phi x_i - \phi x_{i+1}| + C^2 \leqslant 4C^2$$

and so, as

$$\phi(gx_0) = \phi(gx) < \phi(gz) < \phi(gy) = \phi(gx_n),$$

we have $|\phi(gz) - \phi(gx_i)| \leqslant 2C^2$ for some i.

But then

$$C^4 \leqslant |\phi z - \phi x_i| \leqslant d(z, x_i) = d(gz, gx_i)$$
$$\leqslant C|\phi(gz) - \phi(gx_i)| + C^2 \leqslant 2C^3 + C^2,$$

which is absurd since $C > 3$.

The arguments for all other cases are entirely analogous and are left to the reader. $\square$

Lemma 5.72 *Let*

$$H = \{g \in G \mid \phi(gx) < \phi(gy) \text{ whenever } \phi x + C^4 \leqslant \phi y\}.$$

Then H is an open subgroup of G of index at most 2 and, moreover,

$$G \setminus H = \{g \in G \mid \phi(gx) > \phi(gy) \text{ whenever } \phi x + C^4 \leqslant \phi y\}.$$

Proof Observe that each $g \in G$ satisfies exactly one of the following conditions,

(1) $\phi(gx) < \phi(gy)$ whenever $\phi x + C^4 \leqslant \phi y$

or

(2) $\phi(gx) > \phi(gy)$ whenever $\phi x + C^4 \leqslant \phi y$.

If not, there are $g, x_0, x_1, y_0, y_1 \in G$ so that $\phi x_0 + C^4 \leqslant \phi y_0$ and $\phi x_1 + C^4 \leqslant \phi y_1$, while $\phi(gx_0) < \phi(gy_0)$ and $\phi(gx_1) > \phi(gy_1)$. But then we can pick some $u, z \in G$ so that $\phi u \leqslant \min\{\phi x_0, \phi x_1\} - C^4$ and $\max\{\phi y_0, \phi y_1\} + C^4 \leqslant \phi z$, whence by Lemma 5.71 both

$$\phi(gu) < \phi(gx_0) < \phi(gy_0) < \phi(gz)$$

and

$$\phi(gu) > \phi(gx_1) > \phi(gy_1) > \phi(gz),$$

which is absurd.

Now, to see that H is a subgroup, assume that $h, f \in H$, i.e., that $g = h$ and $g = f$ each satisfy condition (1) and pick $x, y \in G$ with $\phi x + C^6 \leqslant \phi y$. Then $\phi(hx) < \phi(hy)$ and

$$|\phi(hx) - \phi(hy)| \geqslant \frac{1}{C} d(x, y) - C \geqslant \frac{1}{C} |\phi x - \phi y| - C \geqslant C^5 - C \geqslant C^4.$$

Thus, $\phi(hx) + C^4 \leqslant \phi(hy)$, whence $\phi(fhx) < \phi(fhy)$. It follows that condition (2) fails for $g = fh$ and hence that, instead, $fh \in H$.

Similarly, if $h \in H$ and $f \notin H$, then $fh \notin H$ and, if $h, f \notin H$, then $fh \in H$. As $1 \in H$, this implies that H is a subgroup of index at most 2. Moreover, because $\phi \colon G \to \mathbb{R}$ is continuous, we see that H is open. $\qquad\square$

Theorem 5.73 *Suppose G is an amenable Polish group coarsely equivalent to $\mathbb{R}$. Then G contains an open subgroup H of index at most 2 that admits coarsely proper continuous homomorphism $H \xrightarrow{\pi} \mathbb{R}$.*

Proof Let ϕ and H be as above and set

$$\rho(h)\xi = \xi(\cdot h)$$

for all $\xi \colon H \to \mathbb{R}$ and $h \in H$. Let also $\mathrm{LUC}(H)$ denote the algebra of bounded real-valued left-uniformly continuous functions on H. Then we may define a cocycle $H \xrightarrow{b} \mathrm{LUC}(H)$ for the right-regular representation

$$\rho \colon H \curvearrowright \mathrm{LUC}(H)$$

by setting

$$b(h) = \phi - \rho(h)\phi.$$

Indeed, to see that $b(h) \in \mathrm{LUC}(H)$, note that

$$\|b(h)\|_\infty = \sup_{x \in H} |\phi(x) - \phi(xh)| \leqslant \sup_{x \in H} d(x, xh) = d(1, h).$$

And, to verify the cocycle equation, observe that

$$b(hf) = \phi - \rho(hf)\phi = (\phi - \rho(h)\phi) + \rho(h)(\phi - \rho(f)\phi)$$
$$= b(h) + \rho(h)b(f).$$

As G is amenable and H has finite index in G, also H is amenable. So pick a ρ-invariant mean $\mathfrak{m} \in \mathrm{LUC}(H)^*$.

To define the homomorphism $\pi \colon H \to \mathbb{R}$, we simply set $\pi(h) = \mathfrak{m}(b(h))$ and note that

$$\pi(hf) = \mathfrak{m}(b(hf))$$
$$= \mathfrak{m}(b(h) + \rho(g)b(f))$$
$$= \mathfrak{m}(b(h)) + \mathfrak{m}(\rho(h)b(f))$$
$$= \mathfrak{m}(b(h)) + \mathfrak{m}(b(f))$$
$$= \pi(h) + \pi(f),$$

i.e., π is a continuous homomorphism.

In order to verify that π is coarsely proper, fix $n \geqslant 4$ and suppose that $h \in H$ satisfies $d(h,1) \geqslant C^{n+2}$. Then, for all $x \in H$, we have $d(xh,x) \geqslant C^{n+2}$ and hence $|\phi(xh) - \phi(x)| \geqslant C^n$. Suppose first that $\phi(h) + C^n \leqslant \phi(1)$. Then, if $x \in H$, we have $\phi(xh) < \phi(x)$ and so

$$\phi(xh) + C^n \leqslant \phi(x)$$

i.e.,

$$C^n \leqslant b(h) \leqslant d(h,1)$$

and

$$C^n \leqslant \pi(h) = \mathfrak{m}(b(h)) \leqslant d(h,1).$$

Similarly, if $\phi(1) + C^n \leqslant \phi(h)$, then $-d(h,1) \leqslant \pi(h) \leqslant -C^n$, which thus shows that π is coarsely proper. $\qquad\square$

Observe that, if G is a locally compact Polish group coarsely equivalent with $\mathbb{R}$, then by the coarse invariance of metric amenability, i.e., Proposition 5.60, we find that also G is metrically amenable and hence amenable by Proposition 5.61. The following well-known fact is thus an immediate corollary of Theorem 5.73.

Corollary 5.74 *Suppose G is a locally compact Polish group coarsely equivalent with $\mathbb{R}$. Then there is an open subgroup H of index at most 2 with a compact normal subgroup $K \trianglelefteq H$ so that H/K is isomorphic either to $\mathbb{R}$ or to $\mathbb{Z}$.*

Proof As noted above, G is amenable. So let H be the subgroup given by Theorem 5.73 and $H \xrightarrow{\pi} \mathbb{R}$ the coarsely proper continuous homomorphism. We note that $\pi[H]$ is closed in $\mathbb{R}$. For if (h_i) is a sequence in H so that $\lim_i \pi(h_i)$ exists in $\mathbb{R}$, then $\{\pi(h_i)\}_i$ is relatively compact in $\mathbb{R}$ and hence $\{h_i\}_i$ is coarsely bounded, that is, relatively compact in H. It follows that, if h is a limit of a subsequence of the h_i, one has $\lim_i \pi(h_i) = \pi(h) \in \pi[H]$.

Set $K = \ker \pi$. Because π is (coarsely) proper, K is a compact normal subgroup of H and π factors through to a proper continuous monomorphism $H/K \xrightarrow{\tilde{\pi}} \mathbb{R}$ with closed image. Therefore, by the open mapping theorem, $\tilde{\pi}$ is a homeomorphism between H/K and a closed subgroup of $\mathbb{R}$. The latter is either $\mathbb{R}$ itself or isomorphic to $\mathbb{Z}$. $\qquad\qquad\square$

Apart from examples such as $\mathbb{Z}/2\mathbb{Z} \ltimes \mathbb{R}$, where $\mathbb{Z}/2\mathbb{Z}$ acts on $\mathbb{R}$ by changing signs, there are other interesting cases. For example, let α be an automorphism of a compact group K and let $\mathbb{Z} \ltimes_\alpha K$ be the corresponding semi-direct product. Again $\mathbb{Z} \ltimes_\alpha K$ is coarsely equivalent with $\mathbb{R}$.

Note also that Corollary 5.74 fails for general Polish groups, because, for example, $\mathrm{Homeo}_{\mathbb{Z}}(\mathbb{R})$ is connected and admits no non-trivial homomorphisms to $\mathbb{R}$. Still we have a weaker statement.

Theorem 5.75 *Let G be a Polish group coarsely equivalent to $\mathbb{R}$. Then there is an open subgroup H of index at most 2 and a coarsely bounded set $A \subseteq H$ so that every $h \in H \setminus A$ generates a cobounded undistorted infinite cyclic subgroup.*

Observe that, by Proposition 5.67, H will be coarsely embedded in G and so, as G is monogenic, the inclusion mapping will be a quasi-isometry between H and G. The last statement of the theorem thus says that, for every $h \in H \setminus A$, the map

$$n \in \mathbb{Z} \mapsto h^n \in H$$

defines a quasi-isometry between $\mathbb{Z}$ and H (and also with G).

Proof Let ϕ and H be as above and suppose that $d(h, 1) \geqslant C^6$ for some $h \in H$. Then either $\phi(1) + C^4 \leqslant \phi(h)$ or $\phi(h) + C^4 \leqslant \phi(1)$ and so, as $\langle h \rangle \subseteq H$, we have either

$$\cdots < \phi(h^{-2}) < \phi(h^{-1}) < \phi(1) < \phi(h) < \phi(h^2) < \cdots$$

or

$$\cdots < \phi(h^2) < \phi(h) < \phi(1) < \phi(h^{-1}) < \phi(h^{-2}) < \cdots.$$

Because also

$$C^4 \leqslant \frac{1}{C}d(h,1) - C \leqslant \frac{1}{C}d(h^{k+1},h^k) - C \leqslant |\phi(h^{k+1}) - \phi(h^k)| \leqslant d(h,1)$$

for all $k \in \mathbb{Z}$, we see that $\{\phi(h^k)\}_{k \in \mathbb{Z}}$ is a linearly ordered bi-infinite sequence in $\mathbb{R}$ whose successive terms have distance between C^4 and $d(h,1)$. In particular, $\{\phi(h^k)\}_{k \in \mathbb{Z}}$ is cobounded in $\mathbb{R}$ and thus $\langle h \rangle$ must be cobounded in H.

Thus, $k \in \mathbb{Z} \mapsto h^k \in H$ is a quasi-isometry of $\mathbb{Z}$ and H and hence $\langle h \rangle$ is a cobounded undistorted infinite cyclic subgroup of H. It therefore suffices to set $A = \{h \in H \mid d(h,1) < C^6\}$. $\square$

In comparison with Corollary 5.74, our Theorem 5.75 feels somewhat raw and might have a crisper algebraic strengthening. For this, one should always keep in mind the three examples D_∞, $\mathbb{Z} \ltimes_\alpha K$, where α is an automorphism of a compact group K, and $\mathrm{Homeo}_\mathbb{Z}(\mathbb{R})$. The following is perhaps too optimistic to hope for.

Problem 5.76 Suppose G is a Polish group coarsely equivalent with $\mathbb{R}$. Can one find a subnormal series of closed subgroups

$$K \trianglelefteq F \trianglelefteq H \trianglelefteq G$$

so that H has index at most 2 in G, the quotient group H/F is coarsely bounded, $F/K \cong \mathbb{Z}$ and K is coarsely bounded in G?

6

Automorphism Groups of Countable Structures

6.1 Non-Archimedean Polish Groups and Large-Scale Geometry

We now turn our attention to the coarse geometry of non-Archimedean Polish groups. Here a Polish group G is said to be *non-Archimedean* if there is a neighbourhood basis at the identity consisting of open subgroups of G.

One particular source of examples of non-Archimedean Polish groups is first-order model theoretical structures. Namely, if $\mathbf{M}$ is a countable first-order structure, e.g., a graph, a group, a field or a lattice, we equip its automorphism group Aut($\mathbf{M}$) with the *permutation group topology*, which is the group topology obtained by declaring the pointwise stabilisers

$$V_A = \{g \in \mathrm{Aut}(\mathbf{M}) \mid \forall x \in A \;\; g(x) = x\}$$

of all finite subsets $A \subseteq \mathbf{M}$ to be open. In this case, one sees that a basis for the topology on Aut($\mathbf{M}$) is given by the family of cosets $f V_A$, where $f \in \mathrm{Aut}(\mathbf{M})$ and $A \subseteq \mathbf{M}$ is finite.

Now, conversely, if G is a non-Archimedean Polish group, then, by considering its action on the left-coset spaces G/V, where V varies over open subgroups of G, one can show that G is topologically isomorphic to the automorphism group Aut($\mathbf{M}$) of some countable first-order structure $\mathbf{M}$.

The investigation of non-Archimedean Polish groups via the interplay between the model theoretical properties of the structure $\mathbf{M}$ and the dynamical and topological properties of the automorphism group Aut($\mathbf{M}$) is currently very active as witnessed, e.g., by the papers [**8, 48, 49, 89**].

In the following, $\mathbf{M}$ denotes a countable first-order structure. We use $\bar{a}, \bar{b}, \bar{c}, \ldots$ as variables for finite tuples of elements of $\mathbf{M}$ and shall write $(\bar{a}, \bar{b})$ to denote the concatenation of the tuples $\bar{a}$ and $\bar{b}$. The automorphism group Aut($\mathbf{M}$) acts naturally on tuples $\bar{a} = (a_1, \ldots, a_n) \in \mathbf{M}^n$ via

$$g \cdot (a_1, \ldots, a_n) = (ga_1, \ldots, ga_n).$$

With this notation, the pointwise stabiliser subgroups

$$V_{\overline{a}} = \{g \in \mathrm{Aut}(\mathbf{M}) \mid g \cdot \overline{a} = \overline{a}\},$$

where $\overline{a}$ ranges over all finite tuples in $\mathbf{M}$, form a neighbourhood basis at the identity in $\mathrm{Aut}(\mathbf{M})$. So, if $A \subseteq \mathbf{M}$ is the finite set enumerated by $\overline{a}$ and $\mathbf{A} \subseteq \mathbf{M}$ is the substructure generated by A, we have $V_{\mathbf{A}} = V_A = V_{\overline{a}}$. An *orbital type* $\mathcal{O}$ in $\mathbf{M}$ is simply the orbit of some tuple $\overline{a}$ under the action of $\mathrm{Aut}(\mathbf{M})$. Also, we let $\mathcal{O}(\overline{a})$ denote the orbital type of $\overline{a}$, i.e., $\mathcal{O}(\overline{a}) = \mathrm{Aut}(\mathbf{M}) \cdot \overline{a}$.

6.2 Orbital Types Formulation

If G is a monogenic non-Archimedean Polish group, there is a completely abstract way of identifying its quasi-isometry type. For the special case of compactly generated non-Archimedean locally compact groups, H. Abels (Beispiel 5.2 in [1]) did this via constructing a vertex transitive and coarsely proper action on a countable connected graph. We may use the same idea in the general case, exercising some caution while dealing with coarsely bounded rather than compact sets.

Namely, fix a symmetric open identity neighbourhood U generating the group and coarsely bounded in G. We construct a vertex transitive and coarsely proper action on a countable connected graph as follows. First, pick an open subgroup V contained in U and let $A \subseteq G$ be a countable set so that $VUV = AV$. Because $a \in VUV$ implies that also $a^{-1} \in (VUV)^{-1} = VUV$, we may assume that A is symmetric, whereby $AV = VUV = (VUV)^{-1} = V^{-1}A^{-1} = VA$ and thus also $(VUV)^k = (AV)^k = A^k V^k = A^k V$ for all $k \geqslant 1$. In particular, we note that $A^k V$ is coarsely bounded in G for all $k \geqslant 1$.

The graph $\mathbb{X}$ is now defined to be the set G/V of left-cosets of V along with the set of edges $\{(gV, gaV) \mid a \in A \text{ and } g \in G\}$. Note that the left-multiplication action of G on G/V is a vertex transitive action of G by automorphisms of $\mathbb{X}$. Moreover, because $G = \bigcup_k (VUV)^k = \bigcup_k A^k V$, one sees that the graph $\mathbb{X}$ is connected and hence the shortest path distance ρ is a well-defined metric on $\mathbb{X}$.

We claim that the action $G \curvearrowright \mathbb{X}$ is coarsely proper. Indeed, note that, if $g_n \to \infty$ in G, then (g_n) eventually leaves every coarsely bounded subset of G and thus, in particular, leaves every $A^k V$. Because, the k-ball around the vertex $1V \in \mathbb{X}$ is contained in the set $A^k V$, one sees that $\rho(g_n \cdot 1V, 1V) \to \infty$, showing that the action is coarsely proper. Therefore, by the Milnor–Schwarz

Lemma (Theorem 2.77), the mapping $g \mapsto gV$ is a quasi-isometry between G and $(\mathbb{X}, \rho)$.

However, this construction neither addresses the question of when G admits a maximal metric nor provides a very informative manner of defining this. For those questions, we need to investigate matters further.

In the following, $\mathbf{M}$ will be a fixed countable first-order structure and $\bar{a}$ will be a finite tuple in $\mathbf{M}$. Observe first that we have a canonical bijective correspondence between the space $\mathrm{Aut}(\mathbf{M})/V_{\bar{a}}$ of left-cosets of the isotropy subgroup $V_{\bar{a}}$ in $\mathrm{Aut}(\mathbf{M})$ and the orbital type $\mathcal{O}(\bar{a})$ given by

$$gV_{\bar{a}} \longmapsto g\bar{a}.$$

Observe also that this correspondence is trivially $\mathrm{Aut}(\mathbf{M})$-equivariant, where the automorphism group $\mathrm{Aut}(\mathbf{M})$ acts on $\mathrm{Aut}(\mathbf{M})/V_{\bar{a}}$ via left-multiplication. What is less immediately transparent is the following correspondence.

Fact 6.1 There is a bijective correspondence between the space

$$V_{\bar{a}} \backslash \mathrm{Aut}(\mathbf{M})/ V_{\bar{a}}$$

of double (or two-sided) cosets $V_{\bar{a}}gV_{\bar{a}}$ of $V_{\bar{a}}$ and the family of orbital types

$$\mathcal{O} \subseteq \mathcal{O}(\bar{a}) \times \mathcal{O}(\bar{a}),$$

which is given by

$$V_{\bar{a}}gV_{\bar{a}} \longmapsto \mathcal{O}(\bar{a}, g\bar{a}).$$

Indeed, to see that the correspondence is well defined and injective, note that

$$V_{\bar{a}}gV_{\bar{a}} = V_{\bar{a}}fV_{\bar{a}} \Leftrightarrow \exists h \in V_{\bar{a}} \;\; hgV_{\bar{a}} = fV_{\bar{a}}$$
$$\Leftrightarrow \exists h \;\; (h\bar{a}, hg\bar{a}) = (\bar{a}, f\bar{a})$$
$$\Leftrightarrow \mathcal{O}(\bar{a}, g\bar{a}) = \mathcal{O}(\bar{a}, f\bar{a}).$$

To see that the correspondence is surjective, note that, because

$$\mathcal{O}(g\bar{a}, h\bar{a}) = \mathcal{O}(\bar{a}, g^{-1}h\bar{a}),$$

the orbital type $\mathcal{O}(g\bar{a}, h\bar{a})$ is the image of the double coset $V_{\bar{a}}g^{-1}hV_{\bar{a}}$.

Observe also that, if S is a finite set of orbital types $\mathcal{O} \subseteq \mathcal{O}(\bar{a}) \times \mathcal{O}(\bar{a})$, then we may write the union of the corresponding collection of double cosets of $V_{\bar{a}}$ as

$$V_{\bar{a}}FV_{\bar{a}} = \bigcup_{\mathcal{O}(g\bar{a}, h\bar{a})\in S} V_{\bar{a}}g^{-1}hV_{\bar{a}} = \bigcup_{\mathcal{O}(\bar{a}, h\bar{a})\in S} V_{\bar{a}}hV_{\bar{a}}$$

for some finite subset $F \subseteq \mathrm{Aut}(\mathbf{M})$. Note, however, that F is not uniquely defined.

If $\mathcal{O} = \mathcal{O}(\overline{b}, \overline{c}) \subseteq \mathcal{O}(\overline{a}) \times \mathcal{O}(\overline{a})$ is an orbital type, then we may define its *transpose* as

$$\mathcal{O}^{\mathsf{T}} = \mathcal{O}(\overline{c}, \overline{b}).$$

We say that a set $\mathcal{S}$ of orbital types $\mathcal{O} \subseteq \mathcal{O}(\overline{a}) \times \mathcal{O}(\overline{a})$ is *symmetric* if it is closed under taking transposes.

Fact 6.2 There is a bijective correspondence between finite symmetric families $\mathcal{S}$ of orbital types $\mathcal{O} \subseteq \mathcal{O}(\overline{a}) \times \mathcal{O}(\overline{a})$ and symmetric finite unions $V_{\overline{a}} F V_{\overline{a}}$ of double cosets, i.e., so that $(V_{\overline{a}} F V_{\overline{a}})^{-1} = V_{\overline{a}} F V_{\overline{a}}$, given by

$$V_{\overline{a}} F V_{\overline{a}} = \bigcup_{\mathcal{O}(g\overline{a}, h\overline{a}) \in \mathcal{S}} V_{\overline{a}} g^{-1} h V_{\overline{a}}$$

and with inverse

$$\mathcal{S} = \left\{ \mathcal{O}(\overline{a}, g\overline{a}) \mid g \in F \right\}.$$

In order to simplify calculations of quasi-isometry types of various automorphism groups and provide a better visualisation of the group structure, we introduce the following graph.

Definition 6.3 Suppose $\overline{a}$ is a tuple in $\mathbf{M}$ and $\mathcal{S}$ is a finite symmetric family of orbital types $\mathcal{O} \subseteq \mathcal{O}(\overline{a}) \times \mathcal{O}(\overline{a})$. We let $\mathbb{X}_{\overline{a}, \mathcal{S}}$ denote the graph whose vertex set is the orbital type $\mathcal{O}(\overline{a})$ and whose edge relation is given by

$$(\overline{b}, \overline{c}) \in \text{Edge } \mathbb{X}_{\overline{a}, \mathcal{S}} \quad \Leftrightarrow \quad \overline{b} \neq \overline{c} \quad \text{and} \quad \mathcal{O}(\overline{b}, \overline{c}) \in \mathcal{S}.$$

Let also $\rho_{\overline{a}, \mathcal{S}}$ denote the corresponding shortest path metric on $\mathbb{X}_{\overline{a}, \mathcal{S}}$, where we stipulate that $\rho_{\overline{a}, \mathcal{S}}(\overline{b}, \overline{c}) = \infty$ whenever $\overline{b}$ and $\overline{c}$ belong to distinct connected components.

We remark that, as the vertex set of $\mathbb{X}_{\overline{a}, \mathcal{S}}$ is just the orbital type of $\overline{a}$, the automorphism group $\text{Aut}(\mathbf{M})$ acts transitively on the vertices of $\mathbb{X}_{\overline{a}, \mathcal{S}}$. Moreover, the edge relation is clearly invariant, meaning that $\text{Aut}(\mathbf{M})$ acts vertex transitively by automorphisms on $\mathbb{X}_{\overline{a}, \mathcal{S}}$. In particular, $\text{Aut}(\mathbf{M})$ preserves the distance $\rho_{\overline{a}, \mathcal{S}}$.

Lemma 6.4 *Suppose $\overline{a}$ is a tuple in $\mathbf{M}$ and $\mathcal{S}$ is a finite symmetric family of orbital types $\mathcal{O} \subseteq \mathcal{O}(\overline{a})$. Then the graph $\mathbb{X}_{\overline{a}, \mathcal{S}}$ is isomorphic to the graph on $\text{Aut}(\mathbf{M}) / V_{\overline{a}}$ defined by the edge relation*

$$(g V_{\overline{a}}, h V_{\overline{a}}) \in \text{Edge} \quad \Leftrightarrow \quad g V_{\overline{a}} \neq h V_{\overline{a}} \quad \text{and} \quad g^{-1} h \in V_{\overline{a}} F V_{\overline{a}},$$

where $V_{\overline{a}} F V_{\overline{a}}$ is the finite union of double cosets corresponding to $\mathcal{S}$.

Proof As above, note that the finite union of double cosets corresponding to S is given by

$$V_{\bar{a}} F V_{\bar{a}} = \bigcup_{\mathcal{O}(g\bar{a}, h\bar{a}) \in S} V_{\bar{a}} g^{-1} h V_{\bar{a}}.$$

Thus, the map $g V_{\bar{a}} \mapsto g\bar{a}$ is a bijection between the vertex sets of the two graphs mapping an edge $(g V_{\bar{a}}, h V_{\bar{a}})$ to the pair $(g\bar{a}, h\bar{a})$. But

$$(g V_{\bar{a}}, h V_{\bar{a}}) \in \text{Edge} \quad \Leftrightarrow \quad g V_{\bar{a}} \neq h V_{\bar{a}} \text{ and } g^{-1} h \in V_{\bar{a}} F V_{\bar{a}}$$
$$\Leftrightarrow \quad g\bar{a} \neq h\bar{a} \text{ and } \mathcal{O}(g\bar{a}, h\bar{a}) \in S$$
$$\Leftrightarrow \quad (g\bar{a}, h\bar{a}) \in \text{Edge } \mathbb{X}_{\bar{a}, S}.$$

In other words, the mapping $g V_{\bar{a}} \mapsto g\bar{a}$ defines an Aut(**M**)-equivariant isomorphism between the two graphs.   □

Lemma 6.5 *Suppose $\bar{a}$ is a fixed tuple in* **M**. *Assume S is a finite symmetric family of orbital types $\mathcal{O} \subseteq \mathcal{O}(\bar{a}) \times \mathcal{O}(\bar{a})$ and $V_{\bar{a}} F V_{\bar{a}}$ the corresponding finite union of double cosets. Then*

$$\rho_{\bar{a}, S}(g\bar{a}, h\bar{a}) \leqslant n \quad \Leftrightarrow \quad g^{-1} h \in (V_{\bar{a}} F V_{\bar{a}})^n$$

*for all $g, h \in$ Aut(**M**).*

Proof Note first that, by Lemma 6.4, we have $\rho_{\bar{a}, S}(g\bar{a}, h\bar{a}) \leqslant n$ if and only if the two cosets $g V_{\bar{a}}$ and $h V_{\bar{a}}$ have distance at most n in the graph on the space of left-cosets of $V_{\bar{a}}$ defined in Lemma 6.4. This latter is equivalent to there being a sequence $f_0 = g, f_1, f_2, \ldots, f_n = h$ so that $f_i^{-1} f_{i+1} \in V_{\bar{a}} F V_{\bar{a}}$ for all $i = 0, \ldots, n - 1$. However, in this case,

$$g^{-1} h = f_0^{-1} f_1 \cdot f_1^{-1} f_2 \cdots f_{n-1}^{-1} f_n \in (V_{\bar{a}} F V_{\bar{a}})^n.$$

Conversely, if $g^{-1} h \in (V_{\bar{a}} F V_{\bar{a}})^n$, then we can write $h = g v_1 v_2 \cdots v_n$ for some $v_i \in V_{\bar{a}} F V_{\bar{a}}$. Setting $f_i = g v_1 \cdots v_i$, we find that $g = f_0$, $h = f_n$ and

$$f_i^{-1} f_{i+1} = v_{i+1} \in V_{\bar{a}} F V_{\bar{a}}$$

for all $i = 0, \ldots, n - 1$.   □

Lemma 6.6 *Suppose $\bar{a}$ is a finite tuple in* **M**. *Then the following are equivalent.*

(1) *The pointwise stabiliser $V_{\bar{a}}$ is coarsely bounded in* Aut(**M**).
(2) *For every tuple $\bar{b}$ in* **M**, *there is a finite symmetric family S of orbital types $\mathcal{O} \subseteq \mathcal{O}(\bar{b}) \times \mathcal{O}(\bar{b})$ so that the orbit*

$$V_{\bar{a}} \cdot \bar{b}$$

has finite $\rho_{\bar{b}, S}$-diameter.

Proof (1)⇒(2) Suppose that $V_{\bar{a}}$ is coarsely bounded in Aut(**M**) and that $\bar{b}$ is a tuple in **M**. Then, as $V_{\bar{b}}$ is open, we may find a finite symmetric set $F \subseteq$ Aut(**M**) and an $n \geqslant 1$ so that

$$V_{\bar{a}} \subseteq \left(V_{\bar{b}} F V_{\bar{b}}\right)^{n}.$$

Now let

$$\mathcal{S} = \left\{\mathcal{O}(\bar{b}, g\bar{b}) \mid g \in F\right\}$$

be the finite symmetric family of orbital types $\mathcal{O} \subseteq \mathcal{O}(\bar{b}) \times \mathcal{O}(\bar{b})$ corresponding to F. Then, for any two $g, h \in V_{\bar{a}}$, we have

$$g^{-1}h \in V_{\bar{a}} \subseteq \left(V_{\bar{b}} F V_{\bar{b}}\right)^{n}$$

and hence

$$\rho_{\bar{b}, \mathcal{S}}\left(g\bar{b}, h\bar{b}\right) \leqslant n$$

by Lemma 6.5. That is, $V_{\bar{a}} \cdot \bar{b}$ has $\rho_{\bar{b}, \mathcal{S}}$-diameter at most n.

(2)⇒(1) Assume that (2) holds. Since the subgroups of the form $V_{\bar{b}}$, with $\bar{b}$ a finite tuple, form a neighbourhood basis at the identity, to show that $V_{\bar{a}}$ is coarsely bounded in Aut(**M**), it is enough to verify that, for all tuples $\bar{b}$ in **M**, there is a finite set $F \subseteq$ Aut(**M**) and an $n \geqslant 1$ so that

$$V_{\bar{a}} \subseteq \left(V_{\bar{b}} F V_{\bar{b}}\right)^{n}.$$

So suppose $\bar{b}$ is given and let $\mathcal{S}$ be a symmetric family of orbital types $\mathcal{O} \subseteq \mathcal{O}(\bar{b}) \times \mathcal{O}(\bar{b})$ so that

$$V_{\bar{a}} \cdot \bar{b}$$

has finite $\rho_{\bar{b}, \mathcal{S}}$-diameter n. Pick then a finite symmetric set $F \subseteq$ Aut(**M**) so that $V_{\bar{b}} F V_{\bar{b}}$ is the union of the double cosets corresponding to $\mathcal{S}$. Thus, for all $g, h \in V_{\bar{a}}$, we have

$$\rho_{\bar{b}, \mathcal{S}}(g\bar{b}, h\bar{b}) \leqslant n$$

and so, by Lemma 6.5,

$$g^{-1}h \in \left(V_{\bar{b}} F V_{\bar{b}}\right)^{n}.$$

In particular, $V_{\bar{a}} \subseteq \left(V_{\bar{b}} F V_{\bar{b}}\right)^{n}$, showing that $V_{\bar{a}}$ is coarsely bounded. □

By using again that the isotropy subgroups $V_{\bar{a}}$ form a neighbourhood basis at the identity in Aut(**M**), we obtain the following criterion for local boundedness and hence the existence of a coarsely proper metric.

Theorem 6.7 *The following are equivalent for the automorphism group* Aut(**M**) *of a countable structure* **M**.

(1) Aut(**M**) *admits a coarsely proper metric.*
(2) Aut(**M**) *is locally bounded.*
(3) *There is a tuple $\bar{a}$ so that, for every tuple $\bar{b}$ in* **M**, *there is a finite symmetric family $\mathcal{S}$ of orbital types $\mathcal{O} \subseteq \mathcal{O}(\bar{b}) \times \mathcal{O}(\bar{b})$ for which the orbit*

$$V_{\bar{a}} \cdot \bar{b}$$

has finite $\rho_{\bar{b}, \mathcal{S}}$-diameter.

Note that $\rho_{\bar{a}, \mathcal{S}}$ is an actual metric exactly when $\mathbb{X}_{\bar{a}, \mathcal{S}}$ is a connected graph. Our next task is to decide when this happens.

Lemma 6.8 *Suppose $\bar{a}$ is a tuple in* **M**. *Then the following are equivalent.*

(1) Aut(**M**) *is finitely generated over $V_{\bar{a}}$,*
(2) *there is a finite symmetric family $\mathcal{S}$ of orbital types $\mathcal{O} \subseteq \mathcal{O}(\bar{a}) \times \mathcal{O}(\bar{a})$ so that $\mathbb{X}_{\bar{a}, \mathcal{S}}$ is connected.*

Proof (1)$\Rightarrow$(2) Suppose that Aut(**M**) is finitely generated over $V_{\bar{a}}$ and pick a finite symmetric set $F \subseteq$ Aut(**M**) containing 1 so that Aut(**M**) $= \langle V_{\bar{a}} \cup F \rangle$. Let also $\mathcal{S} = \{\mathcal{O}(\bar{a}, g\bar{a}) \mid g \in F\}$ be the finite symmetric family of orbital types corresponding to $V_{\bar{a}} F V_{\bar{a}}$. To see that $\mathbb{X}_{\bar{a}, \mathcal{S}}$ is connected, let $g\bar{a}, h\bar{a} \in \mathcal{O}(\bar{a})$ be any two vertices and find $n \geqslant 1$ so that $g^{-1}h \in (V_{\bar{a}} F V_{\bar{a}})^n$. By Lemma 6.5, we find that

$$\rho_{\bar{a}, \mathcal{S}}(g\bar{a}, h\bar{a}) \leqslant n$$

and hence $\mathbb{X}_{\bar{a}, \mathcal{S}}$ is connected.

(2)$\Rightarrow$(1) Assume conversely that $\mathcal{S}$ is a finite symmetric family of orbital types so that $\mathbb{X}_{\bar{a}, \mathcal{S}}$ is connected. Let also $F \subseteq$ Aut(**M**) be a finite set so that $V_{\bar{a}} F V_{\bar{a}}$ is the union of the finitely many double cosets corresponding to $\mathcal{S}$. Then, if $g, h \in$ Aut(**M**) is given, we have

$$n = \rho_{\bar{a}, \mathcal{S}}(g\bar{a}, h\bar{a}) < \infty$$

and so

$$g^{-1}h \in \left(V_{\bar{a}} F V_{\bar{a}}\right)^n.$$

Thus, Aut(**M**) $= \langle V_{\bar{a}} \cup F \rangle$ and Aut(**M**) is finitely generated over $V_{\bar{a}}$. $\square$

With these preliminary results at hand, we can now give a full characteri-sation of when an automorphism group Aut(**M**) carries a well-defined quasi-isometry type and, moreover, provide a direct computation of this.

Theorem 6.9 *Let* **M** *be a countable structure. Then* Aut(**M**) *monogenic if and only if there is a tuple $\bar{a}$ in* **M** *satisfying the following two requirements.*

(1) *For every tuple $\bar{b}$, there is a finite symmetric family $\mathcal{S}$ of orbital types $\mathcal{O} \subseteq \mathcal{O}(\bar{b}) \times \mathcal{O}(\bar{b})$ for which the orbit*

$$V_{\bar{a}} \cdot \bar{b} = \{\bar{c} \mid \mathcal{O}(\bar{c}, \bar{a}) = \mathcal{O}(\bar{b}, \bar{a})\}$$

has finite $\rho_{\bar{b}, \mathcal{S}}$-diameter,
(2) *there is a finite symmetric family $\mathcal{R}$ of orbital types $\mathcal{O} \subseteq \mathcal{O}(\bar{a}) \times \mathcal{O}(\bar{a})$ so that $\mathbb{X}_{\bar{a}, \mathcal{R}}$ is connected.*

Moreover, if $\bar{a}$ and $\mathcal{R}$ are as in (2), then the mapping

$$g \in \text{Aut}(\mathbf{M}) \mapsto g \cdot \bar{a} \in \mathbb{X}_{\bar{a}, \mathcal{R}}$$

is a quasi-isometry between Aut(**M**) *and* $(\mathbb{X}_{\bar{a}, \mathcal{R}}, \rho_{\bar{a}, \mathcal{R}})$.

Proof Note that (1) is simply a restatement of $V_{\bar{a}}$ being coarsely bounded in Aut(**M**), while (2) states that Aut(**M**) is finitely generated over $V_{\bar{a}}$. By Theorem 2.73, these two properties together are equivalent to monogenecity.

For the moreover part, note that, as $\mathbb{X}_{\bar{a}, \mathcal{R}}$ is a connected graph, the metric space $(\mathbb{X}_{\bar{a}, \mathcal{R}}, \rho_{\bar{a}, \mathcal{R}})$ is large-scale geodesic. So let $V_{\bar{a}} F V_{\bar{a}}$ be the finite union of double cosets corresponding to $\mathcal{R}$. Then, by Lemma 6.5, we see that

$$g \in (V_{\bar{a}} F V_{\bar{a}})^n \iff \rho_{\bar{a}, \mathcal{R}}(\bar{a}, g\bar{a}) \leqslant n,$$

which shows that the continuous isometric action Aut(**M**) $\curvearrowright$ $(\mathbb{X}_{\bar{a}, \mathcal{S}}, \rho_{\bar{a}, \mathcal{S}})$ is coarsely proper. Being also transitive, it follows from the Milnor–Schwarz Lemma (Theorem 2.77) that

$$g \in \text{Aut}(\mathbf{M}) \mapsto g \cdot \bar{a} \in \mathbb{X}_{\bar{a}, \mathcal{R}}$$

is a quasi-isometry between Aut(**M**) *and* $(\mathbb{X}_{\bar{a}, \mathcal{R}}, \rho_{\bar{a}, \mathcal{R}})$. $\square$

In cases where Aut(**M**) may not admit a maximal metric, but only a coarsely proper metric, it is still useful to have an explicit calculation of this latter metric. For this, the following lemma will be useful.

Lemma 6.10 *Suppose $\bar{a}$ is finite tuple in a countable structure* **M**. *Let also* $\mathcal{R}_1 \subseteq \mathcal{R}_2 \subseteq \mathcal{R}_3 \subseteq \cdots$ *be an exhaustive sequence of finite symmetric families of orbital types* $\mathcal{O} \subseteq \mathcal{O}(\bar{a}) \times \mathcal{O}(\bar{a})$ *and define a metric* $\rho_{\bar{a},(\mathcal{R}_n)}$ *on* $\mathcal{O}(\bar{a})$ *by*

$$\rho_{\bar{a},(\mathcal{R}_n)}(\bar{b},\bar{c}) = \min \left(\sum_{i=1}^{k} n_i \cdot \rho_{\bar{a},\mathcal{R}_{n_i}}(\bar{d}_{i-1},\bar{d}_i) \,\middle|\, n_i \in \mathbb{N}, \right.$$

$$\left. \bar{d}_i \in \mathcal{O}(\bar{a}), \bar{d}_0 = \bar{b} \text{ and } \bar{d}_k = \bar{c} \right).$$

Assuming that $V_{\bar{a}}$ is coarsely bounded in Aut(**M**), *then the isometric action*

$$\text{Aut}(\mathbf{M}) \curvearrowright \left(\mathcal{O}(\bar{a}), \rho_{\bar{a},(\mathcal{R}_n)} \right)$$

is coarsely proper.

Proof Let us first note that, because the sequence $(\mathcal{R}_n)$ is exhaustive, every orbital type $\mathcal{O}(\bar{b},\bar{c})$ eventually belongs to some $\mathcal{R}_n$, whereby $\rho_{\bar{a},(\mathcal{R}_n)}(\bar{b},\bar{c})$ is finite. Also, $\rho_{\bar{a},(\mathcal{R}_n)}$ satisfies the triangle inequality by definition and hence is a metric.

Note now that, since the $\mathcal{R}_n$ are increasing with n, we have

$$\rho_{\bar{a},\mathcal{R}_m} \leqslant \rho_{\bar{a},\mathcal{R}_n} \leqslant n \cdot \rho_{\bar{a},\mathcal{R}_n}$$

whenever $n \leqslant m$. It thus follows that, for all $m \in \mathbb{N}$,

$$\rho_{\bar{a},(\mathcal{R}_n)}(\bar{b},\bar{c}) \leqslant m \implies \rho_{\bar{a},\mathcal{R}_m}(\bar{b},\bar{c}) \leqslant m$$

and hence

$$\{g \in \text{Aut}(\mathbf{M}) \mid \rho_{\bar{a},(\mathcal{R}_n)}(\bar{a},g\bar{a}) \leqslant m\} \subseteq \{g \in \text{Aut}(\mathbf{M}) \mid \rho_{\bar{a},\mathcal{R}_m}(\bar{a},g\bar{a}) \leqslant m\}.$$

By Lemma 6.5, the latter set is coarsely bounded in Aut(**M**), so the isometric action

$$\text{Aut}(\mathbf{M}) \curvearrowright \left(\mathcal{O}(\bar{a}), \rho_{\bar{a},(\mathcal{R}_n)} \right)$$

is coarsely proper. $\square$

6.3 Homogeneous and Atomic Models

6.3.1 Definability of Metrics

Whereas the preceding sections have largely concentrated on the automorphism group Aut(**M**) of a countable structure **M** without much regard to the actual structure $\mathbf{M}$, its language $\mathcal{L}$ or its theory $T = \text{Th}(\mathbf{M})$, in the present section, we shall study how the theory T may directly influence the large-scale geometry of $\text{Aut}(\mathbf{M})$.

Recall first that, if $\bar{a}$ is a tuple in $\mathbf{M}$, the *type* of $\bar{a}$ is the set

$$\text{tp}^{\mathbf{M}}(\bar{a}) = \{\phi(\bar{x}) \mid \mathbf{M} \models \phi(\bar{a})\}$$

of all parameter-free first-order $\mathcal{L}$-formulas $\phi(\bar{x})$ that hold at $\bar{a}$ in $\mathbf{M}$, where $\bar{x}$ is some fixed tuple of distinct variables of the same length as $\bar{a}$. Similarly, the *quantifier-free type* of $\bar{a}$ is the set

$$\text{qftp}^{\mathbf{M}}(\bar{a}) = \{\phi(\bar{x}) \mid \phi(\bar{x}) \text{ is a quantifier-free formula and } \mathbf{M} \models \phi(\bar{a})\}.$$

A structure $\mathbf{M}$ is said to be ω-*homogeneous* if, for all finite tuples $\bar{a}$ and $\bar{b}$ in $\mathbf{M}$ with the same type $\text{tp}^{\mathbf{M}}(\bar{a}) = \text{tp}^{\mathbf{M}}(\bar{b})$ and all c in $\mathbf{M}$, there is some d in $\mathbf{M}$ so that $\text{tp}^{\mathbf{M}}(\bar{a}, c) = \text{tp}^{\mathbf{M}}(\bar{b}, d)$. By a back-and-forth construction, one sees that, in the case where $\mathbf{M}$ is countable, ω-homogeneity is equivalent to the condition

$$\text{tp}^{\mathbf{M}}(\bar{a}) = \text{tp}^{\mathbf{M}}(\bar{b}) \quad \Leftrightarrow \quad \mathcal{O}(\bar{a}) = \mathcal{O}(\bar{b}).$$

In other words, every orbital type $\mathcal{O}(\bar{a})$ is *type $\emptyset$-definable*, i.e., type definable without parameters.

For a stronger notion, we say that $\mathbf{M}$ is *ultrahomogeneous* if it satisfies

$$\text{qftp}^{\mathbf{M}}(\bar{a}) = \text{qftp}^{\mathbf{M}}(\bar{b}) \quad \Leftrightarrow \quad \mathcal{O}(\bar{a}) = \mathcal{O}(\bar{b}).$$

In other words, every orbital type $\mathcal{O}(\bar{a})$ is defined by the quantifier-free type $\text{qftp}^{\mathbf{M}}(\bar{a})$.

Another requirement is to demand that each individual orbital type $\mathcal{O}(\bar{a})$ is $\emptyset$-*definable* in $\mathbf{M}$, that is, definable by a single formula $\phi(\bar{x})$ without parameters. In other words, for this ϕ, we have

$$\bar{b} \in \mathcal{O}(\bar{a}) \quad \Leftrightarrow \quad \mathbf{M} \models \phi(\bar{b}).$$

We note that such a ϕ necessarily isolates the type $\text{tp}^{\mathbf{M}}(\bar{a})$. Indeed, suppose $\psi \in \text{tp}^{\mathbf{M}}(\bar{a})$. Then, if $\mathbf{M} \models \phi(\bar{b})$, we have $\bar{b} \in \mathcal{O}(\bar{a})$ and thus also $\mathbf{M} \models \psi(\bar{b})$, showing that $\mathbf{M} \models \forall \bar{x} \; (\phi \to \psi)$. Conversely, suppose $\mathbf{M}$ is a countable ω-homogeneous structure and $\phi(\bar{x})$ is a formula without parameters isolating some type $\text{tp}^{\mathbf{M}}(\bar{a})$. Then, if $\mathbf{M} \models \phi(\bar{b})$, we have $\text{tp}^{\mathbf{M}}(\bar{a}) = \text{tp}^{\mathbf{M}}(\bar{b})$ and thus, by ω-homogeneity, $\bar{b} \in \mathcal{O}(\bar{a})$.

We recall that a model $\mathbf{M}$ is *atomic* if every type realised in $\mathbf{M}$ is isolated. As is easy to verify (see Lemma 4.2.14 in [58]), countable atomic models are ω-homogeneous. So, by the discussion above, we see that a countable model $\mathbf{M}$ is atomic if and only if every orbital type $\mathcal{O}(\bar{a})$ is $\emptyset$-definable.

For example, if **M** is a locally finite ultrahomogeneous structure in a finite language $\mathcal{L}$, then **M** is atomic. This follows from the fact that, if **A** is a finite structure in a finite language, then its isomorphism type is described by a single quantifier-free formula.

Lemma 6.11 *Suppose $\bar{a}$ is a finite tuple in a countable atomic model* **M**. *Let S be a finite symmetric collection of orbital types $\mathcal{O} \subseteq \mathcal{O}(\bar{a}) \times \mathcal{O}(\bar{a})$. Then, for every $n \in \mathbb{N}$, the relation*

$$\rho_{\bar{a},S}(\bar{b},\bar{c}) \leqslant n$$

is $\emptyset$-definable in **M**. *Also, if $S_1 \subseteq S_2 \subseteq \cdots$ is an exhaustive sequence of finite symmetric sets of orbital types $\mathcal{O} \subseteq \mathcal{O}(\bar{a}) \times \mathcal{O}(\bar{a})$, then the relation $\rho_{\bar{a},(S_m)}(\bar{b},\bar{c}) \leqslant n$ is $\emptyset$-definable in* **M**.

Proof Let $\phi_1(\bar{x},\bar{y}),\ldots,\phi_k(\bar{x},\bar{y})$ be formulas without parameters defining the orbital types in S. Then

$$\rho_{\bar{a},S}(\bar{b},\bar{c}) \leqslant n \quad \Leftrightarrow$$

$$\mathbf{M} \models \bigvee_{m=0}^{n} \exists \bar{y}_0, \ldots, \bar{y}_m \left(\bigwedge_{j=0}^{m-1} \bigvee_{i=1}^{k} \phi_i(\bar{y}_j, \bar{y}_{j+1}) \text{ and } \bar{b} = \bar{y}_0 \text{ and } \bar{c} = \bar{y}_m \right),$$

showing that $\rho_{\bar{a},S}(\bar{b},\bar{c}) \leqslant n$ is $\emptyset$-definable in **M**.

For the second case, pick formulas $\phi_{m,n}(\bar{x},\bar{y})$ without parameters defining the relations $\rho_{\bar{a},S_m}(\bar{b},\bar{c}) \leqslant n$ in **M**. Then $\rho_{\bar{a},(S_m)}(\bar{b},\bar{c}) \leqslant n$ if and only if the formula

$$\bigvee_{k=0}^{n} \exists \bar{x}_0, \ldots, \bar{x}_k$$

$$\times \left(\bar{x}_0 = \bar{b} \text{ and } \bar{x}_k = \bar{c} \text{ and } \bigvee \left\{ \bigwedge_{i=1}^{k} \phi_{m_i,n_i}(\bar{x}_{i-1},\bar{x}_i) \, \middle| \, \sum_{i=1}^{k} m_i \cdot n_i \leqslant n \right\} \right)$$

is true in **M**. Thus, again, $\rho_{\bar{a},(S_m)}(\bar{b},\bar{c}) \leqslant n$ is $\emptyset$-definable in **M**. $\qquad \square$

6.3.2 Stable Metrics and Theories

We recall the following notion originating in the work of J.-L. Krivine and B. Maurey on stable Banach spaces [53].

Definition 6.12 A metric d on a set X is said to be *stable* if, for all d-bounded sequences (x_n) and (y_m) in X, we have

$$\lim_{n \to \infty} \lim_{m \to \infty} d(x_n, y_m) = \lim_{m \to \infty} \lim_{n \to \infty} d(x_n, y_m),$$

whenever both limits exist.

We mention that stability of the metric is equivalent to requiring that the limit operations $\lim_{n \to \mathcal{U}}$ and $\lim_{m \to \mathcal{V}}$ commute over d for all ultrafilters $\mathcal{U}$ and $\mathcal{V}$ on $\mathbb{N}$.

Now stability of metrics is tightly related to model theoretical stability of which we recall the definition.

Definition 6.13 Let T be a complete theory of a countable language $\mathcal{L}$ and let κ be an infinite cardinal number. We say that T is κ-*stable* if, for all models $\mathbf{M} \models T$ and subsets $A \subseteq \mathbf{M}$ with $|A| \leqslant \kappa$, we have $|S_n^{\mathbf{M}}(A)| \leqslant \kappa$. Also, T is *stable* if it is κ-stable for some infinite cardinal κ.

In the following discussion, we shall always assume that T is a complete theory with infinite models in a countable language $\mathcal{L}$. Of the various consequences of stability of T, the one most closely related to stability of metrics is the fact that, if T is stable and $\mathbf{M}$ is a model of T, then there are no formula $\phi(\bar{x}, \bar{y})$ and tuples $\bar{a}_n, \bar{b}_m, n, m \in \mathbb{N}$, so that

$$n < m \iff \mathbf{M} \models \phi(\bar{a}_n, \bar{b}_m).$$

Knowing this, the following lemma is straightforward.

Lemma 6.14 *Suppose* $\mathbf{M}$ *is a countable atomic model of a stable theory* T *and that* $\bar{a}$ *is a finite tuple in* $\mathbf{M}$. *Let also* ρ *be an integral-valued metric on* $\mathcal{O}(\bar{a})$ *so that, for every* $n \in \mathbb{N}$, *the relation* $\rho(\bar{b}, \bar{c}) \leqslant n$ *is* $\emptyset$-*definable in* $\mathbf{M}$. *Then* ρ *is a stable metric.*

Proof Suppose towards a contradiction that $\bar{a}_n, \bar{b}_m \in \mathcal{O}(\bar{a})$ are bounded sequences in $\mathcal{O}(\bar{a})$ so that

$$r = \lim_{n \to \infty} \lim_{m \to \infty} \rho(\bar{a}_n, \bar{b}_m) \neq \lim_{m \to \infty} \lim_{n \to \infty} \rho(\bar{a}_n, \bar{b}_m)$$

and pick a formula $\phi(\bar{x}, \bar{y})$ so that

$$\rho(\bar{b}, \bar{c}) = r \iff \mathbf{M} \models \phi(\bar{b}, \bar{c}).$$

Then, using $\forall^\infty$ to denote 'for all, but finitely many', we have

$$\forall^\infty n \, \forall^\infty m \quad \mathbf{M} \models \phi(\bar{a}_n, \bar{b}_m),$$

and

$$\forall^\infty m \ \forall^\infty n \quad \mathbf{M} \models \neg\phi(\bar{a}_n, \bar{b}_m).$$

So, upon passing to subsequences of $(\bar{a}_n)$ and $(\bar{b}_m)$, we may suppose that

$$\mathbf{M} \models \phi(\bar{a}_n, \bar{b}_m) \quad \Leftrightarrow \quad n < m.$$

However, the existence of such a formula ϕ and sequences $(\bar{a}_n)$ and $(\bar{b}_m)$ contradicts the stability of T. □

Theorem 6.15 *Suppose $\mathbf{M}$ is a countable atomic model of a stable theory T with $\mathrm{Aut}(\mathbf{M})$ monogenic. Then $\mathrm{Aut}(\mathbf{M})$ admits a stable maximal compatible left-invariant metric.*

Proof By Theorem 6.9, there is a finite tuple $\bar{a}$ and a finite family $\mathcal{S}$ of orbital types so that the mapping

$$g \in \mathrm{Aut}(\mathbf{M}) \mapsto g\bar{a} \in \mathbb{X}_{\bar{a}, \mathcal{S}}$$

is a quasi-isometry of $\mathrm{Aut}(\mathbf{M})$ with $(\mathbb{X}_{\bar{a}, \mathcal{S}}, \rho_{\bar{a}, \mathcal{S}})$. Also, by Lemmas 6.11 and 6.14, $\rho_{\bar{a}, \mathcal{S}}$ is a stable metric on $\mathbb{X}_{\bar{a}, \mathcal{S}}$. Define also a compatible left-invariant stable metric $D \leqslant 1$ on $\mathrm{Aut}(\mathbf{M})$ by

$$D(g, f) = \sum_{n=1}^{\infty} \frac{\chi_{\neq}(g(b_n), f(b_n))}{2^n},$$

where (b_n) is an enumeration of $\mathbf{M}$ and $\chi_{\neq}$ is the characteristic funtion of inequality. The stability of D follows easily from it being an absolutely summable series of the functions $(g, f) \mapsto \frac{\chi_{\neq}(g(b_n), f(b_n))}{2^n}$.

Finally, let

$$d(g, f) = D(g, f) + \rho_{\bar{a}, \mathcal{S}}(g\bar{a}, f\bar{a}).$$

Then d is a maximal and stable compatible left-invariant metric on $\mathrm{Aut}(\mathbf{M})$. □

Similarly, when $\mathrm{Aut}(\mathbf{M})$ is assumed to have only a coarsely proper metric, this can also be taken to be stable. This can be done by working with the metric $\rho_{\bar{a}, (\mathcal{S}_n)}$, where $\mathcal{S}_1 \subseteq \mathcal{S}_2 \subseteq \cdots$ is an exhaustive sequence of finite sets of orbital types on $\mathbf{M}$, instead of $\rho_{\bar{a}, \mathcal{S}}$.

Theorem 6.16 *Suppose $\mathbf{M}$ is a countable atomic model of a stable theory T so that $\mathrm{Aut}(\mathbf{M})$ admits a coarsely proper metric. Then $\mathrm{Aut}(\mathbf{M})$ admits a stable, coarsely proper metric.*

By using the equivalence of the local boundedness and the existence of coarsely proper metrics and that the existence of a stable coarsely proper metric implies coarsely proper actions on reflexive spaces (see Theorem 56 in [**80**]), we have the following corollary.

Corollary 6.17 *Suppose* **M** *is a countable atomic model of a stable theory* T *so that* Aut(**M**) *is locally bounded. Then* Aut(**M**) *admits a coarsely proper continuous affine isometric action on a reflexive Banach space.*

We should briefly review the hypotheses of the preceding theorem. So, in the following, let T be a complete theory with infinite models in a countable language $\mathcal{L}$. We recall that that $\mathbf{M} \models T$ is said to be a *prime model* of T if **M** admits an elementary embedding into every other model of T. By the Omitting Types Theorem, prime models are necessarily atomic. In fact, $\mathbf{M} \models T$ is a prime model of T if and only if **M** is both countable and atomic. Moreover, the theory T admits a countable atomic model if and only if, for every n, the set of isolated types is dense in the type space $S_n(T)$. In particular, this happens if $S_n(T)$ is countable for all n.

Now, by definition, T is ω-stable, if, for every model $\mathbf{M} \models T$, countable subset $A \subseteq \mathbf{M}$ and $n \geqslant 1$, the type space $S_n^{\mathbf{M}}(A)$ is countable. In particular, $S_n(T)$ is countable for every n and hence T has a countable atomic model **M**. Thus, provided that Aut(**M**) is locally bounded, Corollary 6.17 gives a coarsely proper affine isometric action of this automorphism group on a reflexive Banach space.

6.3.3 Fraïssé Classes

A useful tool in the study of ultrahomogeneous countable structures is the theory of R. Fraïssé that allows us to view every such object as a so-called limit of the family of its finitely generated substructures. In the following, we fix a countable language $\mathcal{L}$.

Definition 6.18 A *Fraïssé class* is a class $\mathcal{K}$ of finitely generated $\mathcal{L}$-structures so that

(1) κ contains only countably many isomorphism types,
(2) (hereditary property) if $\mathbf{A} \in \mathcal{K}$ and $\mathbf{B}$ is a finitely generated $\mathcal{L}$-structure embeddable into $\mathbf{A}$, then $\mathbf{B} \in \mathcal{K}$,
(3) (joint embedding property) for all $\mathbf{A}, \mathbf{B} \in \mathcal{K}$, there is some $\mathbf{C} \in \mathcal{K}$ into which both $\mathbf{A}$ and $\mathbf{B}$ embed,
(4) (amalgamation property) if $\mathbf{A}, \mathbf{B}_1, \mathbf{B}_2 \in \mathcal{K}$ and $\eta_i : \mathbf{A} \hookrightarrow \mathbf{B}_i$ are embeddings, then there is some $\mathbf{C} \in \mathcal{K}$ and embeddings $\zeta_i : \mathbf{B}_i \hookrightarrow \mathbf{C}$ so that $\zeta_1 \circ \eta_1 = \zeta_2 \circ \eta_2$.

Also, if **M** is a countable $\mathcal{L}$-structure, we let Age(**M**) denote the class of all finitely generated $\mathcal{L}$-structures embeddable into **M**.

The fundamental theorem of Fraïssé [33] states that, for every Fraïssé class $\mathcal{K}$, there is a unique (up to isomorphism) countable ultrahomogeneous structure **K**, called the *Fraïssé limit* of $\mathcal{K}$, so that Age(**K**) $= \mathcal{K}$ and, conversely, if **M** is a countable ultrahomogeneous structure, then Age(**M**) is a Fraïssé class.

Now, if **K** is the limit of a Fraïssé class $\mathcal{K}$, then **K** is ultrahomogeneous and hence its orbital types correspond to quantifier-free types realised in **K**. Also, as Age(**K**) $= \mathcal{K}$, for every quantifier-free type p realised by some tuple $\bar{a}$ in **K**, we see that the structure **A** $= \langle \bar{a} \rangle$ generated by $\bar{a}$ belongs to $\mathcal{K}$ and that the expansion $\langle \mathbf{A}, \bar{a} \rangle$ of **A** with names for $\bar{a}$ codes p is realised by

$$\phi(\bar{x}) \in p \;\Leftrightarrow\; \langle \mathbf{A}, \bar{a} \rangle \models \phi(\bar{a}).$$

Vice versa, because **A** is generated by $\bar{a}$, the quantifier-free type $\mathrm{qftp}^{\mathbf{A}}(\bar{a})$ fully determines the expanded structure $\langle \mathbf{A}, \bar{a} \rangle$ up to isomorphism. In conclusion, we see that orbital types $\mathcal{O}(\bar{a})$ in **K** correspond to isomorphism types of expanded structures $\langle \mathbf{A}, \bar{a} \rangle$, where $\bar{a}$ is a finite tuple generating some **A** $\in \mathcal{K}$. This also means that Theorem 6.9 may be reformulated by using these isomorphism types in place of orbital types. We leave the details to the reader and instead concentrate on a more restrictive setting.

Suppose now that **K** is the limit of a Fraïssé class $\mathcal{K}$ consisting of finite structures. In other words, **K** is ultrahomogeneous and *locally finite*, meaning that every finitely generated substructure is finite. Then we note that every **A** $\in \mathcal{K}$ can simply be enumerated by some finite tuple $\bar{a}$. Moreover, if **A** is a finite substructure of **K**, then the pointwise stabiliser $V_{\mathbf{A}}$ is a finite index subgroup of the *setwise stabiliser*

$$V_{\{\mathbf{A}\}} = \{ g \in \mathrm{Aut}(\mathbf{K}) \mid g\mathbf{A} = \mathbf{A} \}.$$

In particular, $V_{\{\mathbf{A}\}}$ is coarsely bounded in Aut(**K**) if and only if $V_{\mathbf{A}}$ is. Similarly, if **B** is another finite substructure, then $V_{\mathbf{A}}$ is finitely generated over $V_{\{\mathbf{B}\}}$ if and only if it is finitely generated over $V_{\mathbf{B}}$. Finally, if $(\bar{a}, \bar{c}) \in \mathcal{O}(\bar{a}, \bar{b})$ and **B** and **C** are the substructures of **K** generated by $(\bar{a}, \bar{b})$ and $(\bar{a}, \bar{c})$ respectively, then, for every automorphism $g \in \mathrm{Aut}(\mathbf{K})$ mapping **B** to **C**, there is an $h \in V_{\{\mathbf{B}\}}$ so that $gh(\bar{a}, \bar{b}) = (\bar{a}, \bar{c})$.

By using these observations, one may substitute the orbital types of finite tuples $\bar{a}$ by isomorphism classes of finitely generated substructures of **K** to obtain a reformulation of Theorem 6.9.

Theorem 6.19 *Suppose $\mathcal{K}$ is a Fraïssé class of finite structures with Fraïssé limit **K**. Then Aut(**K**) is monogenic if and only if there is **A** $\in \mathcal{K}$ satisfying the following two conditions.*

(1) *For every* **B** ∈ 𝒦 *containing* **A**, *there are* $n \geqslant 1$ *and an isomorphism invariant family* 𝒮 ⊆ 𝒦, *containing only finitely many isomorphism types, so that, for all* **C** ∈ 𝒦 *and embeddings* $\eta_1, \eta_2 \colon$ **B** ↪ **C** *with* $\eta_1|_{\mathbf{A}} = \eta_2|_{\mathbf{A}}$, *one can find some* **D** ∈ 𝒦 *containing* **C** *and a path* $\mathbf{B}_0 = \eta_1 \mathbf{B}, \mathbf{B}_1, \ldots, \mathbf{B}_n = \eta_2 \mathbf{B}$ *of isomorphic copies of* **B** *inside* **D** *with* ⟨$\mathbf{B}_i \cup \mathbf{B}_{i+1}$⟩ ∈ 𝒮 *for all* i,

(2) *there is an isomorphism invariant family* ℛ ⊆ 𝒦, *containing only finitely many isomorphism types, so that, for all* **B** ∈ 𝒦 *containing* **A** *and isomorphic copies* **A**′ ⊆ **B** *of* **A**, *there is some* **C** ∈ 𝒦 *containing* **B** *and a path* $\mathbf{A}_0, \mathbf{A}_1, \ldots, \mathbf{A}_n$ ⊆ **C** *consisting of isomorphic copies of* **A**, *beginning at* $\mathbf{A}_0 = \mathbf{A}$ *and ending at* $\mathbf{A}_n = \mathbf{A}'$, *satisfying* ⟨$\mathbf{A}_i \cup \mathbf{A}_{i+1}$⟩ ∈ ℛ *for all* i.

6.4 Orbital Independence Relations

The formulation of Theorem 6.9 is rather abstract and it is therefore useful to have some more familiar criteria for being locally bounded or having a well-defined quasi-isometry type. The first such criterion is simply a reformulation of an observation of P. Cameron [15].

Proposition 6.20 (P. Cameron) *Let* **M** *be an* $\aleph_0$-*categorical countable structure. Then, for every tuple* $\bar{a}$ *in* **M**, *there is a finite set* $F \subseteq \mathrm{Aut}(\mathbf{M})$ *so that* $\mathrm{Aut}(\mathbf{M}) = V_{\bar{a}} F V_{\bar{a}}$, *i.e.,* $\mathrm{Aut}(\mathbf{M})$ *is Roelcke pre-compact.*

Proof Since **M** is $\aleph_0$-categorical, the pointwise stabiliser $V_{\bar{a}}$ induces only finitely many orbits on $\mathbf{M}^n$, where n is the length of $\bar{a}$. So let $B \subseteq \mathbf{M}^n$ be a finite set of $V_{\bar{a}}$-orbit representatives. Also, for every $\bar{b} \in B$, pick if possible some $f \in \mathrm{Aut}(\mathbf{M})$ so that $\bar{b} = f\bar{a}$ and let F be the finite set of these f. Then, if $g \in \mathrm{Aut}(\mathbf{M})$, as $g\bar{a} \in \mathbf{M}^n = V_{\bar{a}}B$, there is some $h \in V_{\bar{a}}$ and $\bar{b} \in B$ so that $g\bar{a} = h\bar{b}$. In particular, there is $f \in F$ so that $\bar{b} = f\bar{a}$, whence $g\bar{a} = h\bar{b} = hf\bar{a}$ and thus $g \in hf V_{\bar{a}} \subseteq V_{\bar{a}} F V_{\bar{a}}$. ☐

Thus, for an automorphism group $\mathrm{Aut}(\mathbf{M})$ to have a non-trivial quasi-isometry type, the structure **M** should not be $\aleph_0$-categorical. In this connection, we recall that, if 𝒦 is a Fraïssé class in a finite language ℒ and 𝒦 is *uniformly locally finite*, that is, there is a function $f \colon \mathbb{N} \to \mathbb{N}$ so that every **A** ∈ 𝒦 generated by n elements has size $\leqslant f(n)$, then the Fraïssé limit **K** is $\aleph_0$-categorical. In particular, this applies to Fraïssé classes in finite relational languages.

However, our first concern is to identify locally bounded automorphism groups and for this we consider model theoretical independence relations.

Definition 6.21 Let $\mathbf{M}$ be a countable structure and $A \subseteq \mathbf{M}$ a finite subset. An *orbital A-independence relation* on $\mathbf{M}$ is a binary relation $\underset{A}{\perp}$ defined between finite subsets of $\mathbf{M}$ so that, for all finite $B, C, D \subseteq \mathbf{M}$,

(i) (symmetry) $B \underset{A}{\perp} C \Leftrightarrow C \underset{A}{\perp} B$,
(ii) (monotonicity) $B \underset{A}{\perp} C$ and $D \subseteq C \Rightarrow B \underset{A}{\perp} D$,
(iii) (existence) there is $f \in V_A$ so that $fB \underset{A}{\perp} C$,
(iv) (stationarity) if $B \underset{A}{\perp} C$ and $g \in V_A$ satisfies $gB \underset{A}{\perp} C$, then $g \in V_C V_B$, i.e., there is some $f \in V_C$ agreeing pointwise with g on B.

We read $B \underset{A}{\perp} C$ as 'B is independent from C over A'. Occasionally, it is convenient to let $\underset{A}{\perp}$ be defined between finite tuples rather than sets, which is done by simply viewing a tuple as a name for the set it enumerates. For example, if $\bar{b} = (b_1, \ldots, b_n)$, we let $\bar{b} \underset{A}{\perp} C$ if and only if $\{b_1, \ldots, b_n\} \underset{A}{\perp} C$. Similarly, if $\bar{a}$ enumerates A, we occasionnally denote $\underset{A}{\perp}$ by $\underset{\bar{a}}{\perp}$.

With this convention, the stationarity condition on $\underset{\bar{a}}{\perp}$ can be reformulated as follows. If $\bar{b}$ and $\bar{b}'$ have the same orbital type over $\bar{a}$, i.e., $\mathcal{O}(\bar{b}, \bar{a}) = \mathcal{O}(\bar{b}', \bar{a})$, and are both independent from $\bar{c}$ over $\bar{a}$, then they also have the same orbital type over $\bar{c}$.

Similarly, the existence condition on $\underset{\bar{a}}{\perp}$ can be stated as follows. For all $\bar{b}, \bar{c}$, there is some $\bar{b}'$ independent from $\bar{c}$ over $\bar{a}$ and having the same orbital type over $\bar{a}$ as $\bar{b}$ does.

We should note that, as our interest is in the permutation group Aut($\mathbf{M}$) and not the particular structure $\mathbf{M}$, any two structures $\mathbf{M}$ and $\mathbf{M}'$, possibly of different languages, having the same universe and the exact same automorphism group Aut($\mathbf{M}$) = Aut($\mathbf{M}'$) will essentially be equivalent for our purposes. We also remark that the existence of an orbital A-independence relation does not depend on the exact structure $\mathbf{M}$, but only on its universe and its automorphism group. Thus, in Examples 6.22, 6.23 and 6.24 below, any manner of formalising the mathematical structures as bona fide first-order model theoretical structures of some language with the indicated automorphism group will lead to the same results and hence can safely be left to the reader.

Example 6.22 (Measured Boolean algebras) Let $\mathbf{M}$ denote the Boolean algebra of clopen subsets of Cantor space $\{0, 1\}^{\mathbb{N}}$ equipped with the usual dyadic probability measure μ, i.e., the infinite product of the $\{\frac{1}{2}, \frac{1}{2}\}$-distribution on $\{0, 1\}$. We note that $\mathbf{M}$ is ultrahomogeneous, in the sense that, if $\sigma: A \to B$ is a measure preserving isomorphism between two finite subalgebras of $\mathbf{M}$, then σ extends to a measure preserving automorphism of $\mathbf{M}$.

For two finite subsets A, B, we let $A \perp_{\emptyset} B$ if the Boolean algebras they generate are measure theoretically independent, i.e., if, for all $a_1, \ldots, a_n \in A$ and $b_1, \ldots, b_m \in B$, we have

$$\mu(a_1 \cap \cdots \cap a_n \cap b_1 \cap \cdots \cap b_m) = \mu(a_1 \cap \cdots \cap a_n) \cdot \mu(b_1 \cap \cdots \cap b_m).$$

Remark that, if $\sigma \colon A_1 \to A_2$ and $\eta \colon B_1 \to B_2$ are measure preserving isomorphisms between subalgebras of $\mathbf{M}$ with $A_i \perp_{\emptyset} B_i$, then there is a measure preserving isomorphism $\xi \colon \langle A_1 \cup B_1 \rangle \to \langle A_2 \cup B_2 \rangle$ between the algebras generated extending both σ and η. Namely, $\xi(a \cap b) = \sigma(a) \cap \eta(b)$ for atoms $a \in A_1$ and $b \in B_1$.

Using this and the ultrahomogeneity of $\mathbf{M}$, the stationarity condition (iv) of $\perp_{\emptyset}$ is clear. Also, symmetry and monotonicity are obvious. Finally, for the existence condition (iii), suppose that A and B are given finite subsets of $\mathbf{M}$. Then there is some finite n so that all elements of A and B can be written as unions of basic open sets $N_s = \{x \in \{0,1\}^{\mathbb{N}} \mid s \text{ is an initial segment of } x\}$ for $s \in \{0,1\}^n$. Pick a permutation α of $\mathbb{N}$ so that $\alpha(i) > n$ for all $i \leqslant n$ and note that α induces measure preserving automorphism σ of $\mathbf{M}$ so that $\sigma(A) \perp_{\emptyset} B$.

Thus, $\perp_{\emptyset}$ is an orbital $\emptyset$-independence relation on $\mathbf{M}$. We also note that, by Stone duality, the automorphism group of $\mathbf{M}$ is isomorphic to the group

$$\text{Homeo}(\{0,1\}^{\mathbb{N}}, \mu)$$

of measure-preserving homeomorphisms of Cantor space.

Example 6.23 (The $\aleph_0$-regular tree) Let T_{∞} denote the $\aleph_0$-regular tree. That is, T_{∞} is a countable connected undirected graph without loops in which every vertex has infinite valence. Because T_{∞} is a tree, there is a natural notion of *convex hull*, namely, for a subset $A \subseteq \mathsf{T}_{\infty}$ and a vertex $x \in \mathsf{T}_{\infty}$, we set $x \in \text{conv}(A)$ if there are $a, b \in A$ so that x lies on the unique path from a to b. Now, pick a distinguished vertex $t \in \mathsf{T}_{\infty}$ and, for finite $A, B \subseteq \mathsf{T}_{\infty}$, set

$$A \underset{\{t\}}{\perp} B \iff \text{conv}(A \cup \{t\}) \cap \text{conv}(B \cup \{t\}) = \{t\}.$$

That $\perp_{\{t\}}$ is both symmetric and monotone is obvious. Also, if A and B are finite, then so are $\text{conv}(A \cup \{t\})$ and $\text{conv}(B \cup \{t\})$. This makes it is easy to find a ellitic isometry g with fixed point t, i.e., a rotation of T_{∞} around t, so that $g(\text{conv}(A \cup \{t\})) \cap \text{conv}(B \cup \{t\}) = \{t\}$. Because $g(\text{conv}(A \cup \{t\})) = \text{conv}(gA \cup \{t\})$, one sees that $gA \perp_{\{t\}} B$, verifying the existence condition (iii).

Finally, for the stationarity condition (iv), suppose $B, C \subseteq \mathsf{T}_{\infty}$ are given and g is an elliptic isometry fixing t so that $B \perp_{\{t\}} C$ and $gB \perp_{\{t\}} C$. Then, using again that T_∞ is $\aleph_0$-regular, it is easy to find another elliptic isometry fixing all of $\mathsf{conv}(C \cup \{t\})$ that agrees with g on B.

So $\underset{\{t\}}{\downarrow}$ is an orbital $\{t\}$-independence relation on T_∞.

Example 6.24 (Unitary groups) Fix a countable field $\mathbb{Q} \subseteq \mathfrak{F} \subseteq \mathbb{C}$ closed under complex conjugation and square roots and let $\mathbf{V}$ denote the countable dimensional $\mathfrak{F}$-vector space with basis $(e_i)_{i=1}^\infty$. We define the usual inner product on $\mathbf{V}$ by letting

$$\left\langle \sum_{i=1}^n a_i e_i \ \bigg| \ \sum_{j=1}^m b_j e_j \right\rangle = \sum_i a_i \overline{b_i}$$

and let $\mathcal{U}(\mathbf{V})$ denote the corresponding unitary group, i.e., the group of all invertible linear transformations of $\mathbf{V}$ preserving $\langle \cdot \mid \cdot \rangle$.

For finite subsets $A, B \subseteq \mathbf{V}$, we let

$$A \underset{\emptyset}{\downarrow} B \ \Leftrightarrow \ \mathsf{span}(A) \perp \mathsf{span}(B).$$

That is, A and B are independent whenever they span orthogonal subspaces. Symmetry and monotonicity are clear. Moreover, because we chose our field $\mathfrak{F}$ to be closed under complex conjugation and square roots, the inner product of two vectors lies in $\mathfrak{F}$ and hence so does the norm $\|v\| = \sqrt{\langle v \mid v \rangle}$ of any vector. It follows that the Gram–Schmidt orthonormalisation procedure can be performed within $\mathbf{V}$ and hence every orthonormal set may be extended to an orthonormal basis for $\mathbf{V}$. Using this, one may imitate the details of Example 6.22 to show that $\underset{\emptyset}{\downarrow}$ satisfies conditions (iii) and (iv). (See also Section 6 of [**77**] for additional details.)

Theorem 6.25 *Suppose $\mathbf{M}$ is a countable structure, $A \subseteq \mathbf{M}$ a finite subset and $\underset{A}{\downarrow}$ an orbital A-independence relation. Then the pointwise stabiliser subgroup V_A is coarsely bounded in itself. Thus, if $A = \emptyset$, the automorphism group $\mathrm{Aut}(\mathbf{M})$ is coarsely bounded and, if $A \neq \emptyset$, $\mathrm{Aut}(\mathbf{M})$ is locally bounded.*

Proof Suppose U is an open neighbourhood of 1 in V_A. We will find a finite subset $F \subseteq V_A$ so that $V_A = UFUFU$. By passing to a further subset, we may suppose that U is of the form V_B, where $B \subseteq \mathbf{M}$ is a finite set containing A. We begin by choosing, using property (iii) of the orbital A-independence relation, some $f \in V_A$ so that $fB \underset{A}{\downarrow} B$ and set $F = \{f, f^{-1}\}$.

Now, suppose that $g \in V_A$ is given and choose again, by (iii), some $h \in V_A$ so that $hB \underset{A}{\downarrow} (B \cup gB)$. By (ii), it follows that $hB \underset{A}{\downarrow} B$ and $hB \underset{A}{\downarrow} gB$, whereby, by using (i), we have $B \underset{A}{\downarrow} hB$ and $gB \underset{A}{\downarrow} hB$. Because $g \in V_A$, we can apply (iv) to $C = hB$, whence $g \in V_{hB}V_B = hV_Bh^{-1}V_B$.

However, as $f B \perp_A B$ and $h B \perp_A B$, i.e., $(h f^{-1} \cdot f B) \perp_A B$, and also $h f^{-1} \in V_A$, by (iv) it follows that $h f^{-1} \in V_B V_{fB} = V_B f V_B f^{-1}$. So, finally, $h \in V_B f V_B$ and

$$g \in h V_B h^{-1} V_B \subseteq V_B f V_B \cdot V_B \cdot (V_B f V_B)^{-1} \cdot V_B \subseteq V_B F V_B F V_B$$

as required. $\qquad\qquad\qquad\qquad\qquad\qquad\qquad\qquad\qquad\qquad\qquad\qquad\qquad$ □

By the preceding examples, we see that both the automorphism group of the measured Boolean algebra and the unitary group $\mathcal{U}(\mathbf{V})$ are coarsely bounded, while the automorphism group $\text{Aut}(T_\infty)$ is locally bounded (cf. Theorem 6.20 in [49], Theorem 6.11 in [77], respectively Theorem 6.31 in [49]).

If $\mathbf{M}$ is an $\mathcal{L}$-structure, A is a subset of $\mathbf{M}$ and $\bar{b}$ is a tuple in $\mathbf{M}$, the *type of* $\bar{b}$ *relative to* A is the set

$$\text{tp}^{\mathbf{M}}(\bar{b}/A)$$
$$= \{\phi(\bar{x}) \mid \phi(\bar{x}) \text{ is an } \mathcal{L}\text{-formula with parameters in } A \text{ and } \mathbf{M} \models \phi(\bar{b})\}.$$

We note that, if $\mathbf{M}$ is an ω-homogeneous structure, $\bar{a}, \bar{b}$ are tuples in $\mathbf{M}$ and $A \subseteq \mathbf{M}$ is a finite subset, then

$$\text{tp}^{\mathbf{M}}(\bar{a}/A) = \text{tp}^{\mathbf{M}}(\bar{b}/A) \iff \bar{b} \in V_A \cdot \bar{a}.$$

In this case, we can reformulate conditions (iii) and (iv) of the definition of orbital A-independence relations as follows.

(iii) For all $\bar{a}$ and B, there is $\bar{b}$ with $\text{tp}^{\mathbf{M}}(\bar{b}/A) = \text{tp}^{\mathbf{M}}(\bar{a}/A)$ and $\bar{b} \perp_A B$.
(iv) For all $\bar{a}, \bar{b}$ and B, if $\bar{a} \perp_A B$, $\bar{b} \perp_A B$ and $\text{tp}^{\mathbf{M}}(\bar{a}/A) = \text{tp}^{\mathbf{M}}(\bar{b}/A)$, then $\text{tp}^{\mathbf{M}}(\bar{a}/B) = \text{tp}^{\mathbf{M}}(\bar{b}/B)$.

Also, for the next result, we remark that, if T is a complete theory with infinite models in a countable language $\mathcal{L}$, then T has a countable saturated model if and only if $S_n(T)$ is countable for all n. In particular, this holds if T is ω-stable.

Theorem 6.26 *Suppose that* $\mathbf{M}$ *is a saturated countable model of an* ω-stable *theory. Then* $\text{Aut}(\mathbf{M})$ *is coarsely bounded.*

Proof We note first that, because $\mathbf{M}$ is saturated and countable, it is ω-homogeneous. Now, because $\mathbf{M}$ is the model of an ω-stable theory, there is a corresponding notion of *forking independence* $\bar{a} \perp_A B$ defined by

$$\bar{a} \underset{A}{\perp} B \iff \text{tp}^{\mathbf{M}}(\bar{a}/A \cup B) \text{ is a non-forking extension of } \text{tp}^{\mathbf{M}}(\bar{a}/A)$$

$$\iff \text{RM}(\bar{a}/A \cup B) = \text{RM}(\bar{a}/A),$$

where RM denotes the *Morley rank*. In this case, forking independence $\perp_A$ always satisfies symmetry and monotonicity, i.e., conditions (i) and (ii), for all finite $A \subseteq \mathbf{M}$.

Moreover, by the existence of non-forking extensions, every type $\mathrm{tp}^{\mathbf{M}}(\bar{a}/A)$ has a non-forking extension $q \in S_n(A \cup B)$. Also, as $\mathbf{M}$ is saturated, this extension q is realised by some tuple $\bar{b}$ in $\mathbf{M}$, i.e., $\mathrm{tp}^{\mathbf{M}}(\bar{b}/A \cup B) = q$. Thus, $\mathrm{tp}^{\mathbf{M}}(\bar{b}/A \cup B)$ is a non-forking extension of $\mathrm{tp}^{\mathbf{M}}(\bar{a}/A) = \mathrm{tp}^{\mathbf{M}}(\bar{b}/A)$, which implies that $\bar{b} \perp_A B$. In other words, for all $\bar{a}$ and A, B, there is $\bar{b}$ with $\mathrm{tp}^{\mathbf{M}}(\bar{b}/A) = \mathrm{tp}^{\mathbf{M}}(\bar{a}/A)$ and $\bar{b} \perp_A B$, which verifies the existence condition (iii) for $\perp_A$.

However, forking independence over A, $\perp_A$, may not satisfy the stationarity condition (iv) unless every type $S_n(A)$ is stationary, i.e., unless, for all $B \supseteq A$, every type $p \in S_n(A)$ has a unique non-forking extension in $S_n(B)$. Nevertheless, as we shall show, we can get by with slightly less.

We let $\perp_\emptyset$ denote forking independence over the empty set. Suppose also that $B \subseteq \mathbf{M}$ is a fixed finite subset and let $\bar{a} \in \mathbf{M}^n$ be an enumeration of B. Then there are at most $\deg_M(\mathrm{tp}^{\mathbf{M}}(\bar{a}))$ non-forking extensions of $\mathrm{tp}(\bar{a})$ in $S_n(B)$, where $\deg_M(\mathrm{tp}^{\mathbf{M}}(\bar{a}))$ denotes the *Morley degree* of $\mathrm{tp}^{\mathbf{M}}(\bar{a})$. Choose realisations $\bar{b}_1, \ldots, \bar{b}_k \in \mathbf{M}^n$ for each of these non-forking extensions realised in $\mathbf{M}$. Because $\mathrm{tp}^{\mathbf{M}}(\bar{b}_i) = \mathrm{tp}^{\mathbf{M}}(\bar{a})$, there are $f_1, \ldots, f_k \in \mathrm{Aut}(\mathbf{M})$ so that $\bar{b}_i = f_i \bar{a}$. Let F be the set of these f_i and their inverses. Thus, if $\bar{c} \in \mathbf{M}^n$ satisfies $\mathrm{tp}^{\mathbf{M}}(\bar{c}) = \mathrm{tp}^{\mathbf{M}}(\bar{a})$ and $\bar{c} \perp_\emptyset B$, then there is some i so that $\mathrm{tp}^{\mathbf{M}}(\bar{c}/B) = \mathrm{tp}^{\mathbf{M}}(\bar{b}_i/B)$ and so, for some $h \in V_B$, we have $\bar{c} = h\bar{b}_i = h f_i \bar{a} \in V_B F \cdot \bar{a}$.

Now assume $g \in \mathrm{Aut}(\mathbf{M})$ is given and pick, by condition (iii), some $h \in \mathrm{Aut}(\mathbf{M})$ so that $hB \perp_\emptyset (B \cup gB)$. By monotonicity and symmetry, it follows that $B \perp_\emptyset hB$ and $gB \perp_\emptyset hB$. Also, because Morley rank and hence forking independence are invariant under automorphisms of $\mathbf{M}$, we see that $h^{-1}B \perp_\emptyset B$ and $h^{-1}gB \perp_\emptyset B$. So, as $\bar{a}$ enumerates B, we have $h^{-1}\bar{a} \perp_\emptyset B$ and $h^{-1}g\bar{a} \perp_\emptyset B$, where clearly $\mathrm{tp}^{\mathbf{M}}(h^{-1}\bar{a}) = \mathrm{tp}^{\mathbf{M}}(\bar{a})$ and $\mathrm{tp}^{\mathbf{M}}(h^{-1}g\bar{a}) = \mathrm{tp}^{\mathbf{M}}(\bar{a})$. By our observation above, we deduce that $h^{-1}\bar{a} \in V_B F \cdot \bar{a}$ and $h^{-1}g\bar{a} \in V_B F \cdot \bar{a}$, whence $h^{-1} \in V_B F V_{\bar{a}} = V_B F V_B$ and similarly $h^{-1}g \in V_B F V_B$. Therefore, we finally have that

$$g \in (V_B F V_B)^{-1} V_B F V_B = V_B F V_B F V_B.$$

We have thus shown that, for all finite $B \subseteq \mathbf{M}$, there is a finite subset $F \subseteq \mathrm{Aut}(\mathbf{M})$ so that $\mathrm{Aut}(\mathbf{M}) = V_B F V_B F V_B$, verifying that $\mathrm{Aut}(\mathbf{M})$ is coarsely bounded.   □

Example 6.27 Let us note that Theorem 6.26 fails without the assumption of $\mathbf{M}$ being saturated. To see this, let T be the complete theory of ∞-regular forests, i.e., of undirected graphs without loops in which every vertex has infinite valence, in the language of a single binary edge relation. In all countable models of T, every connected component is then a copy of the $\aleph_0$-regular tree T_∞ and hence the number of connected components is a complete isomorphism invariant for the countable models of T. Moreover, the countable theory T is ω-stable (in fact, T_∞ is also ω-homogeneous). Nevertheless, the automorphism group of the unsaturated structure T_∞ fails to be coarsely bounded, as witnessed by its tautological isometric action on T_∞.

In view of Theorem 6.26 it is natural to wonder if the automorphism group of a countable atomic model of an ω-stable theory is at least locally bounded, and indeed this was mentioned as an open problem in an earlier preprint of this chapter. However, J. Zielinski [98] was able to construct an example of such a structure with an automorphism group isomorphic to the direct product $\mathbb{Q}^{\mathbb{Q}}$, which fails to be locally bounded. It would be useful to isolate other purely model theoretical properties of a countable structure that would ensure local boundedness of its automorphism group.

Example 6.28 Suppose $\mathfrak{F}$ is a countable field and let $\mathcal{L} = \{+, -, 0\} \cup \{\lambda_t \mid t \in \mathfrak{F}\}$ be the language of $\mathfrak{F}$-vector spaces, i.e., $+$ and $-$ are respectively binary and unary function symbols and 0 is a constant symbol representing the underlying abelian group, and λ_t are unary function symbols representing multiplication by the scalar t. Let also T be the theory of infinite $\mathfrak{F}$-vector spaces.

Because $\mathfrak{F}$-vector spaces of size $\aleph_1$ have dimension $\aleph_1$, we see that T is $\aleph_1$-categorical and thus complete and ω-stable. Moreover, provided $\mathfrak{F}$ is infinite, T fails to be $\aleph_0$-categorical, since, for exampe, the one- and two-dimensional $\mathfrak{F}$-vector spaces are non-isomorphic. However, because T is ω-stable, it has a countable saturated model, which is easily seen to be the $\aleph_0$-dimensional $\mathfrak{F}$-vector space denoted $\mathbf{V}$. It thus follows from Theorem 6.26 that the general linear group $\mathrm{GL}(\mathbf{V}) = \mathrm{Aut}(\mathbf{V})$ is coarsely bounded.

Oftentimes, Fraïssé classes admit a canonical form of amalgamation that can be used to define a corresponding notion of independence. One rendering of this is given by K. Tent and M. Ziegler (Example 2.2 in [86]). However, their notion is too weak to ensure that the corresponding independence notion is an orbital independence relation. For this, one needs a stronger form of functoriality, which nevertheless is satisfied in most cases.

Definition 6.29 Suppose $\mathcal{K}$ is a Fraïssé class and that $\mathbf{A} \in \mathcal{K}$. We say that $\mathcal{K}$ admits a *functorial amalgamation* over $\mathbf{A}$ if there is map θ that to all pairs of embeddings $\eta_1: \mathbf{A} \hookrightarrow \mathbf{B}_1$ and $\eta_2: \mathbf{A} \hookrightarrow \mathbf{B}_2$, with $\mathbf{B}_1, \mathbf{B}_2 \in \mathcal{K}$, associates

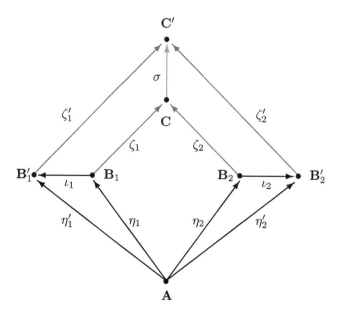

Figure 6.1 Functorial amalgamations.

a pair of embeddings $\zeta_1 : \mathbf{B}_1 \hookrightarrow \mathbf{C}$ and $\zeta_2 : \mathbf{B}_2 \hookrightarrow \mathbf{C}$ into another structure $\mathbf{C} \in \mathcal{K}$ so that $\zeta_1 \circ \eta_1 = \zeta_2 \circ \eta_2$ and, moreover, these satisfy the following conditions.

(1) (symmetry) The pair $\Theta(\eta_2 : \mathbf{A} \hookrightarrow \mathbf{B}_2, \eta_1 : \mathbf{A} \hookrightarrow \mathbf{B}_1)$ is the reverse of the pair $\Theta(\eta_1 : \mathbf{A} \hookrightarrow \mathbf{B}_1, \eta_2 : \mathbf{A} \hookrightarrow \mathbf{B}_2)$.
(2) (functoriality) If $\eta_1 : \mathbf{A} \hookrightarrow \mathbf{B}_1$, $\eta_2 : \mathbf{A} \hookrightarrow \mathbf{B}_2$, $\eta_1' : \mathbf{A} \hookrightarrow \mathbf{B}_1'$ and $\eta_2' : \mathbf{A} \hookrightarrow \mathbf{B}_2'$ are embeddings with $\mathbf{B}_1, \mathbf{B}_2, \mathbf{B}_1', \mathbf{B}_2' \in \mathcal{K}$ and $\iota_1 : \mathbf{B}_1 \hookrightarrow \mathbf{B}_1'$ and $\iota_2 : \mathbf{B}_2 \hookrightarrow \mathbf{B}_2'$ are embeddings with $\iota_i \circ \eta_i = \eta_i'$, then, for

$$\Theta(\eta_1 : \mathbf{A} \hookrightarrow \mathbf{B}_1, \eta_2 : \mathbf{A} \hookrightarrow \mathbf{B}_2) = (\zeta_1 : \mathbf{B}_1 \hookrightarrow \mathbf{C}, \zeta_2 : \mathbf{B}_2 \hookrightarrow \mathbf{C})$$

and

$$\Theta(\eta_1 : \mathbf{A} \hookrightarrow \mathbf{B}_1, \eta_2 : \mathbf{A} \hookrightarrow \mathbf{B}_2) = (\zeta_1' : \mathbf{B}_1' \hookrightarrow \mathbf{C}', \zeta_2' : \mathbf{B}_2' \hookrightarrow \mathbf{C}'),$$

there is an embedding $\sigma : \mathbf{C} \hookrightarrow \mathbf{C}'$ so that $\sigma \circ \zeta_i = \zeta_i' \circ \iota_i$ for $i = 1, 2$.

Please see Figure 6.1 for a graphical depiction.

We note that the pair

$$\Theta(\eta_1 : \mathbf{A} \hookrightarrow \mathbf{B}_1, \eta_2 : \mathbf{A} \hookrightarrow \mathbf{B}_2) = (\zeta_1 : \mathbf{B}_1 \hookrightarrow \mathbf{C}, \zeta_2 : \mathbf{B}_2 \hookrightarrow \mathbf{C})$$

is simply the precise manner of describing the amalgamation $\mathbf{C}$ of the two structures $\mathbf{B}_1$ and $\mathbf{B}_2$ over their common substructure $\mathbf{A}$ (with the additional diagram of embeddings). Thus, symmetry says that the amalgamation should not depend on the order of the structures $\mathbf{B}_1$ and $\mathbf{B}_2$, whereas functoriality states that the amalgamation should commute with embeddings of the $\mathbf{B}_i$ into larger structures $\mathbf{B}'_i$. With this concept at hand, for finite subsets A, B_1, B_2 of the Fraïssé limit $\mathbf{K}$, we may define B_1 and B_2 to be independent over A if $\mathbf{B}_1 = \langle A \cup B_1 \rangle$ and $\mathbf{B}_2 = \langle A \cup B_2 \rangle$ are amalgamated over $\mathbf{A} = \langle A \rangle$ in $\mathbf{K}$ as given by $\Theta(\mathrm{id}_{\mathbf{A}} : \mathbf{A} \hookrightarrow \mathbf{B}_1, \mathrm{id}_{\mathbf{A}} : \mathbf{A} \hookrightarrow \mathbf{B}_2)$. More precisely, we have the following definition.

Definition 6.30 Suppose $\mathcal{K}$ is a Fraïssé class with limit $\mathbf{K}$, $A \subseteq \mathbf{K}$ is a finite subset and Θ is a functorial amalgamation on $\mathcal{K}$ over $\mathbf{A} = \langle A \rangle$. For finite subsets $B_1, B_2 \subseteq \mathbf{K}$ with $\mathbf{B}_i = \langle A \cup B_i \rangle$, $\mathbf{D} = \langle A \cup B_1 \cup B_2 \rangle$ and

$$\Theta\big(\mathrm{id}_{\mathbf{A}} : \mathbf{A} \hookrightarrow \mathbf{B}_1, \mathrm{id}_{\mathbf{A}} : \mathbf{A} \hookrightarrow \mathbf{B}_2\big) = \big(\zeta_1 : \mathbf{B}_1 \hookrightarrow \mathbf{C}, \zeta_2 : \mathbf{B}_2 \hookrightarrow \mathbf{C}\big),$$

we set

$$B_1 \underset{A}{\downarrow} B_2$$

if and only if there is an embedding $\pi : \mathbf{D} \hookrightarrow \mathbf{C}$ so that $\zeta_i = \pi \circ \mathrm{id}_{\mathbf{B}_i}$ for $i = 1, 2$.

With this setup, we readily obtain the following result.

Theorem 6.31 *Suppose $\mathcal{K}$ is a Fraïssé class with limit $\mathbf{K}$, $A \subseteq \mathbf{K}$ is a finite subset and Θ is a functorial amalgamation of $\mathcal{K}$ over $\mathbf{A} = \langle A \rangle$. Let also $\downarrow_A$ be the relation defined from Θ and A as in Definition 6.30. Then $\downarrow_A$ is an orbital A-independence relation on $\mathbf{K}$ and thus V_A is coarsely bounded. In particular, $\mathrm{Aut}(\mathbf{K})$ is locally bounded and hence admits a coarsely proper metric.*

Proof Symmetry and monotonicity of $\downarrow_A$ follow easily from symmetry, respectively functoriality, of Θ. Also, the existence condition on $\downarrow_A$ follows from the ultrahomogeneity of $\mathbf{K}$ and the realisation of the amalgam Θ inside of $\mathbf{K}$.

For stationarity, we use the ultrahomogeneity of $\mathbf{K}$. So, suppose that finite $\bar{a}, \bar{b}$ and $B \subseteq \mathbf{K}$ are given so that $\bar{a} \downarrow_A B$, $\bar{b} \downarrow_A B$ and $\mathrm{tp}^{\mathbf{K}}(\bar{a}/A) = \mathrm{tp}^{\mathbf{K}}(\bar{b}/A)$. We set $\mathbf{B}_1 = \langle \bar{a} \cup A \rangle$, $\mathbf{B}'_1 = \langle \bar{b} \cup A \rangle$, $\mathbf{B}_2 = \langle B \cup A \rangle$, $\mathbf{D} = \langle \bar{a} \cup B \cup A \rangle$ and $\mathbf{D}' = \langle \bar{b} \cup B \cup A \rangle$. Let also

$$\Theta\big(\mathrm{id}_{\mathbf{A}} : \mathbf{A} \hookrightarrow \mathbf{B}_1, \mathrm{id}_{\mathbf{A}} : \mathbf{A} \hookrightarrow \mathbf{B}_2\big) = \big(\zeta_1 : \mathbf{B}_1 \hookrightarrow \mathbf{C}, \zeta_2 : \mathbf{B}_2 \hookrightarrow \mathbf{C}\big),$$

$$\Theta\big(\mathrm{id}_{\mathbf{A}} : \mathbf{A} \hookrightarrow \mathbf{B}'_1, \mathrm{id}_{\mathbf{A}} : \mathbf{A} \hookrightarrow \mathbf{B}_2\big) = \big(\zeta'_1 : \mathbf{B}'_1 \hookrightarrow \mathbf{C}', \zeta'_2 : \mathbf{B}_2 \hookrightarrow \mathbf{C}'\big)$$

and note that there is an isomorphism $\iota\colon \mathbf{B}_1 \hookrightarrow \mathbf{B}_1'$ pointwise fixing A so that $\iota(\overline{a}) = \overline{b}$. By the definition of the independence relation, there are embeddings $\pi\colon \mathbf{D} \hookrightarrow \mathbf{C}$ and $\pi'\colon \mathbf{D}' \hookrightarrow \mathbf{C}'$ so that $\zeta_i = \pi \circ \mathrm{id}_{\mathbf{B}_i}$, $\zeta_1' = \pi' \circ \mathrm{id}_{\mathbf{B}_1'}$ and $\zeta_2' = \pi' \circ \mathrm{id}_{\mathbf{B}_2}$. On the other hand, by the functoriality of Θ, there is an embedding $\sigma\colon \mathbf{C} \hookrightarrow \mathbf{C}'$ so that $\sigma \circ \zeta_1 = \zeta_1' \circ \iota$ and $\sigma \circ \zeta_2 = \zeta_2' \circ \mathrm{id}_{\mathbf{B}_2}$. Thus, $\sigma \circ \pi \circ \mathrm{id}_{\mathbf{B}_1} = \pi' \circ \mathrm{id}_{\mathbf{B}_1'} \circ \iota$ and $\sigma \circ \pi \circ \mathrm{id}_{\mathbf{B}_2} = \pi' \circ \mathrm{id}_{\mathbf{B}_2} \circ \mathrm{id}_{\mathbf{B}_2}$, i.e., $\sigma \circ \pi|_{\mathbf{B}_1} = \pi' \circ \iota$ and $\sigma \circ \pi|_{\mathbf{B}_2} = \pi'|_{\mathbf{B}_2}$. Let now $\rho\colon \pi'[\mathbf{D}'] \hookrightarrow \mathbf{D}'$ be the isomorphism that is inverse to π'. Then $\rho \sigma \pi|_{\mathbf{B}_1} = \iota$ and $\rho \sigma \pi|_{\mathbf{B}_2} = \mathrm{id}_{\mathbf{B}_2}$. So, by ultrahomogeneity of $\mathbf{K}$, there is an automorphism $g \in \mathrm{Aut}(\mathbf{K})$ extending $\rho \sigma \pi$, whence, in particular, $g \in V_{\mathbf{B}_2} \subseteq V_B$, while $g(\overline{a}) = \overline{b}$. It follows that $\mathrm{tp}^{\mathbf{K}}(\overline{a}/B) = \mathrm{tp}^{\mathbf{K}}(\overline{b}/B)$, verifying stationarity. $\qquad\square$

Example 6.32 (Urysohn metric spaces, cf. Example 2.2 (c) in [86]) Suppose $\mathcal{S}$ is a countable additive subsemi-group of the positive reals. Then the class of finite metric spaces with distances in $\mathcal{S}$ forms a Fraïssé class $\mathcal{K}$ with functorial amalgamation over the one-point metric space $\mathbf{P} = \{p\}$. Indeed, if $\mathbf{A}$ and $\mathbf{B}$ belong to $\mathcal{S}$ and intersect exactly in the point p, we can define a metric d on $\mathbf{A} \cup \mathbf{B}$ extending those of $\mathbf{A}$ and $\mathbf{B}$ by letting

$$d(a,b) = d_{\mathbf{A}}(a,p) + d_{\mathbf{B}}(p,b),$$

for $a \in \mathbf{A}$ and $b \in \mathbf{B}$. We thus take this to define the amalgamation of $\mathbf{A}$ and $\mathbf{B}$ over $\mathbf{P}$ and one easily verifies that this provides a functorial amalgamation over $\mathbf{P}$ on the class $\mathcal{K}$.

Two important particular cases are when $\mathcal{S} = \mathbb{Z}_+$, respectively $\mathcal{S} = \mathbb{Q}_+$, in which case the Fraïssé limits are the integral and rational Urysohn metric spaces $\mathbb{Z}\mathbb{U}$ and $\mathbb{Q}\mathbb{U}$. By Theorem 6.31, we see that their isometry groups $\mathrm{Isom}(\mathbb{Z}\mathbb{U})$ and $\mathrm{Isom}(\mathbb{Q}\mathbb{U})$ admit coarsely proper metrics.

We note also that it is vital that $\mathbf{P}$ is non-empty. Indeed, because $\mathrm{Isom}(\mathbb{Z}\mathbb{U})$ and $\mathrm{Isom}(\mathbb{Q}\mathbb{U})$ act transitively on metric spaces of infinite diameter, namely, on $\mathbb{Z}\mathbb{U}$ and $\mathbb{Q}\mathbb{U}$, they are not coarsely bounded and hence the corresponding Fraïssé classes do not admit a functorial amalgamation over the empty space $\emptyset$.

Instead, if, for a given $\mathcal{S}$ and $r \in \mathcal{S}$, we let $\mathcal{K}$ denote the finite metric spaces with distances in $\mathcal{S} \cap [0,r]$, then $\mathcal{K}$ is still a Fraïssé class now admitting functorial amalgamation over the empty space. Namely, to join $\mathbf{A}$ and $\mathbf{B}$, one simply takes the disjoint union and stipulates that $d(a,b) = r$ for all $a \in \mathbf{A}$ and $b \in \mathbf{B}$.

As a particular example, we note that the isometry group $\mathrm{Isom}(\mathbb{Q}\mathbb{U}_1)$ of the rational Urysohn metric space of diameter 1 is coarsely bounded (see Theorem 5.8 in [77]).

Example 6.33 (The ended $\aleph_0$-regular tree) Again, let T_∞ denote the $\aleph_0$-regular tree and fix an *end* $\mathfrak{e}$ of T_∞. That is, $\mathfrak{e}$ is an equivalence class of infinite paths $(v_0, v_1, v_2, \ldots)$ in T_∞ under the equivalence relation

$$(v_0, v_1, v_2, \ldots) \sim (w_0, w_1, w_2, \ldots) \Leftrightarrow \exists k, l \; \forall n \; v_{k+n} = w_{l+n}.$$

So, for every vertex $t \in \mathsf{T}_\infty$, there is a unique path $(v_0, v_1, v_2, \ldots) \in \mathfrak{e}$ beginning at $v_0 = t$. Thus, if r is another vertex in T_∞, we can set $t <_{\mathfrak{e}} r$ if and only if $r = v_n$ for some $n \geqslant 1$. Note that this defines a strict partial ordering $<_{\mathfrak{e}}$ on T_∞ so that every two vertices $t, s \in \mathsf{T}_\infty$ have a least upper bound and, moreover, this least upper bound lies on the geodesic from t to s. Furthermore, we define the function $\vartheta \colon \mathsf{T}_\infty \times \mathsf{T}_\infty \to \mathsf{T}_\infty$ by letting $\vartheta(t, s) = x_1$, where $(x_0, x_1, x_2, \ldots, x_k)$ is the geodesic from $x_0 = t$ to $x_k = s$, for $t \neq s$, and $\vartheta(t, t) = t$.

As is easy to see, the expanded structure $(\mathsf{T}_\infty, <_{\mathfrak{e}}, \vartheta)$ is ultrahomogeneous and locally finite and hence, by Fraïssé's Theorem, is the Fraïssé limit of its age $\mathcal{K} = \mathrm{Age}(\mathsf{T}_\infty, <_{\mathfrak{e}}, \vartheta)$. We also claim that $\mathcal{K}$ admits a functorial amalgamation over the structure on a single vertex t. Indeed, if $\mathbf{A}$ and $\mathbf{B}$ are finite substructures of $(\mathsf{T}, <_{\mathfrak{e}}, \vartheta)$ and we pick a vertex in $t_{\mathbf{A}}$ and $t_{\mathbf{B}}$ in each, then there is a freest amalgamation of $\mathbf{A}$ and $\mathbf{B}$ identifying $t_{\mathbf{A}}$ and $t_{\mathbf{B}}$. Namely, let $t_{\mathbf{A}} = a_0 <_{\mathfrak{e}} a_1 <_{\mathfrak{e}} a_2 <_{\mathfrak{e}} \cdots <_{\mathfrak{e}} a_n$ and $t_{\mathbf{B}} = b_0 <_{\mathfrak{e}} b_1 <_{\mathfrak{e}} b_2 <_{\mathfrak{e}} \cdots <_{\mathfrak{e}} b_m$ be an enumeration of the successors of $t_{\mathbf{A}}$ and $t_{\mathbf{B}}$ in $\mathbf{A}$ and $\mathbf{B}$ respectively. We then take the disjoint union of $\mathbf{A}$ and $\mathbf{B}$ modulo the identifications $a_0 = b_0, \ldots, a_{\min(n,m)} = b_{\min(n,m)}$ and add only the edges from $\mathbf{A}$ and $\mathbf{B}$. There are then unique extensions of $<_{\mathfrak{e}}$ and ϑ to the amalgam making it a member of $\mathcal{K}$. Moreover, this amalgamation is functorial over the single vertex t.

It thus follows that $\mathrm{Aut}(\mathsf{T}_\infty, <_{\mathfrak{e}}, \vartheta)$ is locally bounded as witnessed by the pointwise stabiliser V_t of any fixed vertex $t \in \mathsf{T}_\infty$. Now, as ϑ commutes with automorphisms of T_∞ and $<_{\mathfrak{e}}$ and $\mathfrak{e}$ are interdefinable, we see that $\mathrm{Aut}(\mathsf{T}_\infty, <_{\mathfrak{e}}, \vartheta)$ is simply the group $\mathrm{Aut}(\mathsf{T}_\infty, \mathfrak{e})$ of all automorphisms of T_∞ fixing the end $\mathfrak{e}$.

6.5 Computing Quasi-isometry Types
of Automorphism Groups

Thus far we have been able to show local boundedness of certain automorphism groups. The goal is now to identity their quasi-isometry type insofar as this is well defined.

Example 6.34 (The $\aleph_0$-regular tree) Let T_∞ be the $\aleph_0$-regular tree with automorphism group $\mathrm{Aut}(\mathsf{T}_\infty)$ and fix a vertex $t \in \mathsf{T}_\infty$. By Example 6.23 and Theorem 6.25, we know that subgroup V_t is coarsely bounded. Fix also a neighbour s of t in T_∞ and let $\mathcal{R} = \{\mathcal{O}(t,s)\}$. Now, because $\mathrm{Aut}(\mathsf{T}_\infty)$ acts transitively on the set of oriented edges of T_∞, we see that, if $r \in \mathcal{O}(t) = \mathsf{T}_\infty$ is any vertex and $(v_0, v_1, \ldots, v_m)$ is the geodesic from $v_0 = t$ to $v_m = r$, then $\mathcal{O}(v_i, v_{i+1}) \in \mathcal{R}$ for all i. It thus follows from Theorem 6.9 that $\mathrm{Aut}(\mathsf{T}_\infty)$ admits a maximal metric and, moreover, that

$$g \in \mathrm{Aut}(\mathsf{T}_\infty) \mapsto g(t) \in \mathbb{X}_{t,\mathcal{R}}$$

is a quasi-isometry. However, the graph $\mathbb{X}_{t,\mathcal{R}}$ is simply the tree T_∞ itself, which shows that

$$g \in \mathrm{Aut}(\mathsf{T}_\infty) \mapsto g(t) \in \mathsf{T}_\infty$$

is a quasi-isometry. In other words, the quasi-isometry type of $\mathrm{Aut}(\mathsf{T}_\infty)$ is just the tree T_∞.

Example 6.35 (The ended $\aleph_0$-regular tree) Let $\left(\mathsf{T}_\infty, <_\mathfrak{e}, \vartheta\right)$ be as in Example 6.33. Again, if t is some fixed vertex, the vertex stabiliser V_t is coarsely bounded. Now, $\mathrm{Aut}(\mathsf{T}_\infty, <_\mathfrak{e}, \vartheta)$ acts transitively on the vertices and edges of T_∞, but no longer acts transitively on set of oriented edges. Namely, if (x,y) and (v,w) are edges of T_∞, then $(v,w) \in \mathcal{O}(x,y)$ if and only if $x <_\mathfrak{e} y \leftrightarrow v <_\mathfrak{e} w$. Therefore, let s be any neighbour of t in T and set $\mathcal{R} = \{\mathcal{O}(t,s), \mathcal{O}(s,t)\}$. Then, as in Example 6.34, we see that $\mathbb{X}_{t,\mathcal{R}} = \mathsf{T}_\infty$ and that

$$g \in \mathrm{Aut}(\mathsf{T}_\infty, \mathfrak{e}) \mapsto g(t) \in \mathsf{T}_\infty$$

is a quasi-isometry. So $\mathrm{Aut}(\mathsf{T}_\infty, \mathfrak{e})$ is quasi-isometric to T_∞ and thus also to $\mathrm{Aut}(\mathsf{T}_\infty)$.

Example 6.36 (Urysohn metric spaces) Let $\mathcal{S}$ be a countable additive subsemigroup of the positive reals and let $\mathcal{SU}$ be the limit of the Fraïssé class $\mathcal{K}$ of finite metric spaces with distances in $\mathcal{S}$. As we have seen in Example 6.32, $\mathcal{K}$ admits a functorial amalgamation over the one-point metric space $\mathbf{P} = \{p\}$ and thus the stabiliser V_{x_0} of any chosen point $x_0 \in \mathcal{SU}$ is coarsely bounded.

We remark that $\mathcal{O}(x_0) = \mathcal{SU}$. So fix some point $x_1 \in \mathcal{SU} \setminus \{x_0\}$ and set $s = d(x_0, x_1)$. Let also $\mathcal{R} = \{\mathcal{O}(x_0, x_1)\}$. By the ultrahomogeneity of $\mathcal{SU}$, we see that $\mathcal{O}(y,z) = \mathcal{O}(x_0, x_1) \in \mathcal{R}$ for all $y, z \in \mathcal{SU}$ with $d(y,z) = s$.

Now, for any two points $y, z \in \mathcal{SU}$, let $n_{y,z} = \lceil \frac{d(y,z)}{s} \rceil + 1$. It is then easy to see that there is a finite metric space in $\mathcal{K}$ containing a sequence of points

$v_0, v_1, \ldots, v_{n_{y,z}}$ so that $d(v_i, v_{i+1}) = s$, while $d(v_0, v_{n_{y,z}}) = d(y, z)$. By the ultrahomogeneity of $\mathcal{SU}$, it follows that there is a sequence $w_0, w_1, \ldots, w_{n_{y,z}} \in \mathcal{SU}$ with $w_0 = y$, $w_{n_{y,z}} = z$ and $d(w_i, w_{i+1}) = s$, i.e., $\mathcal{O}(w_i, w_{i+1}) \in \mathcal{R}$ for all i. In other words,

$$\rho_{x_0, \mathcal{R}}(y, z) \leqslant \frac{1}{s} d(y, z) + 2$$

and, in particular, the graph $\mathbb{X}_{x_0, \mathcal{R}}$ is connected. Conversely, if $\rho_{x_0, \mathcal{R}}(y, z) = m$, then there is a finite path $w_0, w_1, \ldots, w_m \in \mathcal{SU}$ with $w_0 = y$, $w_m = z$ and $d(w_i, w_{i+1}) = s$, whereby $d(y, z) \leqslant ms$, showing that

$$\frac{1}{s} d(y, z) \leqslant \rho_{x_0, \mathcal{R}}(y, z) \leqslant \frac{1}{s} d(y, z) + 2.$$

Therefore, the identity map is a quasi-isometry between $\mathbb{X}_{x_0, \mathcal{R}}$ and $\mathcal{SU}$. Because, by Theorem 6.9, the mapping

$$g \in \mathrm{Isom}(\mathcal{SU}) \mapsto g(x_0) \in \mathbb{X}_{x_0, \mathcal{R}}$$

is a quasi-isometry, so is the mapping

$$g \in \mathrm{Isom}(\mathcal{SU}) \mapsto g(x_0) \in \mathcal{SU}.$$

By consequence, the isometry group $\mathrm{Isom}(\mathcal{SU})$ equipped with the permutation group topology is quasi-isometric to the Urysohn space $\mathcal{SU}$.

7

Zappa–Szép Products

7.1 The Topological Structure

A basic result in Banach space theory says that, if X is a Banach space with closed linear subspaces A and B so that $X = A + B$ and $A \cap B = \{0\}$, then X is naturally isomorphic to the direct sum $A \oplus B$ with the product topology. Moreover, the proof of this is a rather straightforward application of the closed graph theorem. Namely, suppose $P \colon X \to A$ is the linear operator defined by $P(x) = a$ if $x = a + b$ for some $b \in B$. Then P has closed graph and hence is bounded. For, if $x_n \xrightarrow{n} x$ and $P(x_n) \xrightarrow{n} a$, then $x - a = \lim_n x_n - P(x_n) \in B$ and so the decomposition $x = a + (x - a)$ shows that $P(x) = a$. Thus both projections $P \colon X \to A$ and $I - P \colon X \to B$ are bounded, whence $X \cong A \oplus B$.

Apart from being formulated specifically for linear spaces, the proof very much depends on the projection P being a morphism and thus gives little hint as to a generalisation to the non-commutative setting. Employing somewhat different ideas, we present this generalisation here.

In the following, we consider a Polish group G with two closed subgroups A and B so that $G = AB$ and $A \cap B = \{1\}$, i.e., so that each element $g \in G$ can be written in a unique manner as $g = ab$ with $a \in A$ and $b \in B$. This situation is expressed by saying that G is the *Zappa–Szép product* of A and B [85, 97] and, of course, includes the familiar cases of internal direct and semi-direct products.

Let $A \times B$ denote the cartesian product with the product topology and define

$$A \times B \xrightarrow{\phi} G$$

by $\phi(a, b) = ab$. By unique decomposition, ϕ is a bijection and, because A and B are homeomorphically embedded in G and multiplication in G is continuous, also ϕ is continuous. It therefore follows that ϕ is a homeomorphism between $A \times B$ and G if and only if the projection maps $G \xrightarrow{\pi_A} A$ and $G \xrightarrow{\pi_B} B$, defined by

$$g = \pi_A(g) \cdot \pi_B(g),$$

are continuous. We verify this by showing that ϕ is an open mapping.

For a subset $D \subseteq G$, let $\mathcal{U}(D)$ denote the largest open subset of G in which D is co-meagre. By continuity of the group operations in G, we have $g\mathcal{U}(D)f = \mathcal{U}(gDf)$ for all $g, f \in G$.

Lemma 7.1 *Suppose $V \subseteq A$ and $W \subseteq B$ are open in A and B respectively. Then*

$$VW \subseteq \mathcal{U}(VW) \quad and \quad \overline{V} \cdot \overline{W} \subseteq \overline{\mathcal{U}(VW)}.$$

Proof Suppose that $a \in V$ and $b \in W$. We choose sets $a \in V_0 \subseteq V$ and $b \in W_0 \subseteq W$, open in A and B respectively, so that $V_0 V_0^{-1} V_0 \subseteq V$ and $W_0 W_0^{-1} W_0 \subseteq W$. Let also $P \subseteq A$ and $Q \subseteq B$ be countable dense subsets, whereby $A = PV_0$, $B = W_0 Q$ and hence also

$$G = \bigcup_{p,q} p V_0 W_0 q.$$

By the Baire Category Theorem, it follows that some $p V_0 W_0 q$ and thus also $V_0 W_0$ is non-meagre in G. Being the continuous image of an open set in $A \times B$ under ϕ, the set $V_0 W_0$ is analytic and hence has the property of Baire. So we have $\mathcal{U}(V_0 W_0) \neq \emptyset$. Now pick some $g \in V_0 W_0 \cap \mathcal{U}(V_0 W_0)$ and write $g = a_0 b_0$ for $a_0 \in V_0$ and $b_0 \in W_0$. Then

$$ab = aa_0^{-1} \cdot a_0 b_0 \cdot b_0^{-1} b$$
$$= aa_0^{-1} \cdot g \cdot b_0^{-1} b$$
$$\in aa_0^{-1} \cdot \mathcal{U}(V_0 W_0) \cdot b_0^{-1} b$$
$$= \mathcal{U}(aa_0^{-1} V_0 W_0 b_0^{-1} b)$$
$$\subseteq \mathcal{U}(V_0 V_0^{-1} V_0 \cdot W_0 W_0^{-1} W_0)$$
$$\subseteq \mathcal{U}(VW).$$

As $a \in V$ and $b \in W$ were arbitrary, this shows that $VW \subseteq \mathcal{U}(VW)$ and hence also $\overline{V} \cdot \overline{W} \subseteq \overline{VW} \subseteq \overline{\mathcal{U}(VW)}$. □

Lemma 7.2 *Suppose $V \subseteq A$ and $W \subseteq B$ are regular open in A and B respectively. Then $VW = \mathcal{U}(VW)$.*

Proof Assume that $V \subseteq A$ and $W \subseteq B$ are regular open, i.e., that $\mathsf{int\,cl}\,V = V$ and hence also $\mathsf{cl\,int}(\sim V) = \,\sim V$, where $\sim V = A \setminus V$, and similarly for W. Here the interior, int, refers to the interior in A and B respectively.

Suppose toward a contradiction that $g \in \mathcal{U}(VW) \setminus VW$ and write $g = ab$ for some $a \in A$ and $b \in B$. Then, either $a \notin V$ or $b \notin W$, say $a \notin V$, the other case being similar. Set $U = \mathsf{int}(\sim V)$, whence $\overline{U} = \mathsf{cl\,int}(\sim V) = \,\sim V$. Then, by Lemma 7.1, we have

$$g = ab \in (\sim V)B = \overline{U} \cdot \overline{B} \subseteq \overline{\mathcal{U}(UB)}.$$

As also $g \in \mathcal{U}(VW)$, it follows that $\mathcal{U}(VW) \cap \overline{\mathcal{U}(UB)} \neq \emptyset$ and thus also that $\mathcal{U}(VW) \cap \mathcal{U}(UB) \neq \emptyset$. Therefore, VW and UB are both co-meagre in the non-empty open set $\mathcal{U}(VW) \cap \mathcal{U}(UB)$, so must intersect, $VW \cap UB \neq \emptyset$. However, as $V \cap U = \emptyset$, this contradicts unique decomposability of elements of G. So $\mathcal{U}(VW) \subseteq VW$ and the reverse inclusion follows directly from Lemma 7.1. □

Theorem 7.3 *Let A and B be closed subgroups of a Polish group G so that $G = AB$ and $A \cap B = \{1\}$. Then the group multiplication, $(a,b) \mapsto ab$, is a homeomorphism from $A \times B$ to G.*

Proof It suffices to note that the sets $V \times W$ with $V \subseteq A$ and $W \subseteq B$ regular open form a basis for the topology on $A \times B$, whence, by Lemma 7.2, the multiplication map is a continuous and open bijection. □

We should point out that the above result can fail entirely when G is no longer assumed to be Polish. Indeed, one could simply take two closed linear subspaces A and B of a separable Banach space X so that $A \cap B = \{0\}$, but not forming a topological direct sum. Then there are unit vectors $a_n \in A$ and $b_n \in B$ so that $a_n - b_n \xrightarrow{n} 0$ and so the topology on the linear subspace $A + B$ is not the product topology.

On the other hand, it is not quite clear whether one needs to assume that G is Polish or if it suffices to assume that it is a *Baire space*, i.e., that the intersection of countable many dense open sets is dense. Examples of these are locally compact Hausdorff spaces and completely metrisable spaces.

Problem 7.4 Does Theorem 7.3 hold for Baire topological groups?

7.2 Examples

Apart from trivial examples such as direct products of Polish groups, there are common instances of the above setup.

Example 7.5 (Internal semidirect products) Another particular case of the Zappa–Szép product is when a Polish group G is the internal semi-direct product of closed subgroups H and N, that is, $G = HN$, $H \cap N = \{1\}$ and N normal in G.

Example 7.6 (Homeomorphism groups of locally compact groups) Suppose H is a locally compact Polish group and consider the group Homeo(H) of homeomorphisms of H equipped with the compact-open topology on the one-point compactification. Then H can be identified with a closed subgroup of Homeo(H) via its left-regular representation $\lambda \colon H \to$ Homeo(H) given by $\lambda_x(y) = xy$. Letting 1_H denote the identity in H and setting

$$K = \{g \in \mathrm{Homeo}(H) \mid g(1_H) = 1_H\},$$

we find that K is closed in Homeo(H), $K \cap H = \{\mathrm{id}\}$ and $KH = \mathrm{Homeo}(H)$. So Homeo(H) is the Zappa–Szép product of the pointwise stabiliser K of 1_H and the group H of translations.

Example 7.7 For a concrete instance of Example 7.6, consider Homeo($\mathbb{T}^2$), where $\mathbb{T} = \mathbb{R}/\mathbb{Z}$. Then K is the group of homeomorphisms of $\mathbb{T}^2$ fixing the point 0 and Homeo($\mathbb{T}^2$) is the Zappa–Szép product of K and $\mathbb{T}^2$ itself.

Example 7.8 For a more general class of examples, suppose H is a locally compact Polish group and G is a subgroup of Homeo(H), which is Polish in some finer group topology and so that G contains the image of H via the left-regular representation, i.e., the group of left-multiplications by elements of H. For example, H could be a Lie group and $G = \mathrm{Diff}^\infty(H)$. Again, H and the pointwise stabiliser K of 1_H are both closed in G and thus G is the Zappa–Szép product of K and H.

Example 7.9 Suppose T_∞ is the countably infinite regular tree, i.e., so that every vertex has denumerable valence. Then T_∞ is isomorphic to the Cayley graph of the free group $\mathbb{F}_\infty$ on a denumerable set of generators and hence the automorphism group Aut(T_∞) can be viewed as a subgroup of the homeomorphism group of the countable discrete group $\mathbb{F}_\infty$. Moreover, under this identification, Aut(T_∞) contains all left-translations by elements of $\mathbb{F}_\infty$ and thus $\mathrm{Aut}(\mathsf{T}_\infty)$ is the Zappa–Szép product of $\mathbb{F}_\infty$ and the pointwise stabiliser

$$K = \{g \in \mathrm{Aut}(\mathsf{T}_\infty) \mid g(r) = r\},$$

where $r \in \mathsf{T}_\infty$ is the vertex corresponding to the identity $1 \in \mathbb{F}_\infty$.

In the preceding examples, the larger group G is decomposed via an action on one of the closed subgroups A and B appearing as a factor in the Zappa–Szép product. As we shall see, this is necessarily so.

Indeed, suppose G is a Polish group with closed subgroups A and B so that $G = AB$ and $A \cap B = \{1\}$. Let also $\pi_A \colon G \to A$ and $\pi_B \colon G \to B$ be the corresponding projection maps, i.e., so that $g = \pi_A(g) \cdot \pi_B(g)$ for all $g \in G$. Then, if $f, g \in G$ and $a \in A$, we have

$$\pi_A(fga)\pi_B(fga) = fga$$
$$= f \cdot \pi_A(ga)\pi_B(ga)$$
$$= \pi_A(f \cdot \pi_A(ga))\pi_B(f \cdot \pi_A(ga))\pi_B(ga),$$

that is,

$$\pi_A(f \cdot \pi_A(ga))^{-1}\pi_A(fga)$$
$$= \pi_B(f \cdot \pi_A(ga))\pi_B(ga)\pi_B(fga)^{-1} \in A \cap B = \{1\}.$$

It follows, in particular, that

$$\pi_A(f \cdot \pi_A(ga)) = \pi_A(fga)$$

for all $f, g \in G$ and $a \in A$. As also $\pi_A(1a) = a$ and as π_A is continuous by Theorem 7.3, we obtain a continuous action $\alpha \colon G \curvearrowright A$ on the topological space A by letting

$$\alpha_g(a) = \pi_A(ga).$$

Observe then that the α-action of A on itself is simply the left-regular representation $\lambda \colon A \curvearrowright A$,

$$\alpha_{a_1}(a_2) = \pi_A(a_1 a_2) = a_1 a_2 = \lambda_{a_1}(a_2),$$

while

$$B = \{g \in G \mid \alpha_g(1) = \pi_A(g) = 1\}.$$

So, in other words, the Zappa–Szép product $G = AB$ arises from a continuous action of G on A, where B is the isotropy subgroup of 1 and A acts by left-multiplication on itself.

By symmetry of this setup, one can of course also see $G = AB$ as arising from a continuous action of G on B, where now A is the isotropy subgroup of 1.

7.3 The Coarse Structure of Zappa–Szép Products

We now turn our attention to the problem of describing the coarse structure of Zappa–Szép products as a function of the coarse structure of the factors in the product. In [38, 39], J. Herndon independently has arrived at other criteria describing the quasi-metric structure provided this exists. One thing that one should be aware of at the outset is that, as opposed to in Section 7.1, here the specific order of the factors becomes vital and the factors will play different roles in the decomposition.

In the following, we fix a Polish group G that is the Zappa–Szép product of two closed subgroups A and B. Define the identification $A \times B \xrightarrow{\phi} G$ by $\phi(a,b) = ab$ and let $G \xrightarrow{\pi_A} A$ and $G \xrightarrow{\pi_B} B$ be the associated projections, i.e., the maps defined by

$$g = \pi_A(g) \cdot \pi_B(g)$$

for all $g \in G$. For $g \in G$ and $X, Y \subseteq G$, we let $g^X = \{xgx^{-1} \mid x \in X\}$ and $Y^X = \{xyx^{-1} \mid x \in X \text{ and } y \in Y\}$.

Lemma 7.10 If $X \subseteq A$, $Y \subseteq B$ and $n \geqslant 1$, we have

$$(XY)^n \subseteq \pi_A[X^B]^n \pi_B[X^B]^{n-1} Y^n.$$

Proof It suffices to show that, for all $a_1, \ldots, a_n \in A$ and $b_1, \ldots, b_n \in B$, we have

$$a_1 b_1 a_2 b_2 \cdots a_n b_n \in a_1 \pi_A(a_2^B) \pi_A(a_3^B) \cdots \pi_A(a_n^B)$$
$$\cdot \pi_B(a_n^B) \pi_B(a_{n-1}^B) \cdots \pi_B(a_2^B) \cdot b_1 b_2 \cdots b_n.$$

We show this by induction on n with the case $n = 1$ being trivial. For the induction step, suppose that

$$a_1 b_1 a_2 b_2 \cdots a_n b_n = ab \cdot b_1 b_2 \cdots b_n$$

for some $a \in a_1 \pi_A(a_2^B) \pi_A(a_3^B) \cdots \pi_A(a_n^B)$ and $b \in \pi_B(a_n^B) \pi_B(a_{n-1}^B) \cdots$
$\pi_B(a_2^B)$. Then, for $a_{n+1} \in A$ and $b_{n+1} \in B$, we have

$$
\begin{aligned}
a_1 b_1 a_2 b_2 & \cdots a_n b_n \cdot a_{n+1} b_{n+1} \\
&= ab \cdot b_1 b_2 \cdots b_n \cdot a_{n+1} b_{n+1} \\
&= a \cdot (bb_1 b_2 \cdots b_n) a_{n+1} (bb_1 b_2 \cdots b_n)^{-1} (bb_1 b_2 \cdots b_n) b_{n+1} \\
&= a \pi_A\big((bb_1 b_2 \cdots b_n) a_{n+1} (bb_1 b_2 \cdots b_n)^{-1}\big) \\
&\quad \cdot \pi_B\big((bb_1 b_2 \cdots b_n) a_{n+1} (bb_1 b_2 \cdots b_n)^{-1}\big) \cdot bb_1 b_2 \cdots b_n b_{n+1} \\
&\in a \pi_A(a_{n+1}^B) \cdot \pi_B(a_{n+1}^B) \cdot bb_1 b_2 \cdots b_n b_{n+1} \\
&\subseteq a_1 \pi_A(a_2^B) \cdots \pi_A(a_n^B) \pi_A(a_{n+1}^B) \\
&\quad \cdot \pi_B(a_{n+1}^B) \pi_B(a_n^B) \cdots \pi_B(a_2^B) \cdot b_1 \cdots b_{n+1},
\end{aligned}
$$

as claimed. $\square$

Lemma 7.11 *The map* $A \times B \xrightarrow{\phi} G$ *is bornologous if and only if* X^B *is coarsely bounded in* G *for every coarsely bounded set* $X \subseteq A$.

Proof Recall that by the results of Section 3.5, the coarse structure on the product group $A \times B$ is simply the product of the coarse structures on A and on B. Thus, a basic entourage in $A \times B$ has the form $E_X \times E_Y$, where X and Y are coarsely bounded subsets of A and B respectively. We observe that

$$
\begin{aligned}
(\phi \times \phi)[E_X \times E_Y] &= \{\big(\phi(a,b), \phi(ax, by)\big) \mid a \in A, x \in X, b \in B, y \in Y\} \\
&= \{(ab, axby) \mid a \in A, x \in X, b \in B, y \in Y\} \\
&= \{(ab, ab \cdot b^{-1} xby) \mid a \in A, x \in X, b \in B, y \in Y\} \\
&\subseteq E_{X^B \cdot Y}.
\end{aligned}
$$

However, because Y is coarsely bounded in B and hence also in G, we find that $X^B \cdot Y$ is coarsely bounded in G if and only if X^B is coarsely bounded in G. The above calculation thus shows that ϕ is bornologous provided that X^B is coarsely bounded in G for every coarsely bounded subset X of A.

For the converse direction, observe that, if ϕ is bornologous and $X \subseteq A$ is coarsely bounded in A, then

$$
(\phi \times \phi)[E_X \times \Delta] = \{(ab, axb) \mid a \in A, x \in X, b \in B\}
$$

is contained in a left-invariant coarse entourage E_C in G. But then $b^{-1} xb = (ab)^{-1} \cdot axb \in C$ for all $x \in X$, $a \in A$ and $b \in B$, i.e., $X^B \subseteq C$ and thus X^B is coarsely bounded in G. $\square$

Lemma 7.12 *Suppose that* A *is locally bounded. Then the following hold.*

(1) *The projection $G \xrightarrow{\pi_A} A$ is bornologous if and only if $\pi_A[X^B]$ is coarsely bounded in A for every coarsely bounded subset $X \subseteq A$.*

(2) *If for every coarsely bounded subset X of A the set $\pi_B[X^B]$ is coarsely bounded in B, then $G \xrightarrow{\pi_B} B$ is modest.*

Proof Fix a coarsely bounded identity neighbourhood $U \subseteq A$. By Theorem 7.3, the sets of the form WV, with W an open identity neighbourhood in A and V an open identity neighbourhood in B, form a neighbourhood basis at the identity in G. Thus, if D is a coarsely bounded subset of G and V an identity neighbourhood in B, there are finite sets $E \subseteq A$ and $F \subseteq B$ and an $n \geqslant 1$ so that $D \subseteq (EUFV)^n$. By Lemma 7.10, it thus follows that

$$D \subseteq (EUFV)^n \subseteq \pi_A\big[(EU)^B\big]^n \pi_B\big[(EU)^B\big]^{n-1}(FV)^n.$$

In particular, for every coarsely bounded set $D \subseteq G$ and every identity neighbourhood $V \subseteq B$, there is a finite set $F \subseteq B$, a coarsely bounded set $X \subseteq A$ and an $n \geqslant 1$, so that

$$\pi_A[D] \subseteq \pi_A\big[X^B\big]^n$$

and

$$\pi_B[D] \subseteq \pi_B\big[X^B\big]^{n-1}(FV)^n.$$

(1) So, assume that $\pi_A[X^B]$ is coarsely bounded in A for all coarsely bounded subsets X of A. To see that π_A is bornologous, let $D \subseteq G$ be a given coarsely bounded set. Then, by the above, there is some coarsely bounded subset X of A so that $\pi_A[D] \subseteq \pi_A\big[X^B\big]^n$ for some n, whence also $\pi_A[D]$ is coarsely bounded in A. Now, for $g \in G$ and $d \in D$, write $g = a_1 b_1$ and $d = a_2 b_2$ for some $a_i \in A$ and $b_i \in B$. Then

$$\pi_A(g)^{-1} \pi_A(gd) = a_1^{-1} a_1 \pi_A(b_1 a_2 b_2) = \pi_A(b_1 a_2 b_1^{-1}) \in \pi_A\big[\pi_A[D]^B\big].$$

Because $\pi_A[D]$ is coarsely bounded in A, also $\pi_A\big[\pi_A[D]^B\big]$ is coarsely bounded in A. So this shows that $G \xrightarrow{\pi_A} A$ is bornologous.

Conversely, assume $G \xrightarrow{\pi_A} A$ is bornologous and that X is a coarsely bounded subset of A. Then X is coarsely bounded in G and so $(\pi_A \times \pi_A)[E_X] \subseteq E_C$ for some coarsely bounded subset C of A. In particular, for $b \in B$ and $x \in X$, we have $(b, bx) \in E_X$ and so

$$\pi_A(bxb^{-1}) = \pi_A(bx) = \pi_A(b)^{-1} \pi_A(bx) \in C,$$

showing that $\pi_A[X^B] \subseteq C$ and thus that $\pi_A[X^B]$ is coarsely bounded in A.

(2) Assume now that $\pi_B[X^B]$ is coarsely bounded in B for all coarsely bounded subsets X of A. To see that $G \xrightarrow{\pi_B} B$ is modest, suppose that $D \subseteq G$

is coarsely bounded and V is an identity neighbourhood in B. Find as above a coarsely bounded subset X of A, a finite set $F \subseteq B$ and an $n \geqslant 1$ so that $\pi_B[D] \subseteq \pi_B[X^B]^{n-1}(FV)^n$. Since $\pi_B[X^B]$ is coarsely bounded in B, this means that also $\pi_B[D]$ is coarsely bounded as required.   □

Theorem 7.13 *Suppose a Polish group G is the Zappa–Szép product of closed subgroups A and B and let*

$$A \times B \xrightarrow{\phi} G$$

be the multiplication map $\phi(a,b) = ab$. Assume also that A is locally bounded. Then ϕ is a coarse equivalence if and only if $\pi_A[X^B]$ is coarsely bounded in A and $\pi_B[X^B]$ is coarsely bounded in B for every coarsely bounded subset $X \subseteq A$.

Proof Observe that the inverse of ϕ is the map that takes $g \in G$ to the pair $\big(\pi_A(g), \pi_B(g)\big) \in A \times B$. Thus, as the coarse structure on $A \times B$ is the product of the coarse structures on A and on B, we see that ϕ^{-1} is bornologous if and only if both $G \xrightarrow{\pi_A} A$ and $G \xrightarrow{\pi_B} B$ are bornologous. It thus follows that ϕ is a coarse equivalence exactly when all of ϕ, π_A and π_B are bornologous.

Suppose first that $A \times B \xrightarrow{\phi} G$ is a coarse equivalence, whence ϕ, π_A and π_B are bornologous, and assume that X is a coarsely bounded subset of A. Then, by Lemma 7.11, X^B is coarsely bounded in G and hence $\pi_A[X^B]$ and $\pi_B[X^B]$ are coarsely bounded in A and B respectively.

For the remainder of the proof, we assume conversely that, for all coarsely bounded subsets X of A, the images $\pi_A[X^B]$ and $\pi_B[X^B]$ are coarsely bounded in A and B respectively.

By Lemma 7.12, π_A is bornologous and π_B is modest. Also, if X is a coarsely bounded subset of A, then $\pi_A[X^B]$ and $\pi_B[X^B]$ are coarsely bounded in G and hence X^B, which satisfies

$$X^B \subseteq \pi_A[X^B] \cdot \pi_B[X^B],$$

is coarsely bounded in G. By Lemma 7.11, this implies that ϕ is bornologous.

Finally, to see that π_B is bornologous, assume $D \subseteq G$ is coarsely bounded. As π_A is bornologous, we have $(\pi_A \times \pi_A)[E_{D^{-1}}] \subseteq E_X$ for some coarsely bounded subset X of A. Also, for $g \in G$ and $d \in D$, we have

$$\pi_B(g)^{-1} \cdot \pi_B(gd) = \pi_B(g)^{-1} \cdot \pi_A(gd)^{-1} gd$$
$$= \pi_B(g)^{-1} \cdot \pi_A(gd)^{-1} \pi_A(g) \cdot \pi_B(g)d$$
$$\in \pi_B(g)^{-1} \cdot X \cdot \pi_B(g) \cdot D$$
$$\subseteq X^B \cdot D.$$

Now, X^B is coarsely bounded in G and hence so is $X^B \cdot D$. Because π_B is modest, $\pi_B[X^B \cdot D]$ is coarsely bounded in B. Thus,

$$\pi_B(g)^{-1}\pi_B(gb) \in B \cap (X^B \cdot D) \subseteq \pi_B[X^B \cdot D]$$

for all $g \in G$ and $d \in D$, showing that π_B is bornologous. □

Remark 7.14 While the two closed subgroups A and B may initially appear to play symmetric roles in the Zappa–Szép product $G = AB$, in light of Theorem 7.13 this is not quite so. Of course, if $G = AB$, then also $G = BA$, but stating that

$$\phi \colon A \times B \to G, \quad \phi(a,b) = ab$$

is a coarse equivalence is not the same as stating that

$$\psi \colon B \times A \to G, \quad \psi(b,a) = ba$$

is a coarse equivalence. This is because we work with the left-coarse structure $\mathcal{E}_L$, which is not in general bi-invariant.

Although case of locally bounded groups may be the most interesting, the assumption that A is locally bounded in Theorem 7.13 might be superfluous.

Problem 7.15 Does Theorem 7.13 hold even without the assumption of A being locally bounded?

Example 7.16 (Internal semi-direct products) Suppose a Polish group G is the internal semi-direct product of two closed subgroups N and H with N locally bounded and normal in G. That is, $G = NH$ with $N \cap H = \{1\}$ and $N \trianglelefteq G$. Let π_N and π_H be the corresponding projections, i.e., chosen so that

$$g = \pi_N(g) \cdot \pi_H(g)$$

for all $g \in G$. As N is normal in G, for any subset $X \subseteq N$, we have $X^H \subseteq N$ and hence $\pi_N(X^H) = X^H$ while $\pi_H(X^H) = \{1\}$.

It thus follows from Theorem 7.13 that the map

$$\phi \colon N \times H \to G, \quad \phi(n,h) = nh$$

is a coarse equivalence if and only if $X^H = \{hxh^{-1} \mid x \in X, h \in H\}$ is coarsely bounded in N for every coarsely bounded set $X \subseteq N$.

Example 7.17 (External semi-direct products) Suppose $\alpha \colon H \curvearrowright N$ is a continuous action of a Polish group H by continuous automorphisms on a locally bounded Polish group N and let $G = N \rtimes_\alpha H$ be the corresponding topological semi-direct product. Thus, G is simply the topological space $N \times H$ equipped with the multiplication

$$(n_1, h_1) \cdot (n_2, h_2) = \big(n_1 \alpha_{h_1}(n_2), h_1 h_2\big).$$

Moreover, N and H can be identified with the subgroups $N \times \{1_H\}$ and $\{1_N\} \times H$ of G with $N \times \{1_H\}$ normal in G.

So G is the Zappa–Szép product of $N \times \{1_H\}$ and $\{1_N\} \times H$. Moreover, as $(n, 1_H) \cdot (1_N, h) = (h, n)$, we see that the projections $\pi_{\{1_H\} \times N}$ and $\pi_{H \times \{1_N\}}$ defined by

$$(h, n) = \pi_{\{1_H\} \times N}(h, n) \cdot \pi_{H \times \{1_N\}}(h, n)$$

are the projection maps to $N \times \{1_H\}$ and $\{1_N\} \times H$ respectively. Therefore,

$$\phi \colon N \times H \to N \rtimes_\alpha H, \quad \phi(n, h) = (n, h)$$

is a coarse equivalence if and only if, for all coarsely bounded subsets X of N, the set

$$\alpha_H(X) = \{\alpha_h(x) \mid h \in H, x \in X\}$$

is coarsely bounded in N.

We sum this up in the following proposition.

Proposition 7.18 *Let* $\alpha \colon H \curvearrowright N$ *be a continuous action of a Polish group* H *by continuous automorphisms on a locally bounded Polish group* N *and let* $N \rtimes_\alpha H$ *be the corresponding topological semi-direct product. Then the formal identity*

$$\phi \colon N \times H \to N \rtimes_\alpha H$$

is a coarse equivalence if and only if, for all coarsely bounded subsets X *of* N, *the set* $\alpha_H(X)$ *is coarsely bounded in* N.

Example 7.19 (Affine isometry groups) Suppose $(X, \|\cdot\|)$ is a separable Banach space. Then the group $\mathrm{Aff}(X)$ of affine isometries of X decomposes as a semi-direct product

$$\mathrm{Aff}(X) = (X, +) \rtimes \mathrm{Isom}(X),$$

where each $A \in \mathrm{Aff}(X)$ is identified with the pair (x, T), so that $A(y) = T(y) + x$ for all $y \in Y$. That is, the projection $\pi_{\mathrm{Isom}(X)}$ associates to $A \in \mathrm{Aff}(X)$ its linear part, whereas π_X is simply the associated cocycle $b \colon \mathrm{Aff}(X) \to X$. Because, if $D \subseteq X$ is norm bounded, also

$$D^{\mathrm{Isom}(X)} = \{T(x) \mid T \in \mathrm{Isom}(X) \text{ and } x \in D\}$$

is norm bounded, we find that $X \times \text{Isom}(X)$ is coarsely equivalent with $\text{Aff}(X)$ via the map that takes (x, T) to the affine isometry $y \mapsto T(y) + x$.

In particular, the cocycle $b \colon \text{Aff}(X) \mapsto X$ is a coarse equivalence (and hence a quasi-isometry) if and only if $\text{Isom}(X)$ is a coarsely bounded group. This re-proves Proposition 3.17.

Example 7.20 (Homeomorphisms of locally compact groups) Suppose G is a subgroup of the group $\text{Homeo}(H)$ of homeomorphisms of a locally compact Polish group H and that G is equipped with a finer Polish group topology. Assume also that G contains the group $\lambda_H \cong H$ of left-translations λ_h by elements $h \in H$ and let $K = \{g \in G \mid g(1) = 1\}$ be the pointwise stabiliser of the identity in H. As observed in Example 7.6, G is then the Zappa–Szép product of λ_H and K. Moreover, if $\pi_{\lambda_H} \colon G \to \lambda_H$ and $\pi_K \colon G \to K$ are the projections associated with the decomposition $G = \lambda_H \cdot K$, we find that

$$\pi_{\lambda_H}(g) = \lambda_{g(1)} \quad \text{and} \quad \pi_K(g) = \lambda_{g(1)^{-1}} \circ g.$$

Indeed, it suffices to note that $\left(\lambda_{g(1)^{-1}} \circ g\right)(1) = \lambda_{g(1)^{-1}}\left(g(1)\right) = 1$ and therefore $\lambda_{g(1)^{-1}} \circ g \in K$. In particular, for $k \in K$ and $h \in H$, we have

$$\pi_{\lambda_H}(k\lambda_h k^{-1}) = \lambda_{k(h)} \quad \text{and} \quad \pi_K(k\lambda_h k^{-1}) = \lambda_{k(h)^{-1}} k\lambda_h k^{-1}.$$

Applying Lemma 7.12 and the fact that coarsely bounded sets in H are simply the relatively compact sets, we find that π_{λ_H} is bornologous if and only if, for every relatively compact open set $U \subseteq H$, the K-invariant open set $K[U] = \bigcup_{k \in K} k[U]$ is relatively compact. It thus follows that π_{λ_H} is bornologous if and only if H admits a covering by K-invariant relatively compact open subsets. Note that this is, in general, stronger than requiring the action $K \curvearrowright H$ to be modest.

Investigating when $\pi_K(X^K)$ is coarsely bounded in K for coarsely bounded subsets $X \subseteq \lambda_H$ ultimately depends on the coarse geometry of K. When K is coarsely bounded as a group or even just as a subset of G, we are led to the following criterion.

Proposition 7.21 *Suppose H is a locally compact Polish groups and $G \leqslant \text{Homeo}(H)$ is a subgroup equipped with a finer Polish group topology. Assume also that G contains the group of left-translations λ_h by elements $h \in H$ and that the pointwise stabiliser $K = \text{stab}_G(1)$ is coarsely bounded in G. Then the inclusion*

$$h \in H \mapsto \lambda_h \in G$$

is a coarse equivalence if and only if H admits a covering by K-invariant relatively compact open subsets.

Proof Suppose first that H admits a covering $\{U_n\}_n$ by K-invariant relatively compact open subsets. Then, by the above discussion, π_{λ_H} is bornologous and hence the inclusion of H into G is bornologous with a bornologous inverse. In particular, H is coarsely embedded in G and is cobounded because $G = HK$ with K coarsely bounded in G. So the inclusion is a coarse equivalence between H and G.

Conversely, suppose that the inclusion $h \in H \mapsto \lambda_h \in G$ is a coarse equivalence. Then λ_H is coarsely embedded in G. Also, because K is coarsely bounded in G, the map $\lambda_h k \in G \mapsto \lambda_h \in G$ is close to the identity map and thus is bornologous. Because λ_H is coarsely embedded, it follows that also $G \xrightarrow{\pi_{\lambda_H}} \lambda_H$ is bornologous and therefore that H is covered by K-invariant relatively compact open sets. □

Instances of this include, for example, $\mathrm{Homeo}_{\mathbb{Z}}(\mathbb{R})$, where $K = \mathsf{stab}(0)$ is isomorphic to $\mathrm{Homeo}_{+}([0, 1])$, and hence is coarsely bounded, and the action of K on $\mathbb{R}$ leaves every open interval $I =] - n, n[$ invariant. So again we see that the inclusion of $\mathbb{R}$ into $\mathrm{Homeo}_{\mathbb{Z}}(\mathbb{R})$ as the group of translations is a coarse equivalence.

Appendix Open Problems

Recall that, if H is a closed subgroup of a Polish group G, then the left-uniform structure of H coincides with that induced from G. That is, the inclusion map $H \hookrightarrow G$ is also a uniform embedding. As we have seen several times, this fails completely when instead we look at the coarse structure; that is, a closed subgroup H of a Polish group G need not be coarsely embedded. This, of course, underlies many of the complications of the theory and it is therefore of significant value to identify cases in which it would be true, which would point to a further robustness of the coarse structure. Many of the problems listed below ask for specific cases of this.

Problem A.1 Which are the Polish groups G so that every closed subgroup $H \leqslant G$ is coarsely embedded?

Suppose G is a Polish group in which the coarsely bounded sets are exactly the relatively compact sets. Then, if H is a closed subgroup and $A \subseteq H$ is coarsely bounded in G, the closure $\overline{A}$ will be compact and thus also coarsely bounded in H. It follows that H is automatically coarsely embedded in G. Therefore, to get instances of Problem A.1, we can look for the following.

Problem A.2 Which are the Polish groups in which the coarsely bounded subsets are exactly the relatively compact sets?

Examples of such groups are, of course, the locally compact second countable groups and direct products $\prod_{n=1}^{\infty} H_n$ of these. Note also that, if G is a non-Archimedean abelian Polish group and A is a coarsely bounded subset, then for every open subgroup $V \leqslant G$ there are a finite set F and some n so that

$$A \subseteq (FV)^n = F^n V^n = F^n V.$$

In other words, every coarsely bounded set in G is covered by finitely many left-translates of every identity neighbourhood and must therefore be relatively compact.

Problem A.3 Suppose H is a closed subgroup of an abelian Polish group G. Must H be coarsely embedded in G?

285

As an indication that the answer to Problem A.3 might be negative, we note that this appears to be open even in the case when $H \cong \mathbb{Z}$. One might still hope for something less.

Problem A.4 Suppose H is a closed subgroup of a locally bounded or monogenic abelian Polish group. Is H also locally bounded, respectively, monogenic?

Recall that, by Proposition 5.67, if H is a closed subgroup of a Polish group G admitting a compact transversal $T \subseteq G$, then H is coarsely embedded in G. The next problem asks if we can avoid the assumption of T being a transversal.

Problem A.5 Suppose H is a co-compact closed subgroup of a Polish group G, i.e., $G = HK$ for some compact set $K \subseteq G$. Is H coarsely embedded in G?

Again, we may aim for something weaker.

Problem A.6 Suppose H is a co-compact closed subgroup of a locally bounded or monogenic Polish group. Is H also locally bounded, respectively, monogenic?

Apart from subgroups being coarsely embedded, the discussion of Chapter 4 shows the importance of determining when group extensions are locally bounded.

Problem A.7 Suppose that K is a closed normal subgroup of a Polish group G and that both K and G/K are locally bounded. Does it follow that G is also locally bounded?

Problem A.8 Suppose that K is a closed normal subgroup of a Polish group G and that K is locally bounded, while G/K is ultralocally bounded. Does it follow that G is also locally bounded?

As the example of $\mathbb{Z} \ltimes \mathbb{Z}^{\mathbb{Z}}$ shows, an open subgroup of a Polish group need not be coarsely embedded. However, the following question still remains open.

Problem A.9 Suppose H is an open subgroup of a coarsely bounded Polish group. Is H also coarsely bounded?

Problem A.10 Let G be a Polish group of finite asymptotic dimension. Is G necessarily locally bounded?

Problem A.11 Let $\mathbb{H}$ be the infinite-dimensional hyperbolic space and G its group of isometries. Is G quasi-isometric to $\mathbb{H}$?

Problem A.12 Let $\mathcal{H}$ be a complex separable infinite-dimensional Hilbert space. Does the general linear group $GL(\mathcal{H})$ have property (PL)?

In relation to this problem, we should note that, by the main result of [73], the unitary group $U(\mathcal{H})$ is coarsely bounded even when viewed as a discrete group and so, in particular, will be coarsely bounded in $GL(\mathcal{H})$. Thus, taking polar decompositions, one only needs to generate the positive invertible operators from any coarsely unbounded set in $GL(\mathcal{H})$.

Problem A.13 Is the relation of having a coarse coupling an equivalence relation on the class of Polish groups?

The class of Polish groups of bounded geometry is clearly one of the best behaved from the perspective of coarse geometry and its development to a large extent mirrors that of the locally compact groups. So, on the one hand, one would like to know how closely these groups are related to the locally compact groups and, on the other hand, discover new interesting examples of such groups to get a better feeling of how they come about.

Problem A.14 Let G be a Polish group of bounded geometry. Must G be coarsely equivalent to a locally compact (second countable) group?

Problem A.15 Find a non-locally compact, topologically simple, Polish group of bounded geometry that is not coarsely bounded.

Problem A.16 More generally, find new interesting Polish groups of bounded geometry.

The next problems, whenever appropriate, are all known to have positive answers for locally compact groups.

Problem A.17 Does every amenable Polish group of bounded geometry have the Haagerup property?

Problem A.18 Suppose G, F and H are Polish groups of bounded geometry and

$$ G \curvearrowright X \curvearrowleft F \curvearrowright Y \curvearrowleft H $$

are two pairs of topological couplings between G and F, respectively, between F and H. Find a direct description of a topological coupling

$$ G \curvearrowright Z \curvearrowleft H $$

between G and H.

Problem A.19 Suppose G is a locally bounded Polish group and assume that $[0,1] \times G \xrightarrow{R} G$ is a continuous contraction of G onto 1_G so that, for every coarsely bounded set A, the restriction $[0,1] \times A \xrightarrow{R} G$ is uniformly continuous. Is R then an efficient contraction, that is, is $R\big[[0,1] \times A\big]$ coarsely bounded in G for every coarsely bounded set A?

Problem A.20 Is the group $\mathrm{Isom}(\mathbb{U})$ efficiently contractible?

Problem A.21 Let G be a Polish group of bounded geometry with a gauge metric d and assume that G is metrically amenable. Does it follow that, for all $\epsilon > 0$ and coarsely bounded C, there is a coarsely bounded set B so that

$$ \mathrm{ent}_d(CBC \setminus B) < \epsilon \cdot \mathrm{ent}_d(B) \ ? $$

Problem A.22 Suppose G is a Polish group coarsely equivalent to $\mathbb{R}$. Can one find a subnormal series of closed subgroups

$$ K \trianglelefteq F \trianglelefteq H \trianglelefteq G $$

so that H has index at most 2 in G, the quotient group H/F is coarsely bounded, $F/K \cong \mathbb{Z}$ and K is coarsely bounded in G?

This conforms with the structure of two-ended locally compact groups and also appears to conform with known examples of two-ended Polish groups.

Problem A.23 Does Theorem 7.3 hold for Baire topological groups?

Problem A.24 Does Theorem 7.13 hold even without the assumption of A being locally bounded?

References

[1] Herbert Abels, *Specker-Kompaktifizierungen von lokal kompakten topologischen Gruppen*, Math. Z., **135** (1973/74), 325–361.

[2] Richard Friederich Arens, *Topologies for homeomorphism groups*, Amer. J. Math., **68** (1946), 593–610.

[3] Uri Bader and Christian Rosendal, *Coarse equivalence and topological couplings of locally compact groups*, Geometriae Dedicata, **196** (2018), 1–9.

[4] Mohammed El Bachir Bekka, Pierre-Alain Chérix and Alain Valette, *Proper Affine Isometric Actions of Amenable Groups, Novikov Conjectures, Index Theorems and Rigidity*, vol. 2 (Oberwolfach, 1993), 1–4, London Mathematical Society Lecture Note Series 227 (Cambridge: Cambridge University Press, 1995).

[5] Mohammed El Bachir Bekka, Pierre de la Harpe and Alain Valette, *Kazhdan's Property (T)*, New Mathematical Monographs, 11 (Cambridge: Cambridge University Press, 2008).

[6] Itaï Ben Yaacov, Alexander Berenstein, C. Ward Henson and Alexander Usvyatsov, Model theory for metric structures. In Z. Chatzidakis, D. Macpherson, A. Pillay and A Wilkie, eds., *Model Theory with Applications to Algebra and Analysis*, vol. 2, London Mathematical Society Lecture Note Series, 350 (Cambridge: Cambridge University Press, 2008), pp. 315–427.

[7] Itaï Ben Yaacov, Julien Melleray and Todor Tsankov, *Metrizable universal minimal flows of Polish groups have a comeagre orbit*, Geom. Funct. Anal., **27**(1) (2017), 67–77.

[8] Itaï Ben Yaacov and Todor Tsankov, *Weakly almost periodic functions, model-theoretic stability, and minimality of topological groups*, Trans. Amer. Math. Soc., **368**(11) (2016), 8267–8294.

[9] Yoav Benyamini and Joram Lindenstrauss, *Geometric Nonlinear Functional Analysis, Volume 1* (Providence, RI: American Mathematical Society, 2000).

[10] George Mark Bergman, *Generating infinite symmetric groups*, Bull. London Math. Soc., **38** (2006), 429–440.

[11] Garrett Birkhoff, *A note on topological groups*, Compositio Math., **3** (1936), 427–430.

[12] Jonathan Block and Shmuel Weinberger, *Aperiodic tilings, positive scalar curvature and amenability of spaces*, J. Amer. Math. Soc., **5** (1992), 907–918.

[13] Bruno de Mendonça Braga, *Topics in the nonlinear geometry of Banach spaces*, Doctoral dissertation, University of Illinois at Chicago (2017).

[14] Nathaniel Brown and Erik Guentner, *Uniform embedding of bounded geometry spaces into reflexive Banach space*, Proc. Amer. Math. Soc. **133**(7) (2005), 2045–2050.

[15] Peter J. Cameron, *Permutation Groups*, London Mathematical Society Student Texts, 45 (Cambridge: Cambridge University Press, 1999).

[16] Peter J. Cameron and Anatoliĭ Moiseevich Vershik, *Some isometry groups of the Urysohn space*, Ann. Pure Appl. Logic, **143**(1–3) (2006), 70–78.

[17] Pierre-Alain Chérix, Michael Cowling, Paul Jolissant, Pierre Julg and Alain Valette, *Groups with the Haagerup Property: Gromov's a-T-menability* (Basel: Birkhäuser, 2001).

[18] Gregory Cherlin, *Homogeneous Ordered Graphs and Metrically Homogeneous Graphs*, book in preparation.

[19] Michael P. Cohen, *On the large scale geometry of diffeomorphism groups of 1-manifolds*, Forum Mathematicum, **30**(1) (2018), 75–86.

[20] Yves de Cornulier, *On lengths on semisimple groups*, J. Topol. Anal., **1**(2), (2009), 113–121.

[21] Yves de Cornulier, On the quasi-isometric classification of locally compact groups. In P.-E. Caprace and N. Monod, eds., *New Directions in Locally Compact Groups*, London Mathematical Society Lecture Notes Series 447 (Cambridge: Cambridge University Press, 2018), pp. 275–342.

[22] Yves de Cornulier and Pierre de la Harpe, *Metric Geometry of Locally Compact Groups*, EMS Tracts in Mathematics, vol. 25 (European Mathematical Society, September 2016).

[23] Yves de Cornulier, Romain Tessera and Alain Valette, *Isometric group actions on Hilbert spaces: growth of cocycles*, Geom. Funct. Anal., **17** (2007), 770–792.

[24] Harry Herbert Corson III and Victor L. Klee Jr, *Topological classification of convex sets*, Proc. Symp. Pure Math. VII, Convexity, Amer. Math. Soc. (1963), pp. 101–181.

[25] Jaques Dixmier, *Dual et quasidual d'une algèbre de Banach involutive*, Trans. Amer. Math. Soc., **104** (1962), 278–283.

[26] Robert D. Edwards and Robion Cromwell Kirby, *Deformations of spaces of imbeddings*, Ann. Math. (2), **93** (1971), 63–88.

[27] Yakov Eliashberg and Tudor Stefan Ratiu, *The diameter of the symplectomorphism group is infinite*, Invent. Math., **103**(2) (1991), 327–340.

[28] Robert Mortimer Ellis, *Universal minimal sets*, Proc. Amer. Math. Soc., **11** (1960), 540–543.

[29] Alex Eskin, David Fisher and Kevin Whyte, *Coarse differentiation of quasi-isometries I: Spaces not quasi-isometric to Cayley graphs*, Ann. Math., **176** (2012), 221–260.

[30] Marián Fabian, Petr Habala, Petr Hájek, Vicente Montesinos Santalucía and Václav Zizler, *Banach Space Theory. The Basis for Linear and Nonlinear Analysis*, CMS Books in Mathematics/Ouvrages de Mathématiques de la SMC (New York: Springer, 2011).

[31] Gordon M. Fisher, *On the group of all homeomorphisms of a manifold*, Trans. Amer. Math. Soc., **97** (1960), 193–212.

[32] Erling Følner, *On groups with full Banach mean value*, Math. Scand., **3** (1955), 243–254.

[33] Roland Fraïssé, *Sur l'extension aux relations de quelques propriétés des ordres*, Ann. Sci. École Norm. Sup., **71** (1954), 363–388.

[34] John M. Franks and Michael Handel, *Distortion elements in group actions on surfaces*, Duke Math. J., **131**(3) (2006), 441–468.

[35] Maxime Gheysens and Nicolas Monod, *Fixed points for bounded orbits in Hilbert spaces*, Annales de l'ENS, **50**(1) (2017), 131–156.

[36] Mikhail Leonidovich Gromov, Asymptotic invariants of infinite groups. In G. A. Niblo and M. A. Roller, eds., *Geometric Group Theory*, vol. 2 (Sussex, 1991), London Mathematical Society Lecture Note Series 182 (Cambridge: Cambridge University Press, 1993), pp. 1–295.

[37] Manuel Herbst, 1-Cocycles of unitary representations of infinite-dimensional unitary groups, Doctoral dissertation, Friedrich-Alexander-Universität Erlangen-Nürnberg (2018).

[38] William Jake Herndon, Absolute continuity and large-scale geometry of Polish groups, arXiv: 1802.10239 (2018).

[39] Willian Jake Herndon, Deformations and Products of Polish Groups, Doctoral dissertation, University of Illinois at Chicago (2019).

[40] Heinz Hopf, *Enden offener Räume und unendliche diskontinuierliche Gruppen*, Comm. Math. Helv., **16** (1943/44), 81–100.

[41] Paul Jaffard, *Traité de topologie générale en vue de ses applications* (Paris: Presses Universitaires de France, 1997).

[42] William Buhmann Johnson, Joram Lindenstrauss and Gideon Schechtman, *Banach spaces determined by their uniform structure*, Geom. Funct. Anal., **6** (1996), 430–470.

[43] Shizuo Kakutani, *Selected Papers, vol. 1*. Edited and with a preface by Robert R. Kallman. With a biographical sketch by Arshag Hajian and Yuji Ito. *Contemporary Mathematicians* (Boston, MA: Birkhäuser Boston, Inc., 1986).

[44] Shizuo Kakutani and Kunihiko Kodaira, *Über das Haarsche Mass in der lokal bikompakten Gruppe*, Proc. Imp. Acad. Tokyo, **20** (1944), 444–450. *Selected Papers, vol. 1*, Robert R. Kallman, ed. (Boston, MA: Birkhäuser Boston, Inc., 1986), pp. 68–74.

[45] Nigel John Kalton, *The non-linear geometry of Banach spaces*, Rev. Mat. Comput., **21**(1) (2008), 7–60.

[46] Miroslav Katětov, *On universal metric spaces*, General Topology and its Relations to Modern Analysis and Algebra, VI (Prague, 1986) (1988), 323–330.

[47] Alexander Sotirios Kechris, *Classical Descriptive Set Theory* (New York: Springer Verlag, 1995).

[48] Alexander Sotirios Kechris, Vladimir Germanovich Pestov and Stevo Todorčević, *Fraïssé limits, Ramsey theory, and topological dynamics of automorphism groups*, Geom. Funct. Anal., **15**(1) (2005), 106–189.

[49] Alexander Sotirios Kechris and Christian Rosendal, *Turbulence, amalgamation and generic automorphisms of homogeneous structures*, Proc. Lond. Math. Soc. (3), **94**(2) (2007), 302–350.

[50] Victor L. Klee Jr, *Invariant metrics in groups (solution of a problem of Banach)*, Proc. Amer. Math. Soc., **3** (1952) 484–487.

[51] Andrey Nikolaevich Kolmogorov, *Asymptotic characteristics of some completely bounded metric spaces*, Dokl. Akad. Nauk SSSR, **108** (1956), 585–589.

[52] Andrey Nikolaevich Kolmogorov and Vladimir Mikhaïlovich Tikhomirov, ϵ-*entropy and ϵ-capacity of sets in function spaces*, Uspehi Mat. Nauk, no. 2 (86), **14** (1959), 3–86; English translation, Amer. Math. Soc. Transl. (2), **17** (1961), 277–364.

[53] Jean-Louis Krivine and Bernard Maurey, *Espaces de Banach stables*, Israel J. Math., **39**(4) (1981), 273–295.

[54] Kazimierz Kuratowski and Czesław Ryll-Nardzewski, *A general theorem on selectors*, Bull. Acad. Pol. Sci. Sér. Sci., Math., Astr. et Phys., **13** (1965), 397–403.

[55] Wolfgang Lusky, *The Gurarij spaces are unique*, Arch. Math. (Basel), **27** (1976), 627–635.

[56] Kathryn Mann, *Automatic continuity for homeomorphism groups and applications. With an appendix on the structure of groups of germs of homeomorphism, written with Frederic Le Roux*. Geometry & Topology, **20**(5) (2016), 3033–3056.

[57] Kathryn Mann and Christian Rosendal, *Large scale geometry of homeomorphism groups*, Ergod. Th. & Dynam. Sys., **38**(7) (2018), 2748–2779.

[58] David Ellis Marker, *Model Theory: An Introduction* (New York: Springer-Verlag, 2002).

[59] Michael G. Megrelishvili, *Every semitopological semigroup compactification of the group $H_+[0, 1]$ is trivial*, Semigroup Forum, **63**(3) (2001), 357–370.

[60] John Meier, *Groups, Graphs and Trees: An Introduction to the Theory of Infinite Groups*, London Math. Soc. Student Texts 73 (Cambridge: Cambridge University Press, 2008).

[61] Julien Melleray, *Topology of the Isometry group of the Urysohn space*, Fundamenta Mathematicae, **207**(3) (2010), 273–287.

[62] Emmanuel Militon, *Distortion elements for surface homeomorphisms*, Geometry & Topology, **18** (2014), 521–614.

[63] Emmanuel Militon, *Conjugacy class of homeomorphisms and distortion elements in groups of homeomorphisms*, Journal de l'École Polytechnique, **5** (2018), 565–604.

[64] John Milnor, *A note on curvature and fundamental group*, J. Diff. Geom., **2** (1968), 1–7.

[65] Jan Mycielski, *Independent sets in topological algebras*, Fund. Math., **55** (1964), 139–147.

[66] Assaf Naor, *Uniform nonextendability from nets*, C. R. Math. Acad. Sci. Paris, **353**(11) (2015), 991–994.

[67] Andrew Nicas and David Rosenthal, *Coarse structures on groups* (English summary), Topology Appl., **159**(14) (2012), 3215–3228.

[68] Piotr W. Nowak and Guoliang Yu, *Large Scale Geometry*, EMS Textbooks in Mathematics (Zürich: European Mathematical Society (EMS), 2012).

[69] Donald Samuel Ornstein and Benjamin Weiss, *Entropy and isomorphism theorems for actions of amenable groups*, J. d'Analyse Mathématique, **48** (1987), 1–141.

[70] Vladimir Germanovich Pestov, *A theorem of Hrushovski–Solecki–Vershik applied to uniform and coarse embeddings of the Urysohn metric space*, Topology Appl., **155**(14) (2008), 1561–1575.

[71] Billy James Pettis, *On continuity and openness of homomorphisms in topological groups*, Ann. Math. (2), **52** (1950), 293–308.

[72] Leonid Polterovich, *Growth of maps, distortion in groups and symplectic geometry*, Invent. Math., **150**(3) (2002), 655–686.

[73] Éric Ricard and Christian Rosendal, *On the algebraic structure of the unitary group*, Collect. Math., **58**(2) (2007), 181–192.

[74] John Roe, *Index Theory, Coarse Geometry, and Topology of Manifolds*, CBMS Regional Conference Series in Mathematics, 90 (published for the Conference Board of the Mathematical Sciences, Washington, DC, by the American Mathematical Society, Providence, RI, 1996).

[75] John Roe, *Lectures on Coarse Geometry*, University Lecture Series, 31 (Providence, RI: American Mathematical Society, 2003).

[76] Walter Roelcke and Susanne Dierolf, *Uniform Structures on Topological Groups and Their Quotients* (New York: McGraw-Hill, 1981).

[77] Christian Rosendal, *A topological version of the Bergman property*, Forum Mathematicum, **21**(2) (2009), 299–332.

[78] Christian Rosendal, *Global and local boundedness of Polish groups*, Indiana Univ. Math. J., **62**(5) (2013), 1621–1678.

[79] Christian Rosendal, Large scale geometry of metrisable groups, arXiv: 1403.3106 (2014).

[80] Christian Rosendal, *Equivariant geometry of Banach spaces and topological groups*, Forum Math., Sigma, **5**(e22) (2017), 62 pages.

[81] Christian Rosendal, *Lipschitz structure and minimal metrics on topological groups*, Arkiv för matematik, **56**(1) (2018), 185–206.

[82] Friedrich Martin Schneider and Andreas Thom, *On Følner sets in topological groups*, Compositio Mathematica, **154**(7) (2018), 1333–1361.

[83] Albert Solomonovich Schwarz, *A volume invariant of coverings* (Russian). Dokl. Akad. Nauk SSSR (N.S.), **105** (1955), 32–34.

[84] Raymond A. Struble, *Metrics in locally compact groups*, Compositio Math., **28** (1974), 217–222.

[85] Jenő Szép, *Über die als Produkt zweier Untergruppen darstellbaren endlichen Gruppen* (German), Comment. Math. Helv., **22** (1949), 31–33.

[86] Katrin Tent and Martin Ziegler, *On the isometry group of the Urysohn space*, J. Lond. Math. Soc. (2), **87** (2013), 289–303.

[87] Romain Tessera, *Large scale Sobolev inequalities on metric measure spaces and applications*, Rev. Mat. Iberoamericana, **24**(3) (2008), 825–865.

[88] Romain Tessera and Alain Valette, *Locally compact groups with every isometric action bounded or proper*, J. Topol. Anal., **12**(02) (2020), 267–292.

[89] Todor Dimitrov Tsankov, *Unitary representations of oligomorphic groups*, Geom. Funct. Anal., **22**(2) (2012), 528–555.

[90] Pavel Samuilovich Urysohn, *Sur un espace métrique universel*, Bull. Sci. Math., **51** (1927), 43–64, 74–90.

[91] Vladimir Vladimirovich Uspenskiĭ, *A universal topological group with a countable base*, Funct. Anal. Appl., **20** (1986), 160–161.

[92] William Veech, *Topological dynamics*, Bull. Amer. Math. Soc, **83**(5) (1977), 775–830.

[93] Nik Weaver, *Lipschitz Algebras* (River Edge, NJ: World Scientific Publishing Co., Inc., 1999).

[94] André Weil, *Sur les espaces à structure uniforme et sur la topologie générale*, Act. Sci. Ind. 551 (Paris: Hermann, 1937).

[95] Kevin Whyte, *Amenability, bi-Lipschitz equivalence, and the von Neumann conjecture*, Duke Math. J., **99** (1999), 93–112.

[96] Kôsaku Yosida, *Mean ergodic theorem in Banach spaces*, Proc. Imp. Acad., **14**(8) (1938), 292–294.

[97] Guido Zappa, Sulla costruzione dei gruppi prodotto di due dati sottogruppi permutabili tra loro (Italian), *Atti Secondo Congresso Un. Mat. Ital., Bologna, 1940*, pp. 119–125. (Rome: Edizioni Cremonense, 1942).

[98] Joseph Zielinski, *An automorphism group of an ω-stable structure that is not locally (OB)*, Math. Log. Q., **62**(6) (2016), 547–551.

[99] Joseph Zielinski, Locally Roelcke precompact groups, arXiv: 1806.03752 (2018).

Index